Methods in Molecular Biology™

Series Editor
John M. Walker
School of Life Sciences
University of Hertfordshire
Hatfield, Hertfordshire, AL10 9AB, UK

For other titles published in this series, go to
www.springer.com/series/7651

Chromatin Immunoprecipitation Assays

Methods and Protocols

Edited by

Philippe Collas

University of Oslo, Oslo, Norway

Humana Press

Editor
Philippe Collas
Department of Biochemistry
University of Oslo
Oslo 0372
Norway
philippe.collas@medisin.uio.no

Series Editor
John Walker
University of Hertfordshire
Halfield, Herts
UK

ISSN 1064-3745 e-ISSN 1940-6029
ISBN 978-1-60327-413-5 e-ISBN 978-1-60327-414-2
DOI 10.1007/978-1-60327-414-2
Springer Dordrecht Heidelberg London New York

Library of Congress Control Number: 2009931091

Cover illustration: The background art is derived from Figure 2 in Chapter 7

Printed on acid-free paper

Springer is part of Springer Science+Business Media (www.springer.com)

Preface

Virtually all aspects of cellular function, such as replication of DNA, separation of chromosomes during cell division, DNA repair, or gene expression, depend on the interaction of proteins with DNA. The nature of DNA-binding proteins is wide and ranges from structural proteins making up the nucleosome, enzymes modulating chromatin structure to enable, facilitate, or repress gene expression, transcription factors, and various cofactors. The biological significance of these associations in the context of gene expression, development, cell differentiation, and disease has immensely been enhanced in the past 20 years by the advent of a technique referred to as *ch*romatin *i*mmuno*p*recipitation, or ChIP. The purpose of the ChIP assay is to identify genomic sequence(s) associated with a protein of interest, for example, your favorite transcription factor, in the genome. ChIP, then, has become the technique of choice to determine the genomic enrichment profiles of transcription factors, post-translationally modified histones, histone variants, or chromatin-modifying enzymes. In the ChIP assay, the protein of interest is immunoprecipitated from a chromatin preparation using specific antibodies. After stringent washes, the DNA is released and the sequences bound by the immunoprecipitated protein are identified. Sequence identification methods have rapidly evolved from dot- or slot-blots in the early days to polymerase chain reaction. Subsequently, the combination of ChIP with DNA microarray or high-throughput sequencing technologies has enabled the profiling of protein occupancy on a genome-wide scale. It has also promoted the appearance of new algorithms for mapping protein binding throughout the genome.

ChIP, therefore, is arguably a power tool. Nevertheless, it has for a long time remained a cumbersome procedure taking several days and requiring very large numbers (several millions) of cells. These limitations have sparked modifications of the assay and variations in DNA detection approaches to shorten the procedure, simplify sample handling, and make ChIP amenable to small cell numbers. As a result, the ChIP assay has become increasingly popular in several areas of molecular and cell biology. To illustrate this point, a PubMed search with the keyword "chromatin immunoprecipitation" brings up four publications in 1988 and a total of over 6,400 to date, including 1,578 publications in 2008 alone (*see* **Fig. 1**).

Release of this volume on *Chromatin Immunoprecipitation Assays* by Humana Press is, therefore, timely. The volume is devoted to recent developments in ChIP and related protocols, which have proven reliable in the literature and which I believe will remain current and of great interest to researchers for many years to come. The chapters describe protocols on subjects such as characterization of ChIP antibodies, ChIP methods for small cell numbers, fast ChIP protocols, and assays adapted to various species and cell types. Several strategies for the analysis of genome-wide data sets are also included. The book also extends beyond ChIP assays per se to include protocols on immunoprecipitation-based DNA methylation analyses, determination

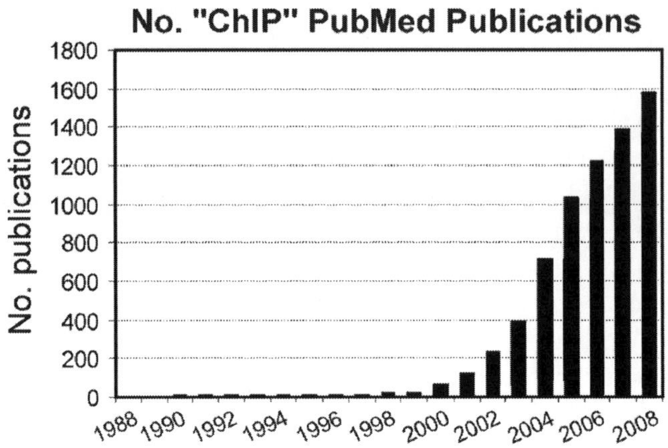

Fig. 1. Yearly number of PubMed publications responding to the search criterion "chromatin immunoprecipitation".

of spatial chromatin organization of large genomic regions, as well as RNA immunoprecipitation.

These protocols have been carefully detailed by researchers deeply involved in their development or improvement. All of the contributors and their teams deserve many thanks for their time, effort, and generosity. It has been fun to work on this project, and I wish to thank John Walker for his invitation to put together this volume, and the entire production team at Humana Press.

Philippe Collas

Contents

Contributors

MONA ABED • *Center for Vascular and Cancer Biology, The Rappaport Faculty of Medicine and Research Institute, Technion-Israel Institute of Technology, Haifa, Israel*
PETER ALESTRÖM • *Department of Basic Sciences and Aquatic Medicine, Norwegian School of Veterinary Science, Oslo, Norway*
MIGUEL A. ANDRADE-NAVARRO • *Max-Delbrück Center for Molecular Medicine, Berlin, Germany*
ZSOLT BACSO • *Department of Biophysics and Cell Biology, University of Debrecen, Debrecen, Hungary*
OLGA BOICO • *Center for Vascular and Cancer Biology, The Rappaport Faculty of Medicine and Research Institute, Technion-Israel Institute of Technology, Haifa, Israel*
KAROL BOMSZTYK • *Molecular and Cellular Biology Program and UW Medicine at Lake Union, University of Washington, Seattle, WA, USA*
LAURENT BULTOT • *Diagenode sa, Sart-Tilman, Liège, Belgium; Université UCL, Bruxelles, Belgium*
JIGUO CHEN • *Department of Biological Sciences, Mississippi State University, Mississippi State, MS, USA*
PHILIPPE COLLAS • *Department of Biochemistry, Institute of Basic Medical Sciences, University of Oslo, Oslo, Norway*
JOHN ARNE DAHL • *Department of Biochemistry, Institute of Basic Medical Sciences, University of Oslo, Oslo, Norway*
JOB DEKKER • *Program in Gene Function and Expression and Department of Biochemistry and Molecular Pharmacology, University of Massachusetts Medical School, Worcester, MA, USA*
OLEG DENISENKO • *UW Medicine at Lake Union, University of Washington, Seattle, WA, USA*
ELVIRA EIVAZOVA • *Vanderbilt University, Nashville, TN, USA*
ALEXEY GAVRILOV • *CNRS UMR-8126, Université Paris-Sud 11, Institut de Cancérologie Gustave Roussy, France; Institute of Gene Biology, Russian Academy of Sciences, Moscow, Russia*
GÉRALDINE GOENS • *Diagenode sa, Sart-Tilman, Liège, Belgium*
RAPHAEL GOTTARDO • *Department of Statistics, University of British Columbia, Vancouver, BC, Canada*
JEAN-JACQUES GOVAL • *Diagenode sa, Sart-Tilman, Liège, Belgium*
EVA HEGEDUS • *Department of Biophysics and Cell Biology, University of Debrecen, Debrecen, Hungary*
DORIT KENYAGIN-KARSENTI • *Center for Vascular and Cancer Biology, The Rappaport Faculty of Medicine and Research Institute, Technion-Israel Institute of Technology, Haifa, Israel*
LASZLO IMRE • *Department of Biophysics and Cell Biology, University of Debrecen, Debrecen, Hungary*

LEIF C. LINDEMAN • *Department of Biochemistry, Institute of Basic Medical Sciences, University of Oslo, Oslo, Norway*

MARC LIPINSKI • *CNRS UMR-8126, Université Paris-Sud 11, Institut de Cancérologie Gustave Roussy, Villejuif, France*

JUANA MAGDALENA • *Diagenode sa, Sart-Tilman, Liège, Belgium*

JENNIFER A. MCCANN • *Department of Medicine and Department of Cellular and Molecular Medicine, Faculty of Medicine, University of Ottawa, Ottawa, ON, Canada*

DOAN XUAN QUANG MINH • *Department of Biophysics and Cell Biology, University of Debrecen, Debrecen, Hungary*

ENRIQUE M. MURO • *Max-Delbrück Center for Molecular Medicine, Berlin, Germany*

JOEL NELSON • *Molecular and Cellular Biology Program, University of Washington, Seattle, WA, USA*

ALES OBRDLIK • *Department of Cell and Molecular Biology, Karolinska Institute, Stockholm, Sweden*

AMIR ORIAN • *Center for Vascular and Cancer Biology, The Rappaport Faculty of Medicine and Research Institute, Technion-Israel Institute of Technology, Haifa, Israel*

PIERGIORGIO PERCIPALLE • *Department of Cell and Molecular Biology, Karolinska Institute, Stockholm, Sweden*

IRYNA PIROZHKOVA • *CNRS UMR-8126, Université Paris-Sud 11, Institut de Cancérologie Gustave Roussy, Villejuif, France*

SERGEY RAZIN • *Institute of Gene Biology, Russian Academy of Sciences, Moscow, Russia*

MICHAEL A. RUDNICKI • *Regenerative Medicine Program and Sprott Centre for Stem Cell Research, Ottawa Health Research Institute, and Department of Medicine and Department of Cellular and Molecular Medicine, Faculty of Medicine, University of Ottawa, Ottawa, ON, Canada*

DORINA RUSU • *Diagenode sa, Sart-Tilman, Liège, Belgium*

ANITA L. SØRENSEN • *Department of Biochemistry, Institute of Basic Medical Sciences, University of Oslo, Oslo, Norway*

GABOR SZABO • *Department of Biophysics and Cell Biology, University of Debrecen, Debrecen, Hungary*

LORANT SZEKVOLGYI • *Institut Curie, Recombinaison et Instabilité Génétique, UMR7147 CNRS, Institut Curie, Université Pierre et Marie Curie, Paris, France*

NYNKE L. VAN BERKUM • *Program in Gene Function and Expression and Department of Biochemistry and Molecular Pharmacology, University of Massachusetts Medical School, Worcester, MA, USA*

YEGOR VASSETZKY • *CNRS UMR-8126, Université Paris-Sud 11, Institut de Cancérologie Gustave Roussy, 39, rue Camille-Desmoulins, 94805 Villejuif CEDEX*

LINN T. VOGT-KIELLAND • *Department of Basic Sciences and Aquatic Medicine, Norwegian School of Veterinary Science, Oslo, Norway*

ZHENGHE WANG • *Department of Genetics and Case Comprehensive Cancer Center, Case Western Reserve University, Cleveland, OH, USA*

Chapter 1

The State-of-the-Art of Chromatin Immunoprecipitation

Philippe Collas

Abstract

The biological significance of interactions of nuclear proteins with DNA in the context of gene expression, cell differentiation, or disease has immensely been enhanced by the advent of chromatin immunoprecipitation (ChIP). ChIP is a technique whereby a protein of interest is selectively immunoprecipitated from a chromatin preparation to determine the DNA sequences associated with it. ChIP has been widely used to map the localization of post-translationally modified histones, histone variants, transcription factors, or chromatin-modifying enzymes on the genome or on a given locus. In spite of its power, ChIP has for a long time remained a cumbersome procedure requiring large number of cells. These limitations have sparked the development of modifications to shorten the procedure, simplify the sample handling, and make the ChIP amenable to small number of cells. In addition, the combination of ChIP with DNA microarray, paired-end ditag, and high-throughput sequencing technologies has in recent years enabled the profiling of histone modifications and transcription factor occupancy on a genome-wide scale. This review highlights the variations on the theme of the ChIP assay, the various detection methods applied downstream of ChIP, and examples of their application.

Key words: Chromatin immunoprecipitation, ChIP, acetylation, methylation, transcription factor, DNA binding, epigenetics.

1. Introduction: Modifications of DNA and Histone Proteins

The interaction between proteins and DNA is essential for many cellular functions such as DNA replication and repair, maintenance of genomic stability, chromosome segregation at mitosis, and regulation of gene expression. Transcription is controlled by the dynamic association of transcription factors and chromatin modifiers with target DNA sequences. These associations take place not only within regulatory regions of genes (promoters and enhancers), but also within coding sequences. They are modulated by

Philippe Collas (ed.), *Chromatin Immunoprecipitation Assays,* Methods in Molecular Biology 567,
DOI 10.1007/978-1-60327-414-2_1, © Humana Press, a part of Springer Science+Business Media, LLC 2009

modifications of DNA such as methylation of CpG dinucleotides (1), by post-translational modifications of histones (2), and by incorporation of histone variants (3–7). These alterations are commonly referred to as epigenetic modifications: they modify the composition of DNA and chromatin without altering genome sequence, and they are passed onto daughter cells (they are heritable).

DNA methylation is generally seen as a hallmark of long-term gene silencing (8, 9). Methyl groups on the cytosine in CpG dinucleotides create target sites for methyl-binding proteins, which induce transcriptional repression by recruiting transcriptional repressors such as histone deacetylases or histone methyltransferases (9). DNA methylation largely contributes to gene repression and as such it is essential for development (10–12), X chromosome inactivation (13), and genomic imprinting (14, 15). The relationship between DNA methylation and gene expression is intricate, and recent evidence based on genome-wide CpG methylation profiling has highlighted CpG content and density of promoters as one component of this complexity (16, 17).

In addition to DNA methylation, post-translational modifications of histone proteins regulate gene expression. The core element of chromatin is the nucleosome, which consists of DNA wrapped around two subunits of histone H2A, H2B, H3, and H4. Nucleosomes are spaced by the linker histone H1. The amino-terminal tails of histones are post-translationally modified to confer physical properties that affect their interactions with DNA. Histone modifications not only influence chromatin packaging, but are also read by adaptor molecules, chromatin-modifying enzymes, transcription factors, and transcriptional repressors, and thereby contribute to the regulation of transcription (2, 18–20).

Histone modifications have been best characterized so far for H3 and H4. They include combinatorial lysine acetylation, lysine methylation, arginine methylation, serine phosphorylation, lysine ubiquitination, lysine SUMOylation, proline isomerization, and glutamate ADP-ribosylation (2) (**Fig. 1.1**). In particular, di- and trimethylation of H3 lysine 9 (H3K9me2, H3K9me3) and trimethylation of H3K27 (H3K27me3) elicit the formation of repressive heterochromatin through the recruitment of heterochromatin protein 1 (21) and polycomb group (PcG) proteins, respectively (22–24). However, whereas H3K9me3 marks constitutive heterochromatin (25), H3K27me3 characterizes facultative heterochromatin, or chromatin domains containing transcriptionally repressed genes that can potentially be activated, for example upon differentiation (26, 27). In contrast, acetylation of histone tails loosens their interaction with DNA and creates a chromatin conformation accessible to targeting of transcriptional activators (28, 29). Thus, acetylation on H3K9 (H3K9ac) and H4K16 (H4K16ac), together with di- or trimethylation of H3K4

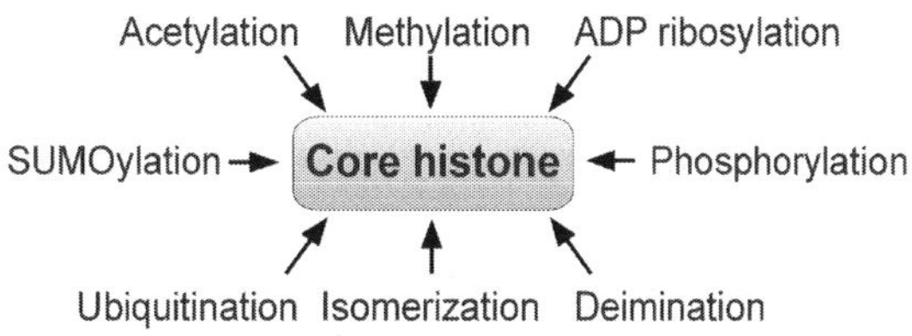

Fig. 1.1. Known post-translational modifications of histones.

(H3K4me2, H3K4me3), is found in euchromatin, often in association with transcriptionally active genes (27, 30–33). The combination of DNA methylation and histone modifications has been proposed to constitute a 'code' read by effector proteins to turn on, turn off, or modulate transcription (20, 34). Increasing evidence also indicates that specific histone modification and DNA methylation patterns mark promoters for potential activation in undifferentiated cells (17, 26, 27, 35).

2. Analysis of DNA-Bound Proteins by Chromatin Immunoprecipitation

Chromatin immunoprecipitation (ChIP) has become the technique of choice to investigate protein–DNA interactions inside the cell (36, 37). ChIP has been used for mapping the localization of post-translationally modified histones and histone variants in the genome and for mapping DNA target sites for transcription factors and other chromosome-associated proteins.

The principle of the ChIP assay is outlined in **Fig. 1.2**. DNA and proteins are commonly reversibly cross-linked with formaldehyde (which is heat-reversible) to covalently attach proteins to target DNA sequences. Formaldehyde cross-links proteins and DNA molecules within ~2 Å of each other, and thus is suitable for looking at proteins which directly bind DNA. The short cross-linking arm of formaldehyde, however, is not suitable for examining proteins that indirectly associate with DNA, such as those found in larger complexes. As a remedy to this limitation, a variety of long-range bifunctional cross-linkers have been used in combination with formaldehyde to detect proteins on target sequences, which could not be detected with formaldehyde alone (38). In contrast to cross-link ChIP, native ChIP (NChIP) omits cross-linking (37, 39). NChIP is well suited for the analysis of histones because of their high affinity for DNA. In both cross-link ChIP and NChIP, chromatin is subsequently fragmented, either by enzymatic digestion with micrococcal nuclease (MNase, which digests DNA at the level of the linker, leaving nucleosomes intact) or by sonication of whole cells or nuclei, into fragments of

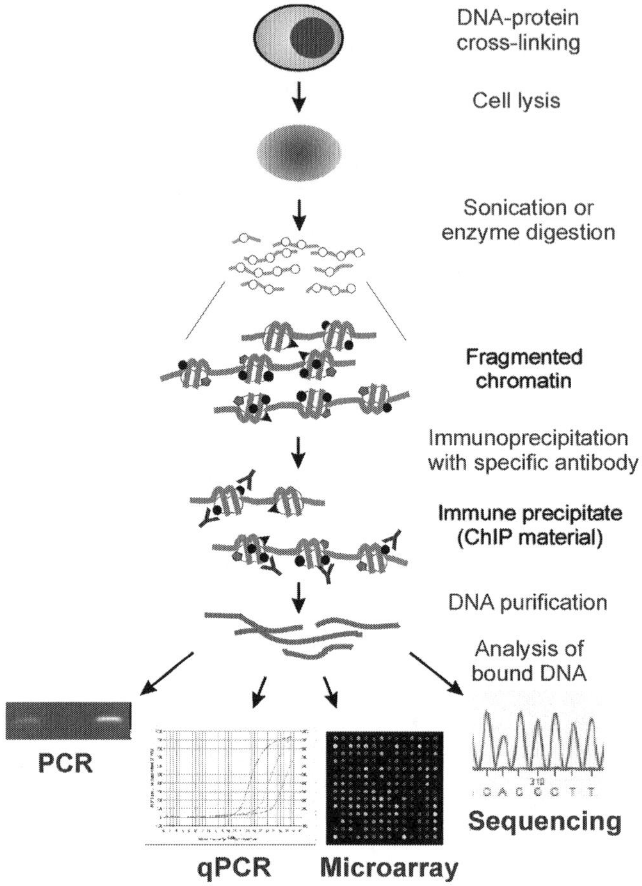

Fig. 1.2. Outline of the chromatin immunoprecipitation (ChIP) assay and various methods of analysis.

200–1,000 base pair (bp), with an average of 500 bp. The lysate is cleared by sedimentation and protein–DNA complexes are immunoprecipitated from the supernatant (chromatin) using antibodies to the protein of interest. Immunoprecipitated complexes are washed under stringent conditions to remove non-specifically bound chromatin, the cross-link is reversed, proteins are digested, and the precipitated ChIP-enriched DNA is purified. DNA sequences associated with the precipitated protein can be identified by end-point polymerase chain reaction (PCR), quantitative (q)PCR, labeling and hybridization to genome-wide or tiling DNA microarrays (ChIP-on-chip) (40–42), molecular cloning and sequencing (43, 44), or direct high-throughput sequencing (ChIP-seq) (45) (**Fig. 1.2**).

Development of techniques leading to the ChIP assay as we know it since the mid-1990s has occurred over many years [reviewed in (46)]. The use of formaldehyde to cross-link proteins with proteins or proteins with DNA, however, was first reported in the 1960s and its application to study histone–DNA interactions within the nucleosome goes back to the mid-late 1970s. The development of anti-histone antibodies 20 years ago, to investigate the association of histones with DNA in relation to transcription, led the path to the ChIP assay (47). Pioneering studies showed that during heat shock, histone H4 remained associated with the *HSP70* gene (47). Subsequent improvements in the procedure enabled the demonstration that the interaction of histone H1 with DNA was altered during changes in transcriptional activity in *Tetrahymena* (48). The availability of antibodies to post-translationally modified histones, in combination with ChIP, has been instrumental in the understanding of transcription regulation in the early 1990s. For instance, antibodies to acetylated histones have been used to show that, using the β-globin locus as a target genomic sequence, core histone acetylation is associated with chromatin that is active or poised for transcription (49–52). The ChIP assays have since been extended to non-histone proteins, including less-abundant protein complexes, and to a wide range of organisms such as protozoa, yeast, sea urchin, flies, fish, and avian and mammalian cells (46).

For well over a decade, ChIP has remained a cumbersome protocol, requiring 3–4 days and large number of cells – in the multi-million range per immunoprecipitation. These limitations have restricted the application of ChIP to large cell samples. Classical ChIP assays also involve extensive sample handling (37, 53), which is a source of loss of material, creates opportunities for technical errors, and enhances inconsistency between replicates. As a remedy to these limitations, modifications have been made to make ChIP protocols shorter, simpler, and allow analysis of small cell samples (39, 54–57).

This introductory review addresses modifications of conventional ChIP assays, which have recently been introduced to simplify and accelerate the procedure and enable the analysis of DNA-bound proteins in small cell samples. Analytical tools that can be combined with ChIP to address the landscape of protein–DNA interactions are also presented.

3. ChIP Assays for Small Cell Numbers

A major drawback of ChIP has for a long time been the requirement for large cell numbers. This has been necessary to compensate for the loss of cells upon recovery after cross-linking, for the overall inefficiency of ChIP, and for the relative insensitivity of detection of ChIP-enriched DNA. The need for elevated cell numbers has hampered the application of ChIP to rare cell samples, such as cells from small tissue biopsies, rare stem cell populations, or cells from embryos. Several recent publications have addressed this issue and reported alterations of conventional ChIP protocols to make the technique applicable to smaller number of cells.

3.1. CChIP

The rationale behind the carrier ChIP, or CChIP, is that the immunoprecipitation of a small amount of chromatin prepared from few mammalian cells (100–1,000) is facilitated by the addition of carrier chromatin from *Drosophila* or any other species sufficiently evolutionarily distant from the species investigated (39). CChIP involves the mixing of cultured *Drosophila* cells with a small number of mammalian cells. Native chromatin fragments are prepared from purified nuclei by partial MNase digestion and immunoprecipitated using antibodies to modified histones. To compensate for the small amount of target DNA precipitated, the ChIP DNA is detected by radioactive PCR and phosphorimaging. Specificity of amplification is monitored for each ChIP by determination of the size of the DNA fragment produced (39).

CChIP has proven to be suitable for the analysis of 100-cell samples. A limitation, however, is that analysis of multiple histone modifications requires multiple aliquots of 100 cells which may or may not be identical. Furthermore, in its published form, CChIP is based on the NChIP procedure (37) and as such is not suited for precipitation of transcription factors. Nonetheless, there is no reason to believe that CChIP is not compatible with cross-linking, and thereby becomes more versatile. Despite these limitations, however, the benefit of CChIP for analyzing small cell samples is already clear.

Using CChIP, O'Neill et al. (39) have reported an analysis of active and repressive histone modifications on a handful of target loci in mouse inner cell mass and trophectoderm cells – the two cell types of the blastocyst. Application of CChIP to embryonic transcription factors in embryos and embryonic stem (ES) cells to unravel common and distinct target genes should enhance our understanding of the molecular basis of pluripotency.

3.2. Q^2ChIP

As an alternative to CChIP, a quick and quantitative (Q^2)ChIP protocol suitable for up to 1,000 histone ChIPs or up to 100 transcription factor ChIPs from as few as 100,000 cells has been developed in our laboratory (56). Q^2ChIP involves a chromatin preparation from a larger number of cells than CChIP, but includes chromatin dilution and aliquoting steps which allow for storage of many identical chromatin aliquots from a single preparation. Because Q^2ChIP involves a cross-linking step, chromatin samples are also suitable for immunoprecipitation of transcription factors or other non-histone DNA-bound proteins. Protein–DNA cross-linking in suspension in the presence of a histone deacetylase inhibitor, elimination of essentially all non-specific background chromatin through a tube-shift after washes of the ChIP material, and combination of cross-linking reversal, protein digestion, and DNA elution into a single 2-h step considerably shorten the procedure and enhance the ChIP efficiency (56). Suitability of Q^2ChIP to small amounts of chromatin has been attributed to the reduction of the number of steps in the procedure and increase in the ratio of antibody-to-target epitope, resulting in an enhanced signal-to-noise ratio. Q^2ChIP has been validated against the conventional ChIP assay from which it was derived (53). It has been used to illustrate changes in histone H3K4, K9, and K27 acetylation and methylation associated with differentiation of embryonal carcinoma cells on developmentally regulated promoters (56).

3.3. μChIP

With the aim of further reducing the number of cells used, we subsequently devised a micro (μ)ChIP protocol suitable for up to nine parallel ChIPs of modified histones and/or RNA polymerase II (RNAPII) from a single batch of 1,000 cells without carrier chromatin (57, 58). The assay can also be downscaled for monitoring the association of one protein with multiple genomic sites in as few as 100 cells and has been adapted for small ($\sim$1 mm^3) tissue biopsies. Modifications of μChIP for analysis of tumor biopsies have been reported recently (58). The assay was validated by assessing several post-translational modifications of histone H3 and binding of RNAPII in embryonal carcinoma cells and in human osteosarcoma biopsies, on developmentally regulated and tissue-specific genes (57).

In μChIP, chromatin is prepared from 1,000 cells and divided into nine aliquots (100-cell ChIP), of which eight can be dedicated to parallel ChIPs, including a negative control, and one serves as an input reference sample. When starting from 100 cells, only one ChIP is possible using the current protocol. Regardless of the starting cell number, the 100-cell ChIP enables the analysis of 3–4 genomic sites by duplicate qPCR without amplification of the ChIP DNA (57). We have since successfully amplified μChIP DNA using whole-genome DNA amplification kits and have been able to apply μChIP to microarrays (J.A. Dahl and P. Collas, unpublished data).

3.4. MicroChIP

Incidentally, at the time our μChIP assay was being evaluated (57), a miniaturized ChIP protocol for 10,000 cells also coincidentally called microChIP was published (54). From batches of 10,000 cells, the assay allows analysis of histone or RNAPII binding throughout the human genome using a ChIP-on-chip approach with high-density oligonucleotide arrays. This microChIP assay

(54) takes approximately 4 days, but presents the main advantage of being applicable to genome-wide studies rather than being restricted to a few genomic regions.

4. Accelerated ChIP Assays: Down to 1 Day

Conventional ChIP protocols are time consuming and limit the number of samples that can be analyzed in parallel. To address this issue, a fast ChIP assay has introduced two modifications which dramatically shorten the procedure (59, 60). First, incubation of antibodies with chromatin in an ultrasonic bath substantially increases the rate of antibody–protein binding, shortening the incubation time to 15 min. Second, in a traditional ChIP assay, elution of the ChIP complex, reversal of cross-linking, and proteinase K digestion of bound proteins require ~9 h, and DNA isolation by phenol:chloroform isoamylalcohol extraction and ethanol precipitation takes almost 1 day. Instead, fast ChIP uses a cation-chelating resin (Chelex-100)-based DNA isolation which reduces the total time for preparation of PCR-ready templates to 1 h (**Fig. 1.3**). We have also reported the shortening of cross-linking reversal, proteinase K digestion, and SDS elution steps into a single 2-h step without loss of ChIP efficiency or specificity (56). It is also possible to purify ChIP DNA with spin columns, but loss of DNA during the procedure limits their application to large ChIP assays.

Using the ChIP material directly as template in the PCR (on-bead PCR) has also been reported in yeast, with results comparable to PCR using purified DNA (61). The possibility of performing the PCR reaction directly on the immunoprecipitated material indicates that the formaldehyde cross-linking reversion step may be omitted, likely because the initial PCR heating step suffices to partially reverse the cross-link. Direct PCR, therefore, holds promises for speeding up the analysis of ChIP products.

Whether end-point or quantitative on-bead PCR can be performed seems, however, to depend on the nature of carrier beads used in ChIP. Direct on-bead PCR is successful with magnetic protein G beads (61) and with agarose-conjugated protein A beads (J.A. Dahl and P. Collas, unpublished data). Furthermore, we have shown that ChIP products precipitated by agarose beads can be directly analyzed by qPCR using SYBR® Green (J.A. Dahl and P. Collas, unpublished data). This is in contrast to magnetic beads which, because of their opacity, interfere with quantification of the SYBR® Green signal during the real-time PCR (**Fig. 1.3**). These observations argue, then, that

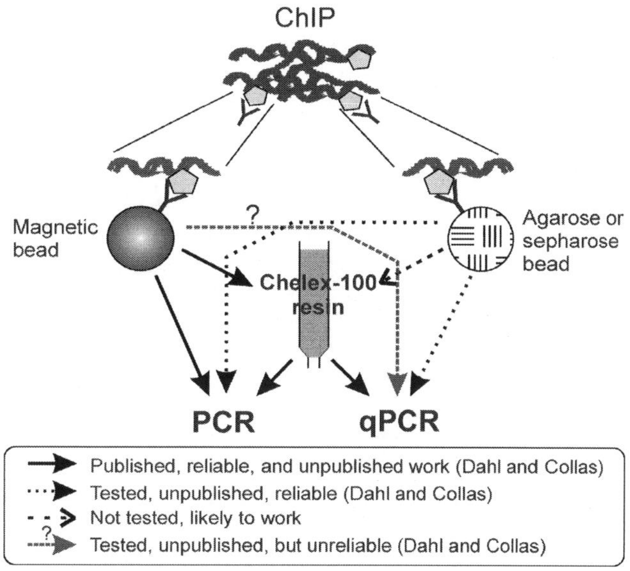

Fig. 1.3. Approaches to accelerate analysis of ChIP DNA fragments. ChIP DNA precipitated using magnetic or paramagnetic beads (*left*) can be directly used as template for PCR or processed through a Chelex-100 DNA purification resin prior to PCR. Chelex-100-purifed DNA can also potentially be used in quantitative (q)PCR assays. Use of DNA in the ChIP complex bound to magnetic bead directly as template for qPCR has proven to be unreliable in our hands (unpublished data), most likely due to the opacity of the magnetic beads which interferes with SYPBR® Green detection. Alternatively, ChIP complexes are precipitated with agarose or sepharose beads (*right*). These are compatible with direct PCR and direct pPCR (our unpublished data).

while direct qPCR is possible with ChIP templates bound to agarose, and most likely sepharose, beads and magnetic beads are currently incompatible with qPCR.

An alternative to Chelex-100 and on-bead PCR has recently been reported in the context of a higher-throughput ChIP assay than those reported till date (62) (**Section 5**). To enable rapid access of the ChIP DNA for PCR with minimal sample handling, the authors have replaced Chelex-100 with a high-pH Tris buffer containing EDTA. PCR-ready DNA recovery is identical to that of Chelex-100, with the advantage that it can be performed in a single tube or in wells without a need for centrifugation (62).

Thus the past 2 years have seen the emergence of creative and attractive variations on the classical ChIP assay, which have enabled a considerable reduction in time, greatly simplified the procedure, and made the ChIP compatible with the analysis of small cell numbers. Notably, the Q^2ChIP and μChIP assays also fit into the 1-day ChIP protocol category.

5. Enhancing Throughput with Matrix ChIP, a Microplate-Based Assay

To increase the throughput of ChIP and simplify the assay, a microplate-based ChIP assay, Matrix ChIP, was recently reported (62). Matrix ChIP takes advantage of antibodies immobilized with protein A coated into each well of a 96-well plate. Besides simplification of sample handling, one rationale for immobilizing antibodies is that they can be maintained in the correct orientation. Such specific orientation can enhance binding capacity to up to 10-fold compared to random-oriented antibodies (63). All steps, from immunoprecipitation to DNA purification, are done in the wells without sample transfers, enabling a potential for automation. As mentioned earlier, recovery of PCR-ready ChIP DNA from the surface-bound antibodies is permitted by the use of simple buffer that facilitates DNA extraction. In its current format, matrix ChIP enables 96 ChIP assays for histone and DNA-bound proteins, including transiently bound protein kinases, in a single day (62).

6. HAP-ChIP: Cleaning Up Nucleosomes for Enhanced Histone ChIP Efficacy

Many modified residues on histone tails serve as docking sites for transcription factors or chromatin-modifying enzymes. In a ChIP assay, binding of these proteins may sterically hinder access of antibodies to a fraction of histone epitopes, resulting in an underestimation of the amount of a given modified histone enriched at a specific locus. To overcome this limitation, a variation on ChIP has been introduced to remove chromatin-bound non-histone proteins prior to immunoprecipitation of nucleosomes (64). This assay takes advantage of high-affinity interaction of DNA with hydroxyapatite (HAP) to wash out chromatin-associated proteins before ChIP under native conditions (HAP-ChIP) (64).

HAP-ChIP consists primarily of five steps. They are purification of nuclei, fragmentation of chromatin with MNase, purification of nucleosomes by HAP chromatography, immunoprecipitation of the nucleosomes, and qPCR analysis of the precipitated DNA. Lysis of nuclei takes place in high concentration of NaCl and is immediately followed by chromatin fragmentation. High-salt lysis is believed to produce an even representation of both euchromatin and heterochromatin, which other NChIP protocols do not necessarily provide (regions of tightly packed heterochromatin are insensitive to MNase under lower salt concentrations). In addition, elution of nucleosomes from HAP occurs with up to 500 mM

$NaPO_4$ at pH 7.2 under low salt conditions. This preserves the interaction of DNA with core histones (histone octamers are eluted from DNA with 2 M NaCl). These procedures result in a preparation of polynucleosomes (1–3 nucleosomes per chromatin fragment), stripped of non-histone proteins (64). HAP-ChIP has been used in combination with qPCR; however, with a few modifications (64), it is speculated to be adaptable to ChIP-on-chip or ChIP-seq.

7. ChIP-on-Beads: Flow Cytometry Analysis of ChIP DNA

Quantitative determination of the amount of DNA associated with an immunoprecipitated protein is commonly done by qPCR (46, 56, 65). A recent protocol, however, calls for the capture of conventional PCR products on microbeads and flow cytometry analysis (66). A standard ChIP is performed, and the ChIP DNA is used as template for end-point PCR in which primers are tagged in their 5′ end with Fam (forward primer) and biotin (reverse primer). The Fam/biotin PCR products are captured and analyzed by flow cytometry. Importantly, labeling must occur in the linear phase of the PCR to ensure reliable quantification. The similarity between the data obtained by qPCR and flow cytometry has been shown for the enrichment of H4 and H3 epitopes on a specific locus in Jurkat cells (66).

The ChIP-on-beads assay has been proposed to be useful for quantitative assessments of ChIP products in a high-throughput manner (66). However, the complexity of the procedure makes it at present difficult to foresee the advantage of ChIP-on-beads over ChIP-qPCR or ChIP-on-chip approaches, especially as long as the qPCR analysis of ChIP products is necessary for evaluation of the linear phase of the PCR-mediated labeling step. Simplification of the ChIP DNA fragment labeling procedure would, however, make ChIP-on-beads amenable for assessing large number of samples for a limited number of genes.

8. Sequential ChIP: Analysis of Histone Modifications or Proteins Co-enriched on Single Chromosome Fragments

An important issue in deciphering the epigenetic code is whether two given histone modifications, transcription factors, or chromatin modifiers are co-enriched on the same locus. Notably, trimethylated H3K4 and H3K27 have been suggested to constitute a ‘bivalent mark’ on genes encoding transcriptional regulators in

ES cells (26, 27, 35) because both modifications could be co-precipitated from the same genomic fragment (27). Indeed, genome-wide approaches in cell types such as ES cells, fibroblasts, and T cells support a view of chromatin domains co-enriched in H3K4me3 and H3K27me3, albeit with distinct profiles and peaks (27, 33, 35, 45, 67). Based on these observations, one may conclude that H3K4me3 and K27me3 may be found on distinct genomic fragments (e.g., two alleles), on the same promoter but on distinct nucleosomes, or may co-exist in a subpopulation of nucleosomes. Similar questions apply to the co-occupancy of two transcription factors on a single locus.

To resolve these issues, a sequential ChIP assay has been developed, whereby one protein is immunoprecipitated from a chromatin sample and a second protein, presumed to be co-enriched on the same genomic fragment, is subsequently immunoprecipitated from chromatin eluted from the first ChIP (68, 69). Sequential ChIP has been used to demonstrate the existence of bivalent histone marks on a single genomic fragment (27). In that study, ES cell chromatin was first immunoprecipitated with antibodies against H3K27me3, and the ChIP chromatin was used for a second immunoprecipitation using antibodies against H3K4me3. Sequential immunoprecipitation, then, retains only chromatin which concomitantly carries both histone modifications. Sequential ChIP has also been used to show the co-occupancy of two or more transcription factors on a genomic site (43, 70–74). The sequential ChIP approach has been detailed and reviewed elsewhere (75, 76). The level of analysis of co-occupancy of two proteins on a locus can potentially be further refined using purified mono-nucleosomes as chromatin templates for ChIP.

9. Methods for Genome-Wide Mapping Protein Binding Sites on DNA

ChIP has for several years been limited to the analysis of predetermined candidate target sequences analyzed by PCR using specific primers. Recently, several strategies have been developed to enable application of ChIP to the discovery of novel target sites for transcriptional regulators and to map the positioning of post-translationally modified histones throughout the genome. These genome-wide approaches have immensely contributed to characterizing the chromatin landscape primarily in the context of pluripotency, differentiation, and disease.

9.1. ChIP-on-Chip

The advent of oligonucleotides microarrays has revolutionized analysis of gene expression and our understanding of transcription profiles. Subsequent development of genomic DNA microarrays (chips) has, when combined with ChIP assays, enabled the mapping of transcription factor binding sites (77, 78) and of histone modifications (79, 80) on large areas in the genome through an approach known as ChIP-on-chip. Despite its relatively recent introduction, ChIP-on-chip has been largely exploited to, for example, map c-myc binding sites in the genome (81, 82), elaborate Oct4, Nanog, and Sox2 transcriptional networks in ES cells (83), identify polycomb target genes (84, 85), or provide a histone modification landscape in T cells (67). Several reviews dedicated to ChIP-on-chip, its variations, and limitations have been published (86–88), thus we only provide here a brief account of the principle.

ChIP-on-chip differs from ChIP-PCR only in the method of analysis of the precipitated DNA (**Fig. 1.4**). ChIP DNA is eluted after cross-link reversal and the ends repaired with a DNA polymerase to generate blunt ends. A linker is applied to each DNA fragment to enable PCR amplification of all fragments. A fluorescent label (usually Cy5) is incorporated during PCR amplification. Similarly, an aliquot of input DNA is labeled with another fluorophore, usually Cy3. The two samples are mixed and hybridized onto a microarray containing oligonucleotide probes covering the whole genome or portions thereof, or probes tiling a region of interest. In this dual-color approach, binding of the

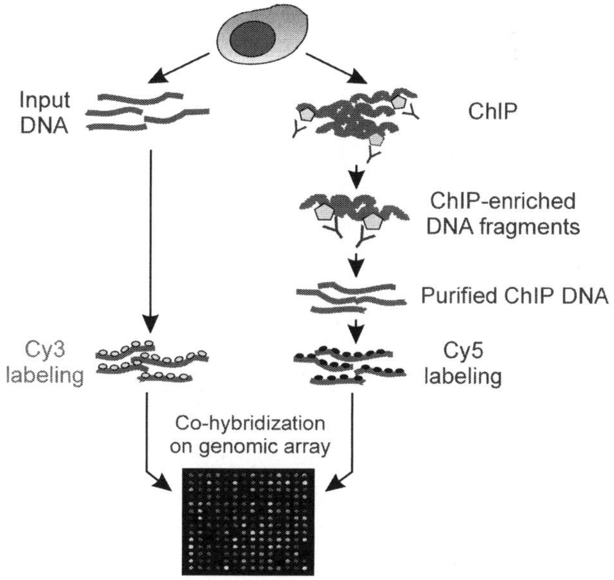

Fig. 1.4. ChIP-on-chip. A protein of interest is selectively immunoprecipitated by ChIP. The ChIP-enriched DNA is amplified by PCR and fluorescently labeled with, e.g., Cy5. An aliquot of purified input DNA is labeled with another fluorophore, e.g., Cy3. The two samples are mixed and hybridized onto a microarray containing genomic probes covering the whole or parts of the genome. Binding of the precipitated protein to a target site is inferred when intensity of the ChIP DNA significantly exceeds that of the input DNA on the array.

immunoprecipitated transcription factor to a genomic site is established when intensity of the ChIP DNA significantly exceeds that of the input DNA on the array. Statistical analysis software and evaluation by the investigator determine the significance of enrichment of the precipitated protein in the region examined. A detailed procedure for ChIP-on-chip has recently been published (42).

9.2. ChIP-Display

ChIP-on-chip is only as informative as the oligonucleotide microarrays onto which the ChIP-enriched DNA is hybridized. This limitation has stimulated the development of methods for unbiased determination of genomic sequences associated with a given protein. Novel transcription factor binding sites can be identified by cloning and sequencing DNA from the ChIP material (89, 90). However, the overwhelming excess of non-specifically precipitated DNA fragments makes ChIP-cloning unpractical. A ChIP-display strategy has been designed and applied to the identification of target genes occupied by the transcription factor Runx2 (91). ChIP-display concentrates DNA fragments containing each target sequence and scatters the remaining, non-specific DNA. Target sequences are concentrated by restriction digestion and electrophoresis, as fragments harboring the same target site acquire the same size. To scatter non-specific fragments, the total pool of restriction fragments is divided into families based on the identity of nucleotides at the ends of these fragments. Because all restriction fragments displaying each given target harbor the same nucleotide ends, they remain in the same family and the family detection signal on gel is not altered. Non-specific background fragments, however, are scattered into many families so that each family detection signal is markedly lower (91).

ChIP-display can unravel transcription factor targets in ChIPs that are enriched for targets by as little as 10- to 20-fold over bulk chromatin (91), and as such shows reasonable sensitivity. Gel electrophoresis display of ChIP DNA products allows a direct comparison of patterns (i.e., targets) obtained from different cell types (91). ChIP-display is also relatively insensitive to background which characterizes ChIP-PCR or ChIP-on-chip approaches. However, ChIP-display is not well suited for a comprehensive analysis of target sequences for proteins with a large number of genomic targets, such as SP1, GATA proteins, histone deacetylases, polycomb proteins, or RNAPII (91), or for the mapping of histone modifications. It is better suited for transcription factors with a more limited number of targets; nonetheless, it lacks quantification of the relative abundance of a transcription factor associated with a given locus, which is enabled by qPCR.

9.3. ChIP-PET

A second strategy developed in response to the limitations of the ChIP-on-chip assay is based on sequencing of portions of the precipitated target DNA. Indeed, with a limited survey of the cloned ChIP DNA fragment library, distinguishing between genuine binding sites and noise without additional molecular validation is challenging. In contrast, with a wide sampling of the ChIP DNA pool, sequencing approaches can identify DNA fragments enriched by ChIP.

ChIP-paired end ditag (PET) exploits the efficiency of sequencing short tags, rather than entire inserts, to enhance information content and increase accuracy of genome mapping (44). ChIP-PET relies on the recently reported gene identification signature strategy in which 5′ and 3′ signatures of full-length cDNAs are extracted into PETs that are concatenated (92, 93). The sequences are subsequently mapped to the genomic sequences to delineate the transcription boundaries of every gene. As in the gene identification signature strategy, a pair of signature sequences (tags) is extracted from the 5′ and 3′ ends of each ChIP DNA fragment, concatenated, and mapped to the genome.

The PET approach has recently been exploited to characterize ChIP DNA fragments in order to achieve unbiased, genome-wide mapping of transcription factor binding sites (43, 44). From a saturated sampling of over 500,000 PET sequences, Wei and colleagues characterized over 65,000 unique p53 ChIP DNA fragments and established overlapping PET clusters to define p53 target sequences with high specificity. The analysis also enabled a refinement of the consensus p53 binding motif and unraveled nearly 100 previously unidentified p53 target genes implicated in p53 function and tumorigenesis (44). In addition, a ChIP-PET analysis of binding sites for Oct4 and Nanog in mouse ES cells has laid out a transcription network regulated by these proteins in these cells (43).

9.4. ChIP-DSL

With the aim of detecting DNA target motifs with higher sensitivity and specificity than through the conventional ChIP-on-chip, a multiplex assay coined as ChIP-DSL was introduced. ChIP-DSL combines ChIP with a DNA ligation and selection (DSL) step (94). The assay involves the pre-determined use, or construction, of a microarray of 40-mer probes onto which the ChIP DNA fragments are to be hybridized. The reason is that a pair of 20-mer 'assay oligonucleotides' is synthesized corresponding to each half of each 40-mer. These 20-mer oligonucleotides are flanked on both sides by a universal primer binding site. These oligonucleotides are mixed into a 'DSL oligo pool'. Following conventional ChIP, the purified ChIP DNA is randomly biotinylated and annealed to the DSL oligo pool. The annealed fragments are captured on streptavidin-conjugated magnetic beads, allowing elimination of the non-annealed 20-mers (the noise). All selected DNA fragments are immobilized onto the beads and those paired by a specific DNA target motif are ligated. Thus, the correctly targeted oligonucleotides are specifically turned into templates for PCR amplification. One of the PCR primers is fluorescently labeled to enable detection after hybridization on the 40-mer probe microarray. The DSL procedure is also carried out for input DNA using PCR primers labeled with a different fluorophore.

ChIP-DSL has been used to identify a large number of novel binding sites for the estrogen receptor alpha in breast cancer-derived MCF7 cells (94). ChIP-DSL has also been used to demonstrate the widespread recruitment of the histone demethylase LSD1 on active promoters, including most estrogen receptor alpha gene targets (95).

ChIP-DSL is claimed to present advantages over ChIP-on-chip (94). Only unique signature motifs are targeted, alleviating potential interference with repetitive and related sequences upon hybridization. Sensitivity of the assay is increased due to the PCR amplification step. Amplification is presumably unbiased because DNA fragments bear the same pair of specific primer binding sites and have the same length.

9.5. ChIP-Sequencing

Perhaps the most powerful strategy to date for identifying protein binding sites across the genome consists of directly and quantitatively sequencing ChIP products. In an ultra high-throughput sequencing approach (35, 45, 96), DNA molecules are arrayed across a surface, locally amplified, subjected to successive cycles of single-base extension (using fluorescently labeled reversible terminators), and imaged after each cycle to determine the inserted base. The length of the reads is short (25–50 nucleotides using the Illumina/Solexa platform); however, millions of DNA fragments can be read simultaneously.

ChIP-Seq has been used to generate 'chromatin-state maps' for ES and lineage-committed cells (35). The data corroborate ChIP-on-chip data on the same cell types reported earlier by the same group (27), as well as results reported independently by ChIP-PET (33). Using the Illumina/Solexa 1G platform, binding sites for the transcription factor STAT1 in HeLa cells (96) and a profiling of histone methylation, histone-variant H2A.Z binding, RNAPII targeting, and CTCF binding throughout the genome (45) have also been reported. All results claim robust overlap between ChIP-seq, ChIP-on-chip, and ChIP-PCR data. Interestingly, the ChIP-seq data illustrate the potential for using ChIP for genome-wide annotation of novel promoters and primary transcripts, active transposable elements, imprinting control regions, and allele-specific transcription (35). Insights into the analysis of large data sets related to array and sequencing data have recently been published (97).

10. Controls, Controls, Controls. . .

In spite of improvements in the ChIP assays to reduce or eliminate background chromatin (56), background does exist and needs to be accounted for using appropriate negative controls. A survey of the ChIP literature reveals the use of various controls, the nature of which seems to mainly depend on the investigator. One classical negative control is the use of no antibodies (also often referred to as a 'bead-only' control). Bead-only controls for unspecific binding of chromatin fragments to the beads used to precipitate the complex of interest. Although it is useful, this control is not as stringent as using an irrelevant antibody, preferably of the same isotype as the experimental antibody, in a parallel chromatin preparation. Enhanced stringency of the control also implies the use of an irrelevant antibody against a nuclear protein. A third negative control consists of comparing, in the same ChIP, protein enrichment on a target sequence relative to enrichment on another, irrelevant, region. This control was performed in our laboratory to demonstrate the specificity of occupancy of Oct4 on the *NANOG* promoter in pluripotent carcinoma cells, whereas it was virtually absent from the *GAPDH* promoter (56). In ChIP-PCR experiments, the negative control may generate a PCR signal that can be used as a reference to express a ChIP-specific enrichment. In ChIP-on-chip or ChIP-cloning-sequencing (such as ChIP-PET) assays, the negative control IP is used in a subtractive approach at the level of array analysis. In addition to a negative control, some investigators use a positive control, such as a high-quality antibody against a well-characterized ubiquitous transcription factor (42). Positive control antibodies are particularly important when setting up new methodologies.

11. Additional Variations on the ChIP Assay

In addition to the techniques reviewed here, various strategies described in this issue have been developed to investigate other aspects of chromatin organization.

11.1. ChIP-BA

Profound understanding of the interplay between histone modifications, DNA methylation, transcription factor binding, and transcription requires the combination of multiple analyses from a single chromatin or DNA sample. The CG content of a transcription factor binding site, thus its methylation state, is likely to affect binding (98). In an attempt to relate transcription factor binding to DNA methylation, ChIP has been combined with bisulfite genomic sequencing analysis in a ChIP-BA approach (99). ChIP DNA fragments are processed for PCR analysis (or array hybridization) and for bisulfite conversion to determine the CpG methylation pattern. ChIP-BA has been used to determine the DNA methylation requirements for binding of a methyl-CpG binding protein (99). The method can also potentially be useful to unravel methylation patterns that are compatible, or incompatible, with the targeting of a specific protein to a genomic region (99). A potential problem with ChIP-BA, however, is noise that is directly turned into a sequence which may be irrelevant. Subtractive strategies may conceivably be utilized provided appropriate controls are performed.

11.2. DamID

An alternative to ChIPing a protein is to label the DNA close to the target site of the protein of interest (100). Labeling consists of a methylation tag put on by a DNA adenine methyltransferase (Dam) fused a DNA binding protein (the protein of interest) (DamID approach) (101). Binding of the transcription factor-Dam protein to DNA elicits adenine methylation in the vicinity of the protein target site. The methylated sites are detected by digestion with a methyl-specific restriction enzyme. The digestion products are purified, amplified using a methylation-specific PCR assay, labeled, and hybridized onto a microarray. DamID has been used to uncover binding sites for transcription factors, DNA methyltransferases, and heterochromatin proteins in *Drosophila*, *Arabidopsis*, and mammalian cells (102–106), and more recently, nuclear lamin B1 (107). Of interest, a comparison of the DamID and ChIP-on-chip approaches has been reported (86).

11.3. MeDIP

A variation of the ChIP assay has been introduced to determine genome-wide profiles of DNA methylation. Methyl-DNA immunoprecipitation (MeDIP) consists of the immunoprecipitation of methylated DNA fragments using an antibody to 5-methyl cytosine (108, 109). Detection of a gene of interest in the methylated DNA fraction can be done by polymerase chain reaction (PCR), hybridization to genomic (promoter or comparative genomic hybridization) arrays (109, 110), or high-throughout sequencing (111). Although MeDIP proves to be a potent method, a constraint of the assay is its limitation to regions with a CpG density of at least 2–3% (108). Below this density, even methylated CpGs will be regarded as unmethylated relative to genome average. MeDIP is being increasingly used to map methylation profiles (the 'methylome') of promoters in a variety of organisms and cell types (16, 109). Reviews on the MeDIP approach have been recently published (111–114).

12. Conclusions and Prospects

ChIP has become the technique of choice for mapping protein–DNA interactions in the cell, identifying novel binding sites for transcription factors or other chromatin-associated proteins, mapping the localization of post-translationally modified histones, and mapping the localization of histone variants. Altogether, these studies unravel an increasingly complex epigenetic landscape in the context of gene expression, definition of gene boundaries, development, differentiation, and disease. Significantly, the advent of ChIP assays for small cell samples has moved ChIP forward into the field of early embryo development and small cancer biopsies. The combination of small-scale ChIP assays with increasingly robust DNA amplification strategies using commercially available kits has also already enabled genome-wide and whole-genome analyses of histone modifications or RNAPII binding in small cell samples. ChIP-on-chip or ChIP-seq analyses of embryos are also much anticipated.

ChIP assays have also in recent years become significantly more user-friendly with fewer steps, reduced sample handling, and faster assays. Efforts have been put into simplifying the isolation of ChIP DNA, for a quicker analysis and minimizing sample loss. Some of the new developments also seem to be suited for automation. In an era which promotes the concept of personalized medicine in a context where epigenetics is increasingly linked to disease, automated whole-genome epigenetic analyses of individual patient material is likely to become a reality.

Acknowledgments

Our work is supported by grants from the Research Council of Norway and from the Norwegian Cancer Society. Thomas Küntziger is thanked for critical reading of the manuscript.

References

1. Antequera, F. (2003) Structure, function and evolution of CpG island promoters. *Cell Mol. Life Sci.* **60**, 1647–1658.
2. Kouzarides, T. (2007) Chromatin modifications and their function. *Cell* **128**, 693–705.
3. Mito, Y., Henikoff, J. G. and Henikoff, S. (2007) Histone replacement marks the boundaries of cis-regulatory domains. *Science* **315**, 1408–1411.
4. Mito, Y., Henikoff, J. G. and Henikoff, S. (2005) Genome-scale profiling of histone H3.3 replacement patterns. *Nat. Genet.* **37**, 1090–1097.
5. Viens, A., Mechold, U., Brouillard, F., Gilbert, C., Leclerc, P. and Ogryzko, V.

(2006) Analysis of human histone H2AZ deposition in vivo argues against its direct role in epigenetic templating mechanisms. *Mol. Cell Biol.* **26**, 5325–5335.

6. Li, B., Pattenden, S. G., Lee, D., Gutierrez, J., Chen, J., Seidel, C., Gerton, J. and Workman, J. L. (2005) Preferential occupancy of histone variant H2AZ at inactive promoters influences local histone modifications and chromatin remodeling. *Proc. Natl. Acad. Sci. U.S.A.* **102**, 18385–18390.
7. Li, A., Eirin-Lopez, J. M. and Ausio, J. (2005) H2AX: tailoring histone H2A for chromatin-dependent genomic integrity. *Biochem. Cell Biol.* **83**, 505–515.
8. Hoffman, A. R. and Hu, J. F. (2006) Directing DNA methylation to inhibit gene expression. *Cell Mol. Neurobiol.* **26**, 425–438.
9. Klose, R. J. and Bird, A. P. (2006) Genomic DNA methylation: the mark and its mediators. *Trends Biochem. Sci.* **31**, 89–97.
10. Morgan, H. D., Santos, F., Green, K., Dean, W. and Reik, W. (2005) Epigenetic reprogramming in mammals. *Hum. Mol. Genet.* **14**, R47–R58.
11. Young, L. E. and Beaujean, N. (2004) DNA methylation in the preimplantation embryo: the differing stories of the mouse and sheep. *Anim. Reprod. Sci.* **82**, 61–78.
12. Razin, A. and Shemer, R. (1995) DNA methylation in early development. *Hum. Mol. Genet.* **4**, 1751–1755.
13. Hellman, A. and Chess, A. (2007) Gene body-specific methylation on the active X chromosome. *Science* **315**, 1141–1143.
14. Tremblay, K. D., Saam, J. R., Ingram, R. S., Tilghman, S. M. and Bartolomei, M. S. (1995) A paternal-specific methylation imprint marks the alleles of the mouse H19 gene. *Nat. Genet.* **9**, 407–413.
15. Reik, W., Collick, A., Norris, M. L., Barton, S. C. and Surani, M. A. (1987) Genomic imprinting determines methylation of parental alleles in transgenic mice. *Nature* **328**, 248–251.
16. Weber, M., Hellmann, I., Stadler, M. B., Ramos, L., Paabo, S., Rebhan, M. and Schubeler, D. (2007) Distribution, silencing potential and evolutionary impact of promoter DNA methylation in the human genome. *Nat. Genet.* **39**, 457–466.
17. Fouse, S. D., Shen, Y., Pellegrini, M., Cole, S., Meissner, A., Van, N. L., Jaenisch, R. and Fan, G. (2008) Promoter CpG methylation contributes to ES cell gene regulation in parallel with Oct4/Nanog, PcG complex, and histone H3 K4/K27 trimethylation. *Cell Stem Cell* **2**, 160–169.
18. Jenuwein, T. and Allis, C. D. (2001) Translating the histone code. *Science* **293**, 1074–1080.
19. Mellor, J. (2006) It takes a PHD to read the histone code. *Cell* **126**, 22–24.
20. Cosgrove, M. S. and Wolberger, C. (2005) How does the histone code work? *Biochem. Cell Biol.* **83**, 468–476.
21. Lachner, M., O'Carroll, D., Rea, S., Mechtler, K. and Jenuwein, T. (2001) Methylation of histone H3 lysine 9 creates a binding site for HP1 proteins. *Nature* **410**, 116–120.
22. Cao, R., Wang, L., Wang, H., Xia, L., Erdjument-Bromage, H., Tempst, P., Jones, R. S. and Zhang, Y. (2002) Role of histone H3 lysine 27 methylation in Polycomb-group silencing. *Science* **298**, 1039–1043.
23. Cao, R. and Zhang, Y. (2004) The functions of E(Z)/EZH2-mediated methylation of lysine 27 in histone H3. *Curr. Opin. Genet. Dev.* **14**, 155–164.
24. Pasini, D., Bracken, A. P., Jensen, M. R., Lazzerini, D. E. and Helin, K. (2004) Suz12 is essential for mouse development and for EZH2 histone methyltransferase activity. *EMBO J.* **23**, 4061–4071.
25. Lachner, M. and Jenuwein, T. (2002) The many faces of histone lysine methylation. *Curr. Opin. Cell Biol.* **14**, 286–298.
26. Azuara, V., Perry, P., Sauer, S., Spivakov, M., Jorgensen, H. F., John, R. M., Gouti, M., Casanova, M., Warnes, G., Merkenschlager, M. and Fisher, A. G. (2006) Chromatin signatures of pluripotent cell lines. *Nat. Cell Biol.* **8**, 532–538.
27. Bernstein, B. E., Mikkelsen, T. S., Xie, X., Kamal, M., Huebert, D. J., Cuff, J., Fry, B., Meissner, A., Wernig, M., Plath, K., Jaenisch, R., Wagschal, A., Feil, R., Schreiber, S. L. and Lander, E. S. (2006) A bivalent chromatin structure marks key developmental genes in embryonic stem cells. *Cell* **125**, 315–326.
28. Kingston, R. E. and Narlikar, G. J. (1999) ATP-dependent remodeling and acetylation as regulators of chromatin fluidity. *Genes Dev.* **13**, 2339–2352.
29. Pray-Grant, M. G., Daniel, J. A., Schieltz, D., Yates, J. R., III and Grant, P. A. (2005) Chd1 chromodomain links histone H3 methylation with SAGA- and SLIK-dependent acetylation. *Nature* **433**, 434–438.

30. Struhl, K. (1998) Histone acetylation and transcriptional regulatory mechanisms. *Genes Dev.* **12**, 599–606.

31. Santos-Rosa, H., Schneider, R., Bannister, A. J., Sherriff, J., Bernstein, B. E., Emre, N. C., Schreiber, S. L., Mellor, J. and Kouzarides, T. (2002) Active genes are tri-methylated at K4 of histone H3. *Nature* **419**, 407–411.

32. Schubeler, D., MacAlpine, D. M., Scalzo, D., Wirbelauer, C., Kooperberg, C., van Leeuwen, F., Gottschling, D. E., O'Neill, L. P., Turner, B. M., Delrow, J., Bell, S. P. and Groudine, M. (2004) The histone modification pattern of active genes revealed through genome-wide chromatin analysis of a higher eukaryote. *Genes Dev.* **18**, 1263–1271.

33. Zhao, X. D., Han, X., Chew, J. L., Liu, J., Chiu, K. P., Choo, A., Orlov, Y. L., Sung, W. K., Shahab, A., Kuznetsov, V. A., Bourque, G., Oh, S., Ruan, Y., Ng, H. H. and Wei, C. L. (2007) Whole-genome mapping of histone H3 Lys4 and 27 trimethylations reveals distinct genomic compartments in human embryonic stem cells. *Cell Stem Cell* **1**, 286–298.

34. Ruthenburg, A. J., Allis, C. D. and Wysocka, J. (2007) Methylation of lysine 4 on histone H3: intricacy of writing and reading a single epigenetic mark. *Mol. Cell* **25**, 15–30.

35. Mikkelsen, T. S., Ku, M., Jaffe, D. B., Issac, B., Lieberman, E., Giannoukos, G., Alvarez, P., Brockman, W., Kim, T. K., Koche, R. P., Lee, W., Mendenhall, E., O'Donovan, A., Presser, A., Russ, C., Xie, X., Meissner, A., Wernig, M., Jaenisch, R., Nusbaum, C., Lander, E. S. and Bernstein, B. E. (2007) Genome-wide maps of chromatin state in pluripotent and lineage-committed cells. *Nature* **448**, 553–560.

36. O'Neill, L. P. and Turner, B. M. (1995) Histone H4 acetylation distinguishes coding regions of the human genome from heterochromatin in a differentiation-dependent but transcription-independent manner. *EMBO J.* **14**, 3946–3957.

37. O'Neill, L. P. and Turner, B. M. (1996) Immunoprecipitation of chromatin. *Methods Enzymol.* **274**, 189–197.

38. Zeng, P. Y., Vakoc, C. R., Chen, Z. C., Blobel, G. A. and Berger, S. L. (2006) In vivo dual cross-linking for identification of indirect DNA-associated proteins by chromatin immunoprecipitation. *Biotechniques* **41**, 694, 696, 698.

39. O'Neill, L. P., Vermilyea, M. D. and Turner, B. M. (2006) Epigenetic characterization of the early embryo with a chromatin immunoprecipitation protocol applicable to small cell populations. *Nat. Genet.* **38**, 835–841.

40. Hanlon, S. E. and Lieb, J. D. (2004) Progress and challenges in profiling the dynamics of chromatin and transcription factor binding with DNA microarrays. *Curr. Opin. Genet. Dev.* **14**, 697–705.

41. Sikder, D. and Kodadek, T. (2005) Genomic studies of transcription factor–DNA interactions. *Curr. Opin. Chem. Biol.* **9**, 38–45.

42. Lee, T. I., Johnstone, S. E. and Young, R. A. (2006) Chromatin immunoprecipitation and microarray-based analysis of protein location. *Nat. Protoc.* **1**, 729–748.

43. Loh, Y. H., Wu, Q., Chew, J. L., Vega, V. B., Zhang, W., Chen, X., Bourque, G., George, J., Leong, B., Liu, J., Wong, K. Y., Sung, K. W., Lee, C. W., Zhao, X. D., Chiu, K. P., Lipovich, L., Kuznetsov, V. A., Robson, P., Stanton, L. W., Wei, C. L., Ruan, Y., Lim, B. and Ng, H. H. (2006) The Oct4 and Nanog transcription network regulates pluripotency in mouse embryonic stem cells. *Nat. Genet.* **38**, 431–440.

44. Wei, C. L., Wu, Q., Vega, V. B., Chiu, K. P., Ng, P., Zhang, T., Shahab, A., Yong, H. C., Fu, Y., Weng, Z., Liu, J., Zhao, X. D., Chew, J. L., Lee, Y. L., Kuznetsov, V. A., Sung, W. K., Miller, L. D., Lim, B., Liu, E. T., Yu, Q., Ng, H. H. and Ruan, Y. (2006) A global map of p53 transcription-factor binding sites in the human genome. *Cell* **124**, 207–219.

45. Barski, A., Cuddapah, S., Cui, K., Roh, T. Y., Schones, D. E., Wang, Z., Wei, G., Chepelev, I. and Zhao, K. (2007) High-resolution profiling of histone methylations in the human genome. *Cell* **129**, 823–837.

46. Kuo, M. H. and Allis, C. D. (1999) In vivo cross-linking and immunoprecipitation for studying dynamic protein:DNA associations in a chromatin environment. *Methods* **19**, 425–433.

47. Solomon, M. J., Larsen, P. L. and Varshavsky, A. (1988) Mapping protein–DNA interactions in vivo with formaldehyde: evidence that histone H4 is retained on a highly transcribed gene. *Cell* **53**, 937–947.

48. Dedon, P. C., Soults, J. A., Allis, C. D. and Gorovsky, M. A. (1991) Formaldehyde cross-linking and immunoprecipitation demonstrate developmental changes in H1 association with transcriptionally active genes. *Mol. Cell Biol.* **11**, 1729–1733.

49. Madisen, L., Krumm, A., Hebbes, T. R. and Groudine, M. (1998) The immunoglobulin heavy chain locus control region increases histone acetylation along linked c-myc genes. *Mol. Cell Biol.* **18**, 6281–6292.
50. Hebbes, T. R., Clayton, A. L., Thorne, A. W. and Crane-Robinson, C. (1994) Core histone hyperacetylation co-maps with generalized DNase I sensitivity in the chicken β-globin chromosomal domain. *EMBO J.* **13**, 1823–1830.
51. Hebbes, T. R., Thorne, A. W., Clayton, A. L. and Crane-Robinson, C. (1992) Histone acetylation and globin gene switching. *Nucleic Acids Res.* **20**, 1017–1022.
52. Hebbes, T. R., Thorne, A. W. and Crane-Robinson, C. (1988) A direct link between core histone acetylation and transcriptionally active chromatin. *EMBO J.* **7**, 1395–1402.
53. Spencer, V. A., Sun, J. M., Li, L. and Davie, J. R. (2003) Chromatin immunoprecipitation: a tool for studying histone acetylation and transcription factor binding. *Methods* **31**, 67–75.
54. Acevedo, L. G., Iniguez, A. L., Holster, H. L., Zhang, X., Green, R. and Farnham, P. J. (2007) Genome-scale ChIP-chip analysis using 10,000 human cells. *Biotechniques* **43**, 791–797.
55. Attema, J. L., Papathanasiou, P., Forsberg, E. C., Xu, J., Smale, S. T. and Weissman, I. L. (2007) Epigenetic characterization of hematopoietic stem cell differentiation using miniChIP and bisulfite sequencing analysis. *Proc. Natl. Acad. Sci. U.S.A.* **104**, 12371–12376.
56. Dahl, J. A. and Collas, P. (2007) Q^2ChIP, a quick and quantitative chromatin immunoprecipitation assay unravels epigenetic dynamics of developmentally regulated genes in human carcinoma cells. *Stem Cells* **25**, 1037–1046.
57. Dahl, J. A. and Collas, P. (2008) MicroChIP – A rapid micro chromatin immunoprecipitation assay for small cell samples and biopsies. *Nucleic Acids Res.* **36**, e15.
58. Dahl, J. A. and Collas, P. (2008) A rapid micro chromatin immunoprecipitation assay (μChIP). *Nat. Protoc.* **3**, 1032–1045.
59. Nelson, J. D., Denisenko, O. and Bomsztyk, K. (2006) Protocol for the fast chromatin immunoprecipitation (ChIP) method. *Nat. Protoc.* **1**, 179–185.
60. Nelson, J. D., Denisenko, O., Sova, P. and Bomsztyk, K. (2006) Fast chromatin immunoprecipitation assay. *Nucleic Acids Res.* **34**, e2.
61. Kohzaki, H. and Murakami, Y. (2007) Faster and easier chromatin immunoprecipitation assay with high sensitivity. *Proteomics* **7**, 10–14.
62. Flanagin, S., Nelson, J. D., Castner, D. G., Denisenko, O. and Bomsztyk, K. (2008) Microplate-based chromatin immunoprecipitation method, Matrix ChIP: a platform to study signaling of complex genomic events. *Nucleic Acids Res.* **36**, e17.
63. Peluso, P., Wilson, D. S., Do, D., Tran, H., Venkatasubbaiah, M., Quincy, D., Heidecker, B., Poindexter, K., Tolani, N., Phelan, M., Witte, K., Jung, L. S., Wagner, P. and Nock, S. (2003) Optimizing antibody immobilization strategies for the construction of protein microarrays. *Anal. Biochem.* **312**, 113–124.
64. Brand, M., Rampalli, S., Chaturvedi, C. P. and Dilworth, F. J. (2008) Analysis of epigenetic modifications of chromatin at specific gene loci by native chromatin immunoprecipitation of nucleosomes isolated using hydroxyapatite chromatography. *Nat. Protoc.* **3**, 398–409.
65. Das, P. M., Ramachandran, K., vanWert, J. and Singal, R. (2004) Chromatin immunoprecipitation assay. *Biotechniques* **37**, 961–969.
66. Szekvolgyi, L., Balint, B. L., Imre, L., Goda, K., Szabo, M., Nagy, L. and Szabo, G. (2006) Chip-on-beads: flow-cytometric evaluation of chromatin immunoprecipitation. *Cytometry* **69**, 1086–1091.
67. Roh, T. Y., Cuddapah, S., Cui, K. and Zhao, K. (2006) The genomic landscape of histone modifications in human T cells. *Proc. Natl. Acad. Sci. U.S.A.* **103**, 15782–15787.
68. Chaya, D., Hayamizu, T., Bustin, M. and Zaret, K. S. (2001) Transcription factor FoxA (HNF3) on a nucleosome at an enhancer complex in liver chromatin. *J. Biol. Chem.* **276**, 44385–44389.
69. Metivier, R., Penot, G., Hubner, M. R., Reid, G., Brand, H., Kos, M. and Gannon, F. (2003) Estrogen receptor-alpha directs ordered, cyclical, and combinatorial recruitment of cofactors on a natural target promoter. *Cell* **115**, 751–763.
70. Wilkinson, D. S., Tsai, W. W., Schumacher, M. A. and Barton, M. C. (2008) Chromatin-bound p53 anchors activated Smads and the mSin3A corepressor to confer transforming-growth-factor-beta-mediated transcription repression. *Mol. Cell Biol.* **28**, 1988–1998.
71. Jalvy, S., Renault, M. A., Lam Shang, L. L., Belloc, I., Reynaud, A., Gadeau, A. P. and

Desgranges, C. (2007) CREB mediates UTP-directed arterial smooth muscle cell migration and expression of the chemotactic protein osteopontin via its interaction with activator protein-1 sites. *Circ. Res.* **100**, 1292–1299.

72. Brunelli, L., Cieslik, K. A., Alcorn, J. L., Vatta, M. and Baldini, A. (2007) Peroxisome proliferator-activated receptor-delta upregulates 14-3-3 epsilon in human endothelial cells via CCAAT/enhancer binding protein-beta. *Circ. Res.* **100**, e59–e71.
73. Kobrossy, L., Rastegar, M. and Featherstone, M. (2006) Interplay between chromatin and trans-acting factors regulating the Hoxd4 promoter during neural differentiation. *J. Biol. Chem.* **281**, 25926–25939.
74. Cui, R., Nguyen, T. T., Taube, J. H., Stratton, S. A., Feuerman, M. H. and Barton, M. C. (2005) Family members p53 and p73 act together in chromatin modification and direct repression of alpha-fetoprotein transcription. *J. Biol. Chem.* **280**, 39152–39160.
75. Geisberg, J. V. and Struhl, K. (2004) Quantitative sequential chromatin immunoprecipitation, a method for analyzing co-occupancy of proteins at genomic regions in vivo. *Nucleic Acids Res.* **32**, e151.
76. Chaya, D. and Zaret, K. S. (2004) Sequential chromatin immunoprecipitation from animal tissues. *Methods Enzymol.* **376**, 361–372.
77. Iyer, V. R., Horak, C. E., Scafe, C. S., Botstein, D., Snyder, M. and Brown, P. O. (2001) Genomic binding sites of the yeast cell-cycle transcription factors SBF and MBF. *Nature* **409**, 533–538.
78. Ren, B., Robert, F., Wyrick, J. J., Aparicio, O., Jennings, E. G., Simon, I., Zeitlinger, J., Schreiber, J., Hannett, N., Kanin, E., Volkert, T. L., Wilson, C. J., Bell, S. P. and Young, R. A. (2000) Genome-wide location and function of DNA binding proteins. *Science* **290**, 2306–2309.
79. Clark, D. J. and Shen, C. H. (2006) Mapping histone modifications by nucleosome immunoprecipitation. *Methods Enzymol.* **410**, 416–430.
80. Loden, M. and van Steensel, B. (2005) Whole-genome views of chromatin structure. *Chromosome. Res.* **13**, 289–298.
81. Dang, C. V., O'Donnell, K. A., Zeller, K. I., Nguyen, T., Osthus, R. C. and Li, F. (2006) The c-Myc target gene network. *Semin. Cancer Biol.* **16**, 253–264.
82. Lee, L. A. and Dang, C. V. (2006) Myc target transcriptomes. *Curr. Top. Microbiol. Immunol.* **302**, 145–167.
83. Boyer, L. A., Lee, T. I., Cole, M. F., Johnstone, S. E., Levine, S. S., Zucker, J. P., Guenther, M. G., Kumar, R. M., Murray, H. L., Jenner, R. G., Gifford, D. K., Melton, D. A., Jaenisch, R. and Young, R. A. (2005) Core transcriptional regulatory circuitry in human embryonic stem cells. *Cell* **122**, 947–956.
84. Boyer, L. A., Plath, K., Zeitlinger, J., Brambrink, T., Medeiros, L. A., Lee, T. I., Levine, S. S., Wernig, M., Tajonar, A., Ray, M. K., Bell, G. W., Otte, A. P., Vidal, M., Gifford, D. K., Young, R. A. and Jaenisch, R. (2006) Polycomb complexes repress developmental regulators in murine embryonic stem cells. *Nature* **441**, 349–353.
85. Lee, T. I., Jenner, R. G., Boyer, L. A., Guenther, M. G., Levine, S. S., Kumar, R. M., Chevalier, B., Johnstone, S. E., Cole, M. F., Isono, K., Koseki, H., Fuchikami, T., Abe, K., Murray, H. L., Zucker, J. P., Yuan, B., Bell, G. W., Herbolsheimer, E., Hannett, N. M., Sun, K., Odom, D. T., Otte, A. P., Volkert, T. L., Bartel, D. P., Melton, D. A., Gifford, D. K., Jaenisch, R. and Young, R. A. (2006) Control of developmental regulators by Polycomb in human embryonic stem cells. *Cell* **125**, 301–313.
86. Bulyk, M. L. (2006) DNA microarray technologies for measuring protein–DNA interactions. *Curr. Opin. Biotechnol.* **17**, 422–430.
87. Hudson, M. E. and Snyder, M. (2006) High-throughput methods of regulatory element discovery. *Biotechniques* **41**, 673, 675, 677.
88. Elnitski, L., Jin, V. X., Farnham, P. J. and Jones, S. J. (2006) Locating mammalian transcription factor binding sites: a survey of computational and experimental techniques. *Genome Res.* **16**, 1455–1464.
89. Weinmann, A. S., Bartley, S. M., Zhang, T., Zhang, M. Q. and Farnham, P. J. (2001) Use of chromatin immunoprecipitation to clone novel E2F target promoters. *Mol. Cell Biol.* **21**, 6820–6832.
90. Hug, B. A., Ahmed, N., Robbins, J. A. and Lazar, M. A. (2004) A chromatin immunoprecipitation screen reveals protein kinase Cbeta as a direct RUNX1 target gene. *J. Biol. Chem.* **279**, 825–830.
91. Barski, A. and Frenkel, B. (2004) ChIP Display: novel method for identification of

genomic targets of transcription factors. *Nucleic Acids Res.* **32**, e104.

92. Ng, P., Wei, C. L., Sung, W. K., Chiu, K. P., Lipovich, L., Ang, C. C., Gupta, S., Shahab, A., Ridwan, A., Wong, C. H., Liu, E. T. and Ruan, Y. (2005) Gene identification signature (GIS) analysis for transcriptome characterization and genome annotation. *Nat. Methods* **2**, 105–111.
93. Ng, P., Tan, J. J., Ooi, H. S., Lee, Y. L., Chiu, K. P., Fullwood, M. J., Srinivasan, K. G., Perbost, C., Du, L., Sung, W. K., Wei, C. L. and Ruan, Y. (2006) Multiplex sequencing of paired-end ditags (MS-PET): a strategy for the ultra-high-throughput analysis of transcriptomes and genomes. *Nucleic Acids Res.* **34**, e84.
94. Kwon, Y. S., Garcia-Bassets, I., Hutt, K. R., Cheng, C. S., Jin, M., Liu, D., Benner, C., Wang, D., Ye, Z., Bibikova, M., Fan, J. B., Duan, L., Glass, C. K., Rosenfeld, M. G. and Fu, X. D. (2007) Sensitive ChIP-DSL technology reveals an extensive estrogen receptor alpha-binding program on human gene promoters. *Proc. Natl. Acad. Sci. U.S.A.* **104**, 4852–4857.
95. Garcia-Bassets, I., Kwon, Y. S., Telese, F., Prefontaine, G. G., Hutt, K. R., Cheng, C. S., Ju, B. G., Ohgi, K. A., Wang, J., Escoubet-Lozach, L., Rose, D. W., Glass, C. K., Fu, X. D. and Rosenfeld, M. G. (2007) Histone methylation-dependent mechanisms impose ligand dependency for gene activation by nuclear receptors. *Cell* **128**, 505–518.
96. Robertson, G., Hirst, M., Bainbridge, M., Bilenky, M., Zhao, Y., Zeng, T., Euskirchen, G., Bernier, B., Varhol, R., Delaney, A., Thiessen, N., Griffith, O. L., He, A., Marra, M., Snyder, M. and Jones, S. (2007) Genome-wide profiles of STAT1 DNA association using chromatin immunoprecipitation and massively parallel sequencing. *Nat. Methods* **4**, 651–657.
97. Bock, C. and Lengauer, T. (2008) Computational epigenetics. *Bioinformatics* **24**, 1–10.
98. Ji, H., Vokes, S. A. and Wong, W. H. (2006) A comparative analysis of genome-wide chromatin immunoprecipitation data for mammalian transcription factors. *Nucleic Acids Res.* **34**, e146.
99. Matarazzo, M. R., Lembo, F., Angrisano, T., Ballestar, E., Ferraro, M., Pero, R., De Bonis, M. L., Bruni, C. B., Esteller, M., D'Esposito, M. and Chiariotti, L. (2004) In vivo analysis of DNA methylation patterns recognized by specific proteins: coupling CHIP and bisulfite analysis. *Biotechniques* **37**, 666–669.
100. Orian, A. (2006) Chromatin profiling, DamID and the emerging landscape of gene expression. *Curr. Opin. Genet. Dev.* **16**, 157–164.
101. van Steensel, B., Delrow, J. and Henikoff, S. (2001) Chromatin profiling using targeted DNA adenine methyltransferase. *Nat. Genet.* **27**, 304–308.
102. Orian, A., van Steensel, B., Delrow, J., Bussemaker, H. J., Li, L., Sawado, T., Williams, E., Loo, L. W., Cowley, S. M., Yost, C., Pierce, S., Edgar, B. A., Parkhurst, S. M. and Eisenman, R. N. (2003) Genomic binding by the Drosophila Myc, Max, Mad/Mnt transcription factor network. *Genes Dev.* **17**, 1101–1114.
103. Tompa, R., McCallum, C. M., Delrow, J., Henikoff, J. G., van Steensel, B. and Henikoff, S. (2002) Genome-wide profiling of DNA methylation reveals transposon targets of CHROMOMETHYLASE3. *Curr. Biol.* **12**, 65–68.
104. de Wit, E., Greil, F. and van Steensel, B. (2005) Genome-wide HP1 binding in Drosophila: developmental plasticity and genomic targeting signals. *Genome Res.* **15**, 1265–1273.
105. van Steensel, B. (2005) Mapping of genetic and epigenetic regulatory networks using microarrays. *Nat. Genet.* **37**, S18–S24.
106. Vogel, M. J., Guelen, L., de Wit, E., Peric-Hupkes, D., Loden, M., Talhout, W., Feenstra, M., Abbas, B., Classen, A. K. and van Steensel, B. (2006) Human heterochromatin proteins form large domains containing KRAB-ZNF genes. *Genome Res.* **16**, 1493–1504.
107. Guelen, L., Pagie, L., Brasset, E., Meuleman, W., Faza, M. B., Talhout, W., Eussen, B. H., de Klein, A., Wessels, L., de Latt, W. and van Steensel, B. (2008) Domain organization of human chromosomes revealed by mapping of nuclear lamina interactions. *Nature* **453**, 948–951.
108. Keshet, I., Schlesinger, Y., Farkash, S., Rand, E., Hecht, M., Segal, E., Pikarski, E., Young, R. A., Niveleau, A., Cedar, H. and Simon, I. (2006) Evidence for an instructive mechanism of de novo methylation in cancer cells. *Nat. Genet.* **38**, 149–153.
109. Weber, M., Davies, J. J., Wittig, D., Oakeley, E. J., Haase, M., Lam, W. L. and Schubeler, D. (2005) Chromosome-wide and promoter-specific analyses identify sites of differential DNA methylation in normal and transformed human cells. *Nat. Genet.* **37**, 853–862.

110. Wilson, I. M., Davies, J. J., Weber, M., Brown, C. J., Alvarez, C. E., Macaulay, C., Schubeler, D. and Lam, W. L. (2006) Epigenomics: mapping the methylome. *Cell Cycle* **5**, 155–158.

111. Taylor, K. H., Kramer, R. S., Davis, J. W., Guo, J., Duff, D. J., Xu, D., Caldwell, C. W. and Shi, H. (2007) Ultradeep bisulfite sequencing analysis of DNA methylation patterns in multiple gene promoters by 454 sequencing. *Cancer Res.* **67**, 8511–8518.

112. Yazaki, J., Gregory, B. D. and Ecker, J. R. (2007) Mapping the genome landscape using tiling array technology. *Curr. Opin. Plant Biol.* **10**, 534–542.

113. Zilberman, D. and Henikoff, S. (2007) Genome-wide analysis of DNA methylation patterns. *Development* **134**, 3959–3965.

114. Jacinto, F. V., Ballestar, E. and Esteller, M. (2008) Methyl-DNA immunoprecipitation (MeDIP): hunting down the DNA methylome. *Biotechniques* **44**, 35, 37, 39.

Chapter 2

Characterization and Quality Control of Antibodies Used in ChIP Assays

Géraldine Goens, Dorina Rusu, Laurent Bultot, Jean-Jacques Goval, and Juana Magdalena

Abstract

We present here the very robust characterization and quality control (QC) process that we have established for our polyclonal antibodies, which are mainly directed against targets relevant to the epigenetics field such as modified histones, modifying enzymes, and chromatin-interacting proteins. The final purpose of the characterization and QC is to label antibodies as chromatin immunoprecipitation (ChIP) grade. Indeed, the ChIP method is extensively used in epigenetics to study gene regulation and relies on the use of antibodies to select the protein of interest and then precipitate and identify the DNA associated to it. We have optimized in-house all protocols and reagents needed from the first to the last step of antibody characterization. First, following immunizations, the rabbit crude serum is tested for immune response. Whether or not the antibody is specific is determined in further characterizations. Then, only specific antibodies are tested in ChIP using an optimized method which is ideal for antibody screening. Once QC is established for one antibody, it is used to similarly characterize each antibody batch in order to supply researchers in a reproducible manner with validated antibodies. All in all, this demonstrates that we develop epigenetics research tools based on everyday's researcher's needs by providing batch-specific fully characterized ChIP-grade antibodies.

Key words: Antibody, characterization, quality control, specificity, chromatin immunoprecipitation.

1. Introduction

Extensive characterization of antibodies represents a real need in the research field (1–4). A defined quality control (QC) for each antibody is also of extreme importance due to possible batch to batch variation. Moreover, the use of chromatin immunoprecipitation (ChIP) grade antibodies is essential in any experiment aiming to study protein–DNA interactions. We present here the

Philippe Collas (ed.), *Chromatin Immunoprecipitation Assays,* Methods in Molecular Biology 567,
DOI 10.1007/978-1-60327-414-2_2, © Humana Press, a part of Springer Science+Business Media, LLC 2009

very robust QC that we established for our antibodies, which are mainly directed against modified histones, chromatin-modifying enzymes, and chromatin-interacting proteins. As ChIP is a major method used to study gene regulation in epigenetics looking at in vivo protein–DNA interactions, we focus all our attention on the ChIP-grade antibody characterization using a variety of methods in sequentially ordered steps.

We first design immunogenic peptides in order to produce polyclonal antibodies directed against the target of interest, using preferentially N-terminal and C-terminal regions, including modifications when applicable (e.g., modified histone tails). We use the Lasergene software by DNASTAR (Madison, USA) to design the peptides, looking for large regions with high hydrophobicity. Selected regions are then checked for high surface probability and high antigenicity index. We choose peptides of about 16 amino acids or less, avoiding alpha helices and repeats. We use maximum two peptides for one target per rabbit immunization. Two rabbits are injected with the chosen peptide(s), which is conjugated to KLH to boost the antibody production (*see* **Note 2**). Although both crude sera and purified antibodies are submitted to a similar step-by-step QC, we focus first on crude sera before undertaking any purification (**Fig. 2.1A,B**).

Step 1: As soon as bleeds are available, the crude serum is first tested in ELISA side by side with the pre-immune for immune response assessment. Antibodies from crude sera can be affinity purified and tested in ELISA before and after purification (*see* **Section 3.1**). Step 2: Whether or not the antibody is specific is determined during further characterization. We use dot blot and western blot when applicable (note that at this stage, it is also possible to perform immunoprecipitations (IP) and immunofluorescence (IF) assays) (*see* **Sections 3.2** and **3.3**). Step 3: Then specific antibodies are tested in ChIP (*see* **Section 3.4**). Our LowCell# ChIP kit which was proven to give reproducible results is used for antibody screening. It is an ideal tool as it also ensures the use of low amount of reagents per reaction (not only cells but also antibodies, inhibitors, and buffers), the number of steps is greatly reduced, and handling is much easier. Finally, it is crucial to characterize each antibody batch with an established QC to validate the antibodies in a reproducible manner. An example of QC strategy is given below and results are shown for an antibody raised against one modified histone (H3K9me3; **Figs. 2.1**, **2.2**, and **2.3**).

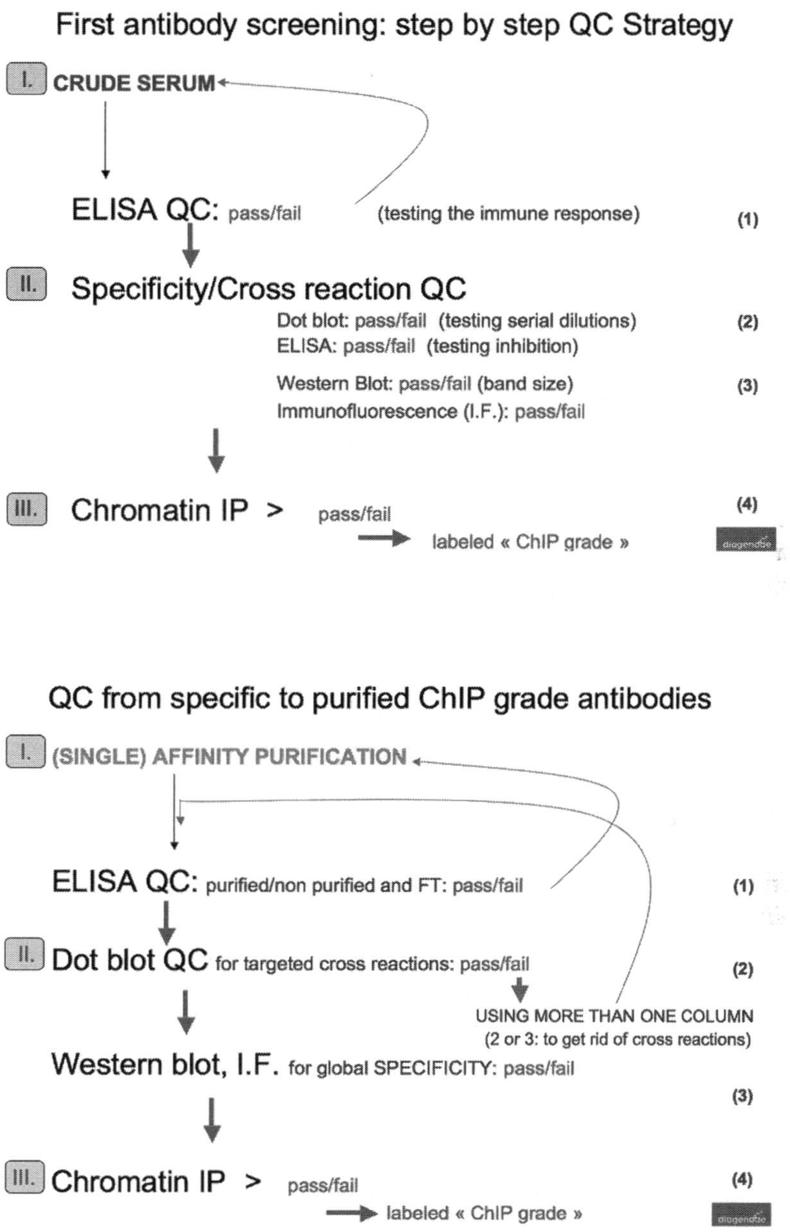

Fig. 2.1. In order to validate our antibodies, we go step by step. We start by designing immunogenic peptides. After immunization, we analyze the rabbit crude sera for immune response and antibody specificity (**A**), this corresponds to Steps 1 and 2, respectively. Affinity purified antibodies undergo a similar QC (**B**). The specific antibodies undergo ChIP validation (Step 3). Once ChIP graded, other tests can be performed such as ChIP-chip and ChIP-seq to validate the antibody further.

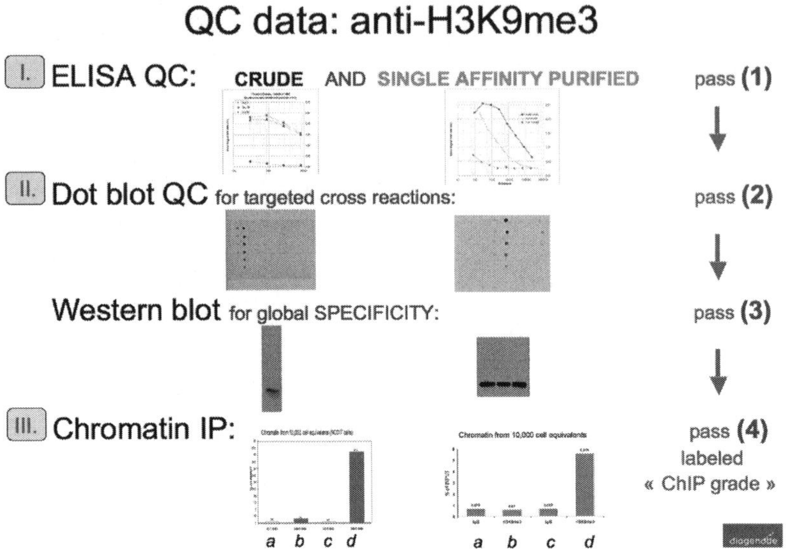

Fig. 2.2. Here is an example of antibody QC data obtained with crude serum and corresponding affinity purified antibody. In order to validate our antibodies directed against histone H3K9me3 (cat. no. CS-056 and pAb-056, Diagenode), we go step by step from Steps 1–3. After immunization, we analyze the rabbit crude sera for immune response and antibody specificity. Affinity purified antibodies undergo similar QC. In ELISA, pre-immune and flow-through after purification do not give any signal, while crude sera and purified antibody fraction give a positive signal. In dot blot, the specificity was tested using mono-, di-, and tri-methylated peptide sequences containing H3K9me1,2,3, H4K20me1,me2,me3, H3K27me1,me2,me3, and H3K36me1,me2,me3 (from *right* to *left*). Specific antibodies are then further validated in ChIP. We use the pre-serum as negative IP control (*a, c*), which gave no ChIP signal. We also use one positive (*a, b*) and one negative (*c, d*) PCR target for each antibody being tested. A good ChIP signal was obtained with the positive PCR target used after the IP of chromatin with the antibody anti-H3K9me3 (*d*). Note that optimal dilutions of both crude serum and purified antibodies to be used in each assay are determined by titration. Here, in dot blot, western blot, and ChIP, the dilutions are 1:10,000, 1:750, and 1:5,000, respectively, using crude serum and 1:1,000, 1:500, and 1 μg/IP using purified antibody.

2. Materials

2.1. ELISA

1. Strips F8 BioOne, High Binding (cat. no. 762.061, Greiner) or 96-well microplate BioOne, High Binding (cat. no. 655.081, Greiner).
2. Peptide solution stock: 10 mg/mL in 50 m*M* Tris-HCl, pH 8.0.
3. Coating buffer: 0.1 *M* carbonate–bicarbonate, pH 9.6.
4. Phosphate-buffered saline with Tween (PBS-T): 0.05% Tween 20 (v/v) in PBS.

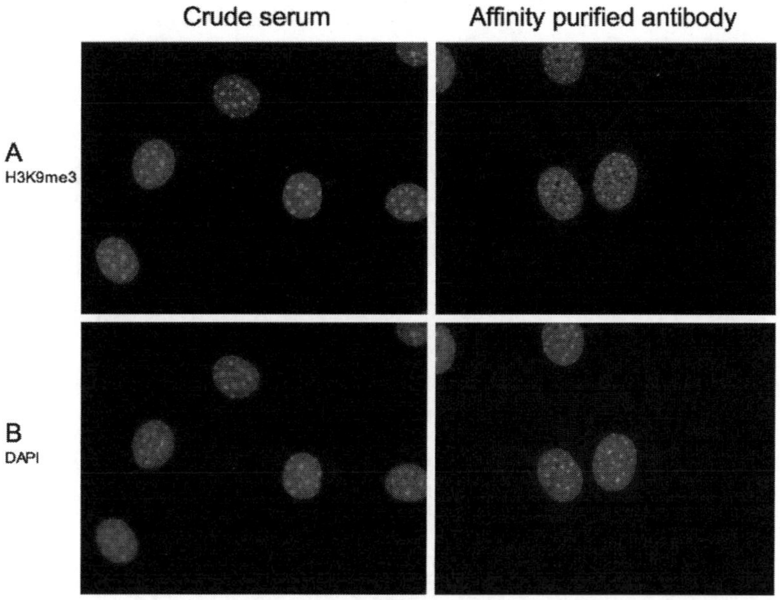

Fig. 2.3. Here is an additional antibody characterization that has been done to show antibody-specific binding to its target localized in the nucleus. Indirect immunofluorescence results obtained with the antibody anti-H3K9me3 (cat. no. CS-056 and pAb-056, Diagenode). NIH3T3 cells are stained with the antibody directed against H3K9me3 and with DAPI. Cells are formaldehyde fixed, permeabilized with Triton X-100, and then blocked with BSA containing PBS. (**A**) Cells are immunofluorescently labeled with the rabbit polyclonal antibody anti-H3K9me3 (both pAb-056 and CS-056 at dilution 1:200, and incubated for 1 h at RT) followed by goat anti-rabbit antibody conjugated to FITC. (**B**) Nuclei were DAPI stained to label specifically the DNA. Note the presence of more intense spots showing the distribution pattern of this modified histone. Both, antibody and DAPI staining are restricted to the nucleus.

5. ELISA saturation buffer: 3% (w/v) BSA in PBS-T.
6. ELISA dilution buffer: 1% (w/v) BSA in PBS-T.
7. ProClin 300 (Sigma).
8. ELISA wash buffer: 0.01% (v/v) ProClin 300 in PBS-T.
9. Primary antibody (rabbit pre-immune and crude serum, reference antibody).
10. HRP-conjugated goat antibody anti-rabbit IgG.
11. ELISA substrate: tetramethyl benzidine (TMB).
12. ELISA stop solution: 1 M H_2SO_4(3X, 3 M Rectapur).
13. Keyhole limpet hemocyanin (KLH).
14. Microplate reader.

2.2. Dot Blot

1. Plate of 96-wells F (None or low binding; cat. no. 269620, NUNC).

2. PVDF membrane (cat. no. 162-0176, Bio-Rad).
3. Aliquot of 10 μL of 5 m*M* peptide stock.
4. Dot blot buffers: 50 m*M* Tris-HCl, pH 7.5 (sterile, filtered on 0.2 μm); TBS (20 m*M* Tris-HCl, pH 7.5, 150 m*M* NaCl); TBS-T (0.05% (v/v) Tween in TBS); DB blocking buffer (2% (w/v) BSA in TBS-T); primary antibody dilution buffer (3% (w/v) BSA in TBS-T); secondary antibody dilution buffer (5% (w/v) low-fat dry milk in TBS-T).
5. Ponceau S solution (cat. no. 33427, Serva) used to double-check spotting efficiency.
6. Primary antibody (crude serum, pre-immune, and/or purified antibody).
7. Secondary antibody (enhanced chemiluminescent (ECL) peroxidase labeled anti-rabbit; cat. no. NA934VS, GE Healthcare).
8. Peroxidase substrate (ECL Advance western blotting detection kit; cat. no. RPN2135, GE Healthcare).
9. Imaging system (chemiluminescence detection; Kodak Gel Logic 1500).

2.3. Western Blot

2.3.1. Histone Extraction

1. Cultured cells and tissue-culture grade PBS (cat. no. 14190, Gibco).
2. Triton extraction buffer (TEB; 0.5% (v/v) Triton X-100 in PBS).
3. Protease inhibitors (100X solution; P8340, Sigma). Add to TEB before use.
4. 0.2 *N* HCl.
5. Bradford reagent (Sigma).

2.3.2. Nuclear Extract Preparation

1. Cultured cells, tissue-culture grade PBS, and tissue culture scrapers.
2. Igepal-CA630. Prepare 10% (w/v) Igepal-CA630 in H_2O.
3. Protease inhibitors (100X solution; P8340, Sigma) to be added to buffers before use.
4. Membrane lysis buffer: 10 m*M* Hepes, pH 8.0, 1.5 m*M* $MgCl_2$,10 m*M* KCl, 1 m*M* DTT.
5. Nuclear envelope lysis buffer: 20 m*M* Hepes, pH 8.0, 1.5 m*M* $MgCl_2$, 25% glycerol, 420 m*M* NaCl, 0.2 m*M* EDTA, 1 m*M* DTT.

2.3.3. Western Blot

1. SDS-PAGE: 40% acrylamide solution and 2% bis solution; SDS-PAGE migration buffer (10X) and broad range protein molecular weight marker.

2. Laemmli sample buffer (2X); beta-mercaptoethanol. Complete Laemmli sample buffer (Laemmli sample buffer supplemented with 5% beta-mercapto-ethanol).
3. Transfer buffers: 10X Tris/glycine/SDS, 10X Tris/glycine, and methanol for transfer from gel to PVDF 0.45 μm membrane. Mini-trans blot electrophoretic transfer cell.
4. Ponceau S solution used to double-check transfer efficiency.
5. Western blot buffers: TBS (20 m*M* Tris-HCl, pH 7.4, 150 m*M* NaCl); TBS-T (0.05% (v/v) Tween in TBS); WB buffer (5% (w/v) low-fat dry milk in TBS-T).
6. Streptavidin peroxidase polymer used to detect the molecular weight marker.
7. Primary antibody (crude serum, pre-immune, and/or purified antibody).
8. Secondary antibody (enhanced chemiluminescent (ECL) peroxidase labeled anti-rabbit).
9. ECL Western blotting detection kit.
10. Gel imaging system.

2.4. Chromatin Immunoprecipitation

1. Cultured cells. Trypsin–EDTA. Formaldehyde to fix the cells. Consider that you need chromatin from 10,000 cells per IP.
2. Bioruptor™ (cat. no. UCD-200, Diagenode) to prepare sheared chromatin.
3. LowCell# ChIP kit (cat. no. kch-maglow-016, Diagenode).
4. Magnetic rack (cat. no. kch-816-001, Diagenode).
5. Antibody (crude serum, pre-immune, and/or purified antibody).
6. Phosphate buffered saline (PBS).
7. 1 *M* sodium butyrate (1 *M* NaBu).
8. RNAse/DNase-free 1.5 mL tubes.
9. Galaxy Mini with strip rotor.
10. Centrifuge for 1.5 mL tubes (4°C), rotating wheel (4°C), and vortex.
11. Floating rack for 1.5 mL tubes, tube claps, and boiling water.
12. Thermomixer (50 and 65°C).
13. Quantitative PCR facilities and reagents.

3. Methods

3.1. ELISA Test (Characterization and QC Step 1)

As soon as bleeds are available, the crude serum is first tested for immune response assessment (*see* **Note 1**). The crude and pre-immune sera are tested side by side in ELISA, the pre-immune being used as negative control (*see* **Note 2**). The peptide that has been used for the rabbit immunizations to raise the antibody is coated on a 96-well plate. When recognition of the peptide by the crude serum is observed, the serum can be tested further (**Fig. 2.2**). Antibodies from crude sera can also be affinity purified and the ELISA method is then used again to compare purified antibody fractions to initial crude serum (*see* **Note 7**; **Fig. 2.2**). We also include in our standardized protocol the use of a reference antibody to enable comparison of data from experiment to experiment.

1. Prepare solutions of peptide and KLH in carbonate buffer (100 ng/100 μL).
2. Coat the wells in duplicate; adding 100 ng/100 μL of peptide per well in two eight-well strips (total of 16 wells); and 100 ng/100 μL of KLH per well in another two eight-well strips. In addition, in another eight-well strip, add the ELISA negative control (carbonate buffer alone, in four wells) and the ELISA peptide positive control (peptide to be tested with the serum of reference or ELISA antibody positive control, in four wells) (*see* **Note 3**).
3. Incubate overnight at 4°C.
4. Wash twice with ELISA wash buffer and dry on paper.
5. Add ELISA saturation buffer (125 μL/well) and incubate 1 h at room temperature.
6. Wash once with ELISA wash buffer and dry on paper.
7. Each antibody sample is tested in duplicate (in two eight-well strips) and at different dilutions (in eight wells, from wells A to G). Using ELISA dilution buffer, prepare serial dilutions of both crude serum and pre-immune (for two strips each, prepare 250 μL of each diluted antibody sample). From wells A to G, dilutions are: 1:50; 1:150; 1:450; 1:1,350; 1:4,050; 1:12,150, and 1:36,450.
8. Add 100 μL of each dilution of antibody in duplicate and incubate overnight at 4°C. Add 100 μL of ELISA dilution buffer in two wells as negative ELISA control. Also, add 100 μL positive antibody control in another two wells.
9. Wash four times with deionized water and dry on paper.
10. Dilute the HRP-conjugated goat antibody anti-rabbit IgG (1:100,000) in ELISA dilution buffer.

11. Add 100 μL/well of diluted HRP-conjugated secondary antibody.
12. Incubate 1.5 h at room temperature.
13. Wash four times with deionized water and dry on paper.
14. Add 100 μL/well of TMB.
15. Incubate 30 min at room temperature.
16. Add 100 μL/well of ELISA stop solution.
17. Read at 450 nm on an ELISA plate reader.

3.2. Dot Blot (Characterization and QC Step 2)

When a crude serum is shown by ELISA to recognize the peptide used for immunizations, the crude serum undergoes more characterization. The antibody cross-reactivity can be tested against several other peptides. The crude serum directed against a determined histone modification is tested against other histone modifications by dot blot using corresponding peptides spotted on membrane (e.g., for H3K9me3, other histone modifications include mono- and di-methylation of the same lysine and mono-, di-, and tri-methylation of other lysines in the same and other histones). It should be pointed out that some lysines are contained in very similar amino acid sequence, e.g., H3K9 and H3K27 (2). Dot blot analysis to check antibody specificity was reported earlier (2–3). Based on previous publications and optimization in-house of our protocols, we set up a standardized QC method. A good antibody only recognizes the peptide used to generate the immune response (**Fig. 2.2**).

3.2.1. Peptide Dilution in an Eight-Well Strip

1. Prepare aliquots of 10 μL of 5 m*M* peptide stock.
2. Add 990 μL of 50 m*M* Tris-HCl, pH 7.5, in each 10 μL of 5 m*M* peptide stock to obtain a peptide concentrated at 50 pmol/μL (peptide solutions can be aliquoted and kept at −20°C).
3. In a 96-well plate, per eight-well strip, add 50 m*M* Tris-HCl, pH 7.5, in the successive wells as follows: B (100 μL), C (100 μL), D (240 μL), E (240 μL), F (240 μL), and G (100 μL). Prepare one row per peptide.
4. Add 200 μL of each diluted peptide in the well A of one row.
5. Make a serial dilution of the peptide as follows: transfer 100 μL of peptide solution from well A to B, then from well B to C. Transfer 60 μL from well C to D, D to E, and then E to F.

3.2.2. Spotting Membranes with Serially Diluted Peptides

1. Cut a PVDF membrane (size: *X* cm/7 cm – *X* is the number of peptides to spot).
2. Wet a filter paper with TBS.
3. Re-hydrate the PVDF membrane 1 min in methanol 100%.

4. Wash the membrane 5 min in deionized or distilled water.
5. Wash the membrane 10 s in TBS.
6. Place the wet filter paper on a plane surface.
7. Place PVDF membrane on the wet filter paper.
8. Spot each dilution of peptide on the membrane in drops of 2 μL: spot 1 (100 p*M*), spot 2 (50 p*M*), spot 3 (25 p*M*), spot 4 (5 p*M*), spot 5 (1 p*M*), spot 6 (0.2 p*M*), and spot 6 (50 m*M* Tris) (*see* **Note 4**).

3.2.3. Incubation of Peptide Blots with the Antibody and Detection

1. Incubate the membrane 1 h at room temperature with DB blocking buffer.
2. Incubate the membrane overnight at 4°C with primary antibody diluted in primary antibody dilution buffer (*see* **Note 5**).
3. Wash the membrane four times 10 min with TBS-T.
4. Incubate the membrane 1 h at room temperature with the secondary antibody at the dilution 1:20,000 in secondary antibody dilution buffer.
5. Wash the membrane four times 10 min with TBS-T.
6. Proceed to detection by incubating the membrane with the appropriate substrate as follows. Prepare the detection solution (ECL Advance western blotting detection kit: 750 μL solution A and 750 μL solution B gives 1.5 mL for two membranes).
7. Incubate the membrane for 5 min with the freshly prepared detection solution.
8. Visualize and take pictures.

3.3. Western Blot (Characterization and QC Step 3)

When a crude serum is shown by ELISA to recognize the peptide used for immunizations, the crude serum undergoes more characterization. For antibody directed against modified histones, the antibody cross-reactivity is assessed by dot blot as described above and by western blot using histone extracts. For any other antibody, cross-reactivity and specificity are observed by using the western blot method on nuclear extracts. Use cellular extracts, if the protein target is strictly cytoplasmic. By western blot, the specific antibody detects a single protein band of expected molecular weight (**Fig. 2.2**). At this stage, it is also possible to perform immunoprecipitations (IP) and immunofluorescence (IF) assays to determine further antibody specificity (*see* **Fig. 2.3**).

3.3.1. Histone Extraction

1. Harvest 10 million cells and wash with PBS.
2. Resuspend cells in TEB freshly supplemented with protease inhibitors at a cell density of 10 million cells per milliliter.
3. Lyse cells on ice for 10 min with gentle stirring.
4. Centrifuge at 380*g* for 10 min at 4°C. Discard the supernatant.

5. Wash the cells in half the volume of TEB (0.5 mL) and centrifuge as above.
6. Resuspend the pellet in 250 μL of 0.2 *N* HCl (cell density of 4×10^6 cells per mL).
7. Incubate 1 h at 4°C. This is the step for acid extraction of histones.
8. Centrifuge samples at 380*g* for 10 min at 4°C.
9. Removed the supernatant and determine protein concentration using the Bradford assay reagents. The protein content should be about 500–1000 μg of protein/mL.
10. Dilute histones to 0.5 μg/μL, add equal volume of 2X complete Laemmli sample buffer (final histone concentration: 0.25 μg/μL) and store at –20°C or directly load on gel.

3.3.2. Nuclear Extract Preparation

1. Aspirate culture medium and wash the cells twice with ice-cold PBS.
2. Add 3 mL ice-cold PBS and scrape cells gently into a 15 mL tube.
3. Centrifuge for 5 min at 380*g* at 4°C.
4. Carefully aspirate supernatant and keep the pellet.
5. For each culture flask resuspend the pellet in 4 mL of ice-cold membrane lysis buffer freshly supplemented with protease inhibitors.
6. Transfer to 1.5 mL tubes, and add 1 mL of cell suspension per tube.
7. Incubate 15 min on ice to allow cells to swell.
8. Add 100 μL of 10% Igepal-CA630 per tube and vortex for 10 s.
9. Centrifuge 2–3 min at 14,000*g*.
10. Carefully aspirate supernatant; this is the cytoplasmic fraction. Keep the pellet.
11. Resuspend the pellet in 200 μL ice-cold nuclear envelope lysis buffer freshly supplemented with protease inhibitors.
12. Vortex 30 s; rotate vigorously for 30 min at 4°C.
13. Centrifuge 15 min at maximum speed. Keep the supernatants, and transfer all the supernatant fractions (see Step 6 above) in a single new ice-cold tube.
14. Aliquot and store at –80°C until use. Do not freeze/thaw.
15. Determine protein concentration using the Bradford reagent.

3.3.3. Immunoblotting

1. Perform an SDS-PAGE electrophoresis using a standard protocol and instructions from the buffer supplier (Bio-Rad). For histone analysis, we use a stacking gel of 4% acrylamide and

running gel of 12% polyacrylamide. For nuclear extracts analysis, use a running gel according to the expected molecular weight of the target of interest.

2. Cut and treat a piece of PVDF membrane as described in **Section 3.2.2** (Steps 1–5).
3. Transfer the proteins from gel to membrane using a standard protocol and instructions from the buffer and apparatus supplier (*see* **Note 4**). For histone analysis, the transfer buffer 1X contains 0.05% SDS and 20% methanol final (mix both transfer buffers: Tris/glycine/SDS and Tris/glycine, and add methanol). For nuclear extracts analysis, the Tris/glycine transfer buffer is used supplemented with 20% methanol. Transfer for 1 h at 100 V.
4. Incubate the PVDF membrane in WB buffer during 1 h at room temperature.
5. Dilute the primary antibody in WB buffer (for dilutions to use and titration to perform, *see* **Note 5**).
6. Add the diluted antibody solution to the membrane and incubate overnight at 4°C.
7. Wash the membrane in WB buffer 5 min twice, and wash 10 min twice again.
8. Add to WB buffer both secondary antibody (1:50,000) and S-HRP (1:3,000).
9. Incubate the membrane 1 h in WB buffer supplemented with secondary antibody and S-HRP.
10. Wash the membrane in TBS-T 5 min twice, and wash 10 min twice again.
11. Prepare the detection solution (ECL Advance western blotting detection kit: 750 μL solution A and 750 μL solution B gives you 1.5 mL for two membranes).
12. Incubate the membrane 5 min with the freshly prepared detection solution.
13. Visualize and take pictures.

3.4. Chromatin Immunoprecipitation (Characterization and QC Step 4)

Antibodies that have been shown to be specific in the previous two steps of the characterization and QC are submitted to the ChIP assay. It is essential to use a standardized protocol such as in a kit, including IP controls and to analyze by qPCR the isolated DNA looking at two loci: a locus that is positive for the target of interest and a locus that is negative (**Fig. 2.2**). We use the LowCell# ChIP method, which enables the immunoprecipitation of up to 14 parallel histone ChIP reactions plus two controls from a total of as few as 16,000 cells in a day's work. It requires low amounts of reagents per assay, the number of steps is reduced, and rapid handling at constant temperature is enabled by the use of our magnetic rack (*see* **Note 8**). It is, therefore, a valuable tool for antibody characterization and QC, which involves titration and batch testing.

3.4.1. Binding Antibodies to Magnetic Beads

1. Wash twice the protein A-coated paramagnetic beads with ice-cold Buffer A as follows: add Buffer A, suspend the beads in Buffer A, then centrifuge for 5 min at 1,300 rpm, discard the supernatant, and keep the bead pellet. 10 μL of beads are needed per IP. Scale accordingly.
2. After washing, resuspend in Buffer A to the same bead concentration as the stock.
3. Aliquot 90 μL of Buffer A per 200-μL PCR tube for each magnetic ChIP reaction.
4. Add 10 μL of pre-washed protein A-beads per IP tube.
5. Add the specific antibody and positive and negative control antibodies (*see* **Note 6**).
6. Incubate the IP tubes at 40 rpm on a rotating wheel for at least 2 h at 4°C.

3.4.2. Cell Collection and Protein–DNA Cross-Linking

1. Immediately before harvesting the cells, add inhibitors, if needed, to the culture medium and mix gently.
2. Prepare cells as described in section "4. Kit Assay Protocol". Count the cells.
3. Label new 1.5 mL tube(s), add PBS (including inhibitors) to a final volume of 500 μL after cells have been added. Transfer cells and wash the pipette tip thoroughly.
4. Add 13.5 μL of 36.6% formaldehyde per 500 μL sample.
5. Mix by gentle vortexing. Incubate for 8 min at room temperature to allow fixation to take place.
6. Add 57 μL of 1.25 *M* glycine to the sample.
7. Mix by gentle vortexing. Incubate for 5 min at room temperature. This is to stop the fixation.
8. Centrifuge at 470*g* for 10 min at 4°C.
9. Aspirate the supernatant. Take care not to remove the cells. Aspirate slowly and leave approximately 30 μL of the solution behind.

3.4.3. Cell Lysis and Bioruptor™ Chromatin Shearing

1. Wash the cross-linked cells twice with 0.5 mL ice-cold PBS (adding NaBu and/or any other inhibitor of choice). Add the solution, gently vortex, and centrifuge at 470*g* (in a swing-out rotor with soft settings for deceleration) for 10 min at 4°C.

2. After the last wash, aspirate the supernatant. Leave about 10–20 μL behind.
3. Add protease inhibitor and NaBu to Buffer B at RT. This is the complete Buffer B. Keep the buffer at room temperature until use, discard what is not used during the day.
4. Add 130 μL of complete Buffer B (RT) to the cells. Vortex until resuspension. Incubate for 5 min on ice.
5. Submit the samples to sonication to shear the chromatin using the Bioruptor™ for 12 cycles of 30s "ON", 30s "OFF" each.
6. Use the sheared chromatin directly in ChIP.
7. Add 5 μL of protease inhibitor mix per milliliter of Buffer A. Add NaBu (20 m*M* final) or any other inhibitor to Buffer A.
8. Add 870 μL complete Buffer A to the 130 μL of sheared chromatin.
9. Once shearing efficiency is assessed, proceed to the next step.

3.4.4. Magnetic Immunoprecipitation

1. Briefly spin the 0.2 mL tubes containing the antibody-coated beads to bring down liquid caught in the lid.
2. Place tubes in the ice-cold magnetic rack (cooled by placing on ice), and wait for 1 min.
3. Discard the supernatant. Keep the pellet of antibody-coated beads.
4. Use 100 μL of diluted sheared chromatin per IP. Transfer 100 μL to each 0.2 mL IP tube. Keep 100 μL as input sample; keep at 4°C.
5. Close the tube caps and remove tubes from magnetic field.
6. Incubate under constant rotation on a rotator at 40 rpm for 2 h up to overnight, at 4°C.

3.4.5. Washes After Magnetic Immunoprecipitation

1. Wash three times using 100 μL ice-cold Buffer A. Each wash is done as follows: add buffer, invert to mix, incubate for 4 min at 4°C on a rotating wheel (40 rpm), spin, place in the magnetic rack, wait for 1 min, and discard the buffer. Keep the captured beads.
2. Wash one time with Buffer C. Add 100 μl Buffer C to the beads and invert to mix. Incubate on a rotating wheel for 4 min at 4°C (40 rpm). Spin and place the clean tubes now containing the beads in the magnetic rack after washing; capture the beads and remove Buffer C.

3.4.6. DNA Purification

1. Put water to boil.
2. Label new 1.5 mL tubes. IP# 1–8 (one row), IP# 1–8, and # 9–16 (two rows).

3. Add 100 μL of DNA purifying slurry directly to the washed beads and remove the eight-tube strips from the Diagenode Magnetic Rack. Mix by pipetting up and down and transfer the ChIP sample (beads and DNA purifying slurry) into the newly labeled 1.5 mL tubes.
4. Add 100 μL of input sample in a clean 1.5 mL tube and supplement with 100 μL of DNA purifying slurry.
5. Invert the tubes and lock the tubes with tube claps.
6. Incubate the samples for 10 min in boiling water.
7. Turn on the thermomixer, set the temperature at 55°C.
8. Thaw the provided proteinase K on ice.
9. Label new 1.5 mL tubes. IP#1–8 (one row), IP# 1–8, and # 9–16 (two rows).
10. Take the tubes out of the boiling water (boiling water will be needed again) and spin briefly to bring down the liquid caught in the lid.
11. Take off the tube claps. Wait for samples to cool down.
12. Add 1 μL of proteinase K to each sample and 2 μL for the input.
13. Vortex for 2s at medium power.
14. Shake all the samples for 30 min at 1,000 rpm in the thermomixer at 55°C.
15. Spin briefly and lock the tubes with tube claps before boiling.
16. Incubate the samples for 10 min in boiling water.
17. Centrifuge 1 min at 14,000*g* at 4°C.
18. Do not disturb the pellet. Transfer 50 μL of the IP sample supernatant and 150 μL of the input sample supernatant to the newly labeled 1.5 mL tubes. The pellet of the input sample can be discarded.
19. Add 100 μL of water to the pellet of the IP sample.
20. Vortex for 10 s at medium power.
21. Centrifuge for 1 min at 14,000*g* at 4°C.
22. Collect 100 μL of supernatant and pool with the previous supernatant; mix; the DNA sample can be tested in qPCR.

4. Notes

1. The ELISA is a quantitative method used to determine the concentration of a primary antibody using a series of dilutions of crude sera in antigen-coated wells. We plot the absorbance versus antibody dilution to estimate the antibody titer.

2. From starting rabbit immunization at day 0, we obtain bleeds at day 66, day 87, 4 months and then the final bleed at month 4.5. Volumes of bleeds are of 20, 20, 20 and 50 mL, respectively. We start by testing bleeds from day 66.
3. Keyhole limpet hemocyanin (KLH) is the most common protein carriers, and KLH is preferred since it is more antigenic in the majority of animals. Carrier molecule is critical since peptide molecules alone often fail to initiate an immune response. In ELISA, it is essential to test the sera against KLH. In addition, a known peptide is coated on the wells and used as peptide positive control; it is to be tested with a serum of reference (or ELISA antibody positive control), which was already shown to recognize the peptide.
4. Several dot blot membranes can be spotted and stored (dried between two filter papers) during several weeks (one aliquot of 10 μL of 5 m*M* peptide stock is enough for about 200 membranes). For regular spotting use multichannel pipette and/or draw on the membrane a grid (1 cm^2) with a pencil. You can use Red Ponceau to color and double-check the spotting, but do not use it to quantify between peptides of different sequences as they will be stained differently based on their sequences (2). Ponceau S solution can also be used to double-check SDS-PAGE transfer efficiency. Incubate membranes in Ponceau solution for 5 min and wash twice in deionized water.
5. In dot blots and western blots, the dilution of the primary antibody depends of antibody titer: 1:1,000 could be the starting dilution, but a titration should be done (depending on results) to determine an optimum concentration for each antibody.
6. In ChIP, the amount of the antibody to use is about 1–5 μg/IP. It is advised to perform a titration of the antibody, e.g., use in ChIP: 1, 2, and 5 μg of antibody to determine the best ChIP conditions. Crude serum dilutions depend on titration as well; dilute the crude serum at 1:1,000 and 1:5,000 if the corresponding titer is high in ELISA and dot blots. Dilute the crude sera 10 times less, if otherwise. Note that antibodies with high titers are the best (4).
7. Affinity purification must be performed with the antigen that was used for generating the immune response. Antibody purification method used is affinity chromatography with coupled peptide on a pre-packed HiTrapTMNHS-activated HP column (#17-07-01, GE HealthCare) followed by a buffer exchange by Gel Filtration on G-25 fine (HiPrepTMTM 26/10 Desalting, #17-5087-01, GE HealthCare) on Äkta-Prime™ System (#11-0013-13, GE HealthCare). After

peptide affinity purification, the antibody specificity must be checked since the antibody preference substrate might have been altered as well as its titer greatly reduced.

8. A magnetic rack from Diagenode has been specially designed for simple sample handling with the LowCell# ChIP kit. It can hold up to 16X 0.2 mL tubes simultaneously in a chilled environment even on the bench top, and enables efficient and fast magnetic separation. Note that the LowCell# ChIP kit also allows immunoprecipitation of transcription factors as well as histones.

Acknowledgments

We would like to thank Thomas Jenuwein and Laura O'Neill for frequent and very helpful discussions on antibody characterization. We also acknowledge Henk Stunnenberg and lab members for exchange of critical comments on antibody testing. This work was supported by a grant from the European Union called HEROIC. We are indebted to all the partners of HEROIC for their contribution and help in testing antibodies.

References

1. Harlow, Ed. and Lane, D. (1998) Antibodies: a laboratory manual. Cold Spring Harbor Laboratory Press, New York.
2. Sarma, K., Nishioka, K. and Reinberg, D. (2004) Tips in analyzing antibodies directed against specific histone tail modifications. *Methods Enzymol.* **376**, 255–269.
3. Burgos, L., Peters, A.H., Opravil, S., Kauer, M., Mechtler, K. and Jenuwein, T. (2004) Generation and characterization of methyl-lysine histone antibodies. *Methods Enzymol.* **376**, 234–254.
4. Cheung, P. (2004) Generation and characterization of antibodies directed against di-modified histones, and comments on antibody and epitope recognition. *Methods Enzymol.* **376**, 221–234.

Chapter 3

The Fast Chromatin Immunoprecipitation Method

Joel Nelson, Oleg Denisenko, and Karol Bomsztyk

Abstract

The chromatin immunoprecipitation assay (ChIP assay) has greatly facilitated the recent, dramatic expansion of our knowledge of the protein–DNA interactions involved in regulating gene expression, DNA repair, and cell division. The power of the assay is that it gives a researcher the ability to not only detect a specific protein–DNA interaction in vivo but also determine the relative density of factors along genes or the entire genome. Though powerful, the traditional assay is time consuming (involving 2 days or more) and laborious. With Fast ChIP, we simplified the assay to greatly reduce the time and labor involved. The improved assay is especially useful for studies which involve many samples, including the probing of multiple chromatin factors simultaneously and/or looking at genomic events over several time points. Using Fast ChIP, 24 sheared chromatin samples can be processed to yield PCR-ready DNA in 5 h.

Key words: Chromatin immunoprecipitation, ChIP-chip, tissue ChIP, transcription, DNA repair.

1. Introduction

DNA in the eukaryotic nucleus is complexed with proteins and RNAs in chromatin, one of the most intensely studied structures in biology today (1–3). Chromatin is complex, dynamic, responsive to intra- and extra-cellular signals and is involved in regulating most aspects of DNA metabolism including transcription, DNA repair, DNA replication, and chromosome condensation (3–5). Chromatin immunoprecipitation (ChIP) is a powerful method used to study the interactions of proteins (or specific modified forms of proteins) with DNA in vivo (6, 7). ChIP can be used not only to detect the interaction of a protein with a specific region of the genome but also to estimate the relative density of this interaction.

Philippe Collas (ed.), *Chromatin Immunoprecipitation Assays,* Methods in Molecular Biology 567,
DOI 10.1007/978-1-60327-414-2_3, © Humana Press, a part of Springer Science+Business Media, LLC 2009

The ChIP assay represents a major advancement in the study of chromatin processes and its use has increased dramatically over the last few years.

The ChIP assay begins with the cross-linking of protein–DNA complexes by the fixation of cells/tissues with formaldehyde (6–8). After lysing the cells, the nuclei are disrupted and the chromatin is sheared either by sonication (6, 7) or by digestion with micrococcal nuclease (9). The chromatin fragments, typically between 500 and 1,000 base pairs in length, are immunoprecipitated using an antibody specific to the protein of interest (6, 7). After reversing the cross-links, the DNA is isolated and used in one of the several detection methods including dot/slot blot (10), PCR or qPCR (11), hybridization to a DNA microarray (ChIP-chip) (12), or sequenced using a rapid sequencing technology (ChIP-seq) (13). Enrichment of a particular DNA region over other sites where the factor is not expected to bind indicates that the protein interacts with this region.

The traditional ChIP assay, though it has proved to be powerful, is time consuming and laborious. The slowest step of the traditional ChIP assay is the 5 h reversal of cross-linking (8) and the most laborious step is the DNA cleanup, which involves phenol:chloroform extractions and ethanol precipitation (6). In Fast ChIP, cross-links are reversed during a 10 min incubation at 100°C in the presence of Chelex-100. In addition, since Fast ChIP does not require the addition of sodium bicarbonate/SDS buffer to elute the chromatin from the beads (the high temperature is sufficient), the DNA cleanup step is not necessary. After the 100°C incubation, the DNA-containing supernatant is directly used in PCR (11). Thus, several hours and a great deal of labor in the traditional assay are replaced with a 10 min incubation in Fast ChIP.

Another improvement in Fast ChIP is the use of an ultrasonic bath to increase the rate of antibody–chromatin interaction (14). In the traditional ChIP assay, the antibody–chromatin incubation can take anywhere from 1 h to overnight (6). With the use of the ultrasonic bath, this incubation is decreased to 15 min (11). The combination of these two improvements in Fast ChIP not only allows the assay to be easily completed in 1 day, starting with sonicated chromatin extracts, but also gives enough time for the products to be analyzed by qPCR in the same day (*see* **Fig.** 3.1 for an outline of the method).

Due to its simplicity and reduced labor, Fast ChIP facilitates studies which involve multiple chromatin samples, multiple antibodies, or both. These include studies where (i) multiple proteins or protein modifications (e.g., histone modifications) are observed simultaneously; (ii) multiple time points are observed;

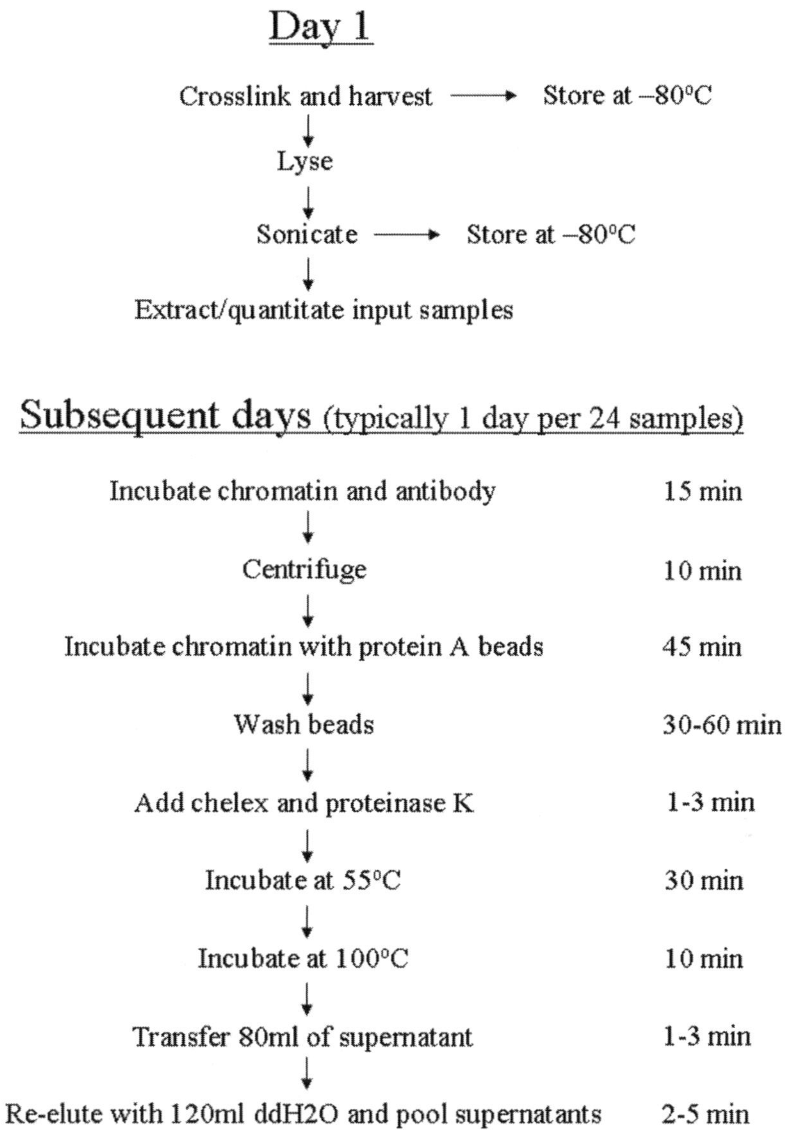

Fig. 3.1. A suggested outline for Fast ChIP. This outline assumes that sonication conditions have not been optimized for the cell/tissue type used and/or that tissue is being used as the chromatin input, requiring quantitation of the input DNA to adjust the input chromatin (**Section 3.4**, Steps 1–14). If neither of these cases applies, the method can be condensed into 1 day.

or (iii) antibodies and chromatin extracts are being screened for their suitability in ChIP. Beginning with sheared chromatin, 24 ChIP samples can be easily processed to yield PCR-ready DNA in 5 h. Also, the short time required for completion of the assay is helpful when optimizing conditions for a particular antibody or when learning the assay for the first time. We have used Fast ChIP with chromatin from tissue culture (15), mammalian tissues (16), and yeast cultures (17), and it is likely that it is compatible with most other sources of chromatin. Though we have designed Fast ChIP for analysis by PCR or qPCR, it has also been used, with the addition of a column cleanup step, in ChIP-chip studies (12). Thus, it is likely that Fast ChIP may be used for most ChIP applications, including ChIP-seq.

2. Materials

2.1. Reagents

See also under Buffers and Solutions.

1. Protein A–Sepharose (Amersham, cat. no. 17-5280-01).
2. Phosphate buffered saline (PBS).
3. SYBR Green PCR Master Mix.

2.2. Buffers and Solutions

1. 1 *M* Glycine.
2. IP buffer: 150 m*M* NaCl, 50 m*M* Tris–HCl, pH 7.8, 5 m*M* EDTA, pH 8.0, 0.5% (v/v) NP-40, 1% (v/v) Triton X-100.
3. Lysis/sonication buffer: make it fresh before each use. Per 1 mL of IP buffer, add the following protease inhibitors: 5 μL PMSF (0.1 *M* in isopropanol; stored at –20°C; re-dissolve at room temperature before pipetting) and 1 μL leupeptin (10 μg/μL; aliquoted and stored at –20°C) and keep on ice. In addition, the following phosphatase inhibitors may be added if required for ChIP with phosphospecific antibodies: 10 μL β-glycerophosphate (1 *M*; stored at 4°C), 10 μL sodium fluoride (1 *M*; stored at 4°C; resuspend before pipetting), 10 μL sodium molybdate dihydrate (10 m*M*; stored at 4°C), 1 μL sodium orthovanadate (100 m*M*; stored at –20°C), and 13.84 mg *p*-nitrophenylphosphate (stored at 4°C).
4. 10% Chelex-100 in ddH_2O (Bio-Rad, cat. no. 142-1253).
5. 20 μg/μL proteinase K in ddH_2O.
6. TE, pH 9.0: 10 m*M* Tris–HCl, 1 m*M* EDTA, bring to pH 9.0 with 5 *M* NaOH.

2.3. Equipment

1. Sonicator with microtip (e.g., Misonix Sonicator 3000).
2. Refrigerated microcentrifuge.
3. Heat blocks and hot plate (for 55°C incubation and boiling water incubation).
4. Tube rotator or tumbler at 4°C.

5. Set up for quantitative PCR (e.g., ABI 7900 real-time PCR system).
6. Ultrasonic bath (optional).

3. Methods

The steps which make Fast ChIP unique compared to other ChIP methods are immunoprecipitation and preparation of PCR-ready DNA. Therefore, the following methods for cross-linking, lysis, and sonication are based on what has worked in our laboratory, but are certainly not the only methods compatible with Fast ChIP. If a researcher has previously established his/her own chromatin preparation method for ChIP, they should continue to use this method with Fast ChIP.

To ensure equal loading of different chromatin samples, especially necessary when tissue fragments are used, we suggest extracting total DNA from each chromatin sample (**Section 3.4**, Steps 1–14) and measuring the amount of DNA for each by qPCR. If the samples differ by more than 25%, the amount of chromatin loaded (**Section 3.5**, Step 1) should be adjusted based on this measurement. If the amount of chromatin is adjusted, remember to use an average of the input samples while calculating the percent of input (**Section 3.6**).

If extracting the input DNA for quantitation to adjust chromatin loading for ChIP (especially if using tissue samples) or for analyzing the chromatin fragmentation (optimizing the sonication conditions), we suggest doing the cross-linking, lysis, and sonication steps on a separate day from the ChIP. If using cells from tissue culture, equal chromatin loading can be more easily controlled than in tissue samples by ensuring equal density on plates. Therefore, if sonication conditions have already been optimized, for tissue culture the entire assay can be completed in 1 day with the input DNA extraction and the ChIP being processed simultaneously.

3.1. Cross-Linking

3.1.1. Tissue Culture

1. Keep in mind that approximately 4×10^5–10^6 cells are required per IP sample.
2. Add 40 μL 37% formaldehyde per milliliter of tissue culture medium directly to the dish/flask (1.42% final concentration), swirl, and incubate at room temperature for 15 min (*see* **Note 1**).
3. Quench formaldehyde by adding 141 μL of 1 *M* glycine per milliliter of medium (125 m*M* final concentration) and incubate for 5 min at room temperature.

4. Harvest cells by scraping and centrifuging at 2,000*g* for 5 min (4°C).
5. Keep cells on ice and wash twice with ice-cold IP buffer. After aspirating the PBS, the cell pellet can be stored at –80°C for at least a year.

3.1.2. Fresh or Frozen Tissue

This method has been used in our laboratory for ChIP on both kidney and liver tissue and is likely to be effective in other tissues which have similar numbers of cells per volume of tissue.

1. Place approximately 0.1 cm^3 piece of fresh or frozen (–80°C) tissue in 1 mL of PBS containing 1% formaldehyde at room temperature and quickly mince with forceps into 1–2 mm^3 fragments.
2. Incubate tissue fragments at room temperature for 20 min (*see* **Note 1**).
3. Centrifuge at 2,000–3,000*g* for 1 min (4°C) and discard the supernatant.
4. Suspend pellet in 1 mL PBS with 125 m*M* glycine and incubate for 5 min at room temperature.
5. Centrifuge tissue fragments, and discard the supernatant.
6. Wash twice with PBS and place on ice for the lysis/sonication step (**Section 3.2**, Step 4).

3.1.3. Yeast Culture

For both cross-linking and lysis of yeast cells, we use the method described by Kuo and Allis (6) up to the point where whole cell lysate is obtained (*see* **Note 1**). At this point, Fast ChIP can be used, beginning at the sonication steps (**Section 3.3**).

3.2. Lysis

1. Lyse approximately 10^7 cells by resuspending in 1 mL ice-cold lysis/sonication buffer (*see* **Note 2**) and pipetting up and down several times.
2. Collect the insoluble material, which includes the nuclei, by centrifuging at 12,000*g* for 1 min (4°C), and aspirate the supernatant.
3. Resuspend the pellet once more in 1 mL lysis/sonication buffer, collect the pellet by centrifugation, and aspirate the supernatant. This washes away residual soluble proteins from the pellet leaving insoluble chromatin, nuclear matrix, and associated cytoskeleton.
4. For tissues, resuspend cross-linked fragments (**Section 3.1.2**, Step 6) in 1 mL lysis buffer and proceed to the sonication step (**Section 3.3**).

3.3. Sonication

1. Resuspend the pellet in 1 mL ice-cold lysis/sonication buffer, and split into two 500 μL fractions. At this point, both the fractions should be in 1.5 mL microcentrifuge tubes. Both the volume of buffer and the geometry of the tube used for sonication affect fragmentation efficiency with volumes of 500 μL or less and 1.5 mL microcentrifuge tubes (for tissues 1 mL buffer and 2 mL tubes) being optimal.
2. The protocol used for sonication can vary widely and must be optimized for each cell or tissue type and sonicator set-up. Optimal fragment sizes are typically between 0.5 and 1 kb as determined by running sonicated chromatin on 1% agarose after DNA extraction and reversal of cross-links (**Section** 3.4, Steps 1–14). The following are suggestions for optimizing sonication using a microtip:
   a. Sonication can cause heating of the sample; so the tube should be immersed in an ice-water bath during sonication.
   b. Foaming can occur if the microtip gets too close to the surface of the sample during sonication. The tip should remain no more than a few millimeters from the bottom of the tube during sonication. If foaming does occur, stop sonication and wait till the majority of bubbles rise to the surface before continuing sonication.
   c. The two variables to optimize are the total amount of sonication time and the power output of the sonicator.
   d. To avoid excessive heating, the total sonication time should be broken up into rounds of 10–20 s each, with at least 2 min of rest on ice between each round. In addition, sonication is more efficient if each round is broken up into approximately 1 s pulses rather than continuous sonication, since the power of sonication decreases gradually after the beginning of each pulse.
   e. The higher the power output of the sonicator the faster the fragmentation of the chromatin and the more heating the sample is exposed to. Start with a power output 50% or less of the total power output for the sonicator and increase as needed such that the samples are not overheated by the end of each round of sonication, but the amount of time required for sonication is not prohibitive considering the number of samples to be sonicated.
   f. Other factors which affect sonication efficiency are the cell concentration and the extent of cross-linking of the chromatin. Diluting the chromatin and/or reducing the cross-linking time or concentration of formaldehyde can increase sonication efficiency.

3. After sonication, the chromatin should be cleared by centrifugation at 12,000*g* for 10 min (4°C).
4. Transfer the supernatant to a new tube and aliquot for storage at –80°C (*see* **Note 3**). Save one aliquot of 10 μL for extracting total DNA for the 'input' sample.

3.4. Isolating Total DNA (Input Sample)

Unless otherwise stated, Steps 1–14 can be performed at room temperature.

1. Precipitate DNA from the 10 μl aliquot from **Section** 3.3 Step 4 for 10 min at room temperature with 30 μL absolute or 96% ethanol.
2. Pellet the DNA by centrifugation at 12,000*g* for 3 min (4°C).
3. Aspirate or decant the supernatant and add 50 μL 75% ethanol.
4. Centrifuge at 12,000*g* for 1 min (4°C), and remove as much of the supernatant as possible.
5. Dry the pellets to completion (they should become transparent after drying).
6. Add 100 μL of 10% Chelex-100 slurry to the dried pellets (*see* **Note 10**).
7. Boil for 10 min and cool by centrifuging for 1 min (4°C).
8. Add 1 μL of 20 μg/μL proteinase K to each tube and vortex. Briefly centrifuge to bring contents to the bottom of the tube.
9. Incubate at 55°C for 30 min, gently resuspending the Chelex once or twice during the incubation.
10. Boil for 10 min and centrifuge the condensate to the bottom of the tube at 10,000*g* for 1 min (4°C).
11. Transfer 80 μL of the supernatant to a new tube.
12. Add 120 μL ddH_2O to each tube containing Chelex slurry, vortex, and centrifuge the contents to the bottom of the tube.
13. Remove 120 μL of the supernatant and pool with the 80 μL supernatant from Step 10 (*see* **Note 11**).
14. The DNA can be run undiluted on 1% agarose. For PCR, use no less than a 1:20 dilution in TE, since some of the remaining contaminants can be inhibitory to PCR.

3.5. Immunoprecipitation

1. For each IP sample, dilute the equivalent of 1×10^6 cells of chromatin to 200 μL with ice-cold lysis/sonication buffer (*see* **Notes 4, 5**).
2. Add specific or mock antibodies to each sample and mix by inverting (*see* **Notes 6,** 7).
3. Turn the ultrasonic bath on and float samples in the bath for 15 min at 4°C (*see* **Note 8**).

4. Clear the solution by centrifugation at 12,000*g* for 10 min (4°C). This step is essential to remove non-specific insoluble chromatin aggregates which may contaminate the final product.
5. While the chromatin and antibodies are incubating, transfer approximately 20 μL per IP sample of protein A agarose slurry to a clean tube (*see* **Notes 9, 10**). Wash 1–3 times with IP buffer to remove ethanol.
6. Resuspend beads in 180 μL IP buffer for every 20 μL of beads (*see* **Note 5**). Dispense 200 μL of the diluted slurry to new tubes, 1 tube for each IP sample (*see* **Note 10**). Centrifuge and aspirate buffer. Visually inspect tubes to make sure each one has the same amount of beads.
7. Transfer no more than the top 90% of each cleared chromatin sample from Step 4 (avoiding the pellet at the bottom of the tube) to the tubes with the beads.
8. Rotate tubes at 4°C for 45 min with a rotating platform or tumbler. The rotation should be fast enough to keep the beads suspended.
9. Centrifuge the tubes at 10,000*g* for 1 min (4°C) and aspirate the supernatant.
10. Wash the beads (resuspend with buffer, centrifuge, and aspirate the supernatant) five times with 1 mL ice-cold IP buffer. After the last wash, remove as much supernatant as possible without removing the beads.
11. Add 100 μL of 10% Chelex-100 slurry to the washed beads (*see* **Note 10**).
12. Add 1 μL of 20 μg/μL proteinase K to each tube and vortex. Briefly centrifuge contents to the bottom of the tube.
13. Incubate at 55°C for 30 min. Gently resuspend beads and Chelex-100 once or twice during the incubation.
14. Boil samples for 10 min.
15. Centrifuge samples at 10,000*g* for 1 min (4°C) to cool samples and bring condensate to the bottom of the tube.
16. Transfer 80 μL of supernatant to new tubes.
17. Add 120 μL ddH_2O to each tube containing Chelex/protein A beads slurry, vortex, and centrifuge contents to the bottom of the tube (*see* **Note 11**).
18. Remove 120 μL of the supernatant and pool with the 80 μL supernatant from Step 16.
19. The PCR-ready DNA can be stored at –20°C and repeatedly thawed and frozen over several months without loss of PCR signal.

3.6. PCR and Calculation of Enrichment

We use 2.35 μL of IP DNA or diluted input DNA in 5 μL reactions with 0.15 μL of primer pair (each primer at 10 μ*M*), and 2.5 μL of master mix (SensiMix containing SYBR green and ROX) in 384-well PCR plates. The reactions are run in triplicate in 384-well PCR plates on the ABI 7900 for 40 cycles with the default two-step method. Data are acquired and analyzed using the SDS 2.2.1 software. The threshold is set manually and Cts are imported to EXCEL for calculations.

We express enrichment of the immunoprecipitated region of the genome as the percent of input DNA. To eliminate the differences in amplification efficiencies of different primers, relative amounts of DNA for the IP, mock, and input samples are calculated for each primer using a standard curve. The standard curve consists of serial dilutions of total DNA from the same cell type or tissue used in the experiment and is run each time a primer pair is used. We suggest making up a large amount of each dilution in TE buffer and aliquoting them for multiple uses so that the standard curve can be run repeatedly without error due to degradation of the DNA.

PCR-primer efficiency curves are fit to the natural log of concentration vs. Ct for each dilution using an *r*-squared best fit. The relative amount for each ChIP and input DNA sample is calculated from their respective averaged Ct values using the formula:

$$[\mathrm{DNA}] = \frac{b \times e^{m\times \mathrm{AvgCt}}}{\mathrm{Dilution}} \qquad ([1])$$

where *b* and *m* are the curve fit parameters from the primer calibration curve that is generated for each PCR experiment. Dilution is the cumulative dilution of ChIP DNA as compared to the input DNA sample. Final results are expressed either as a fraction or percent of input using the following equation:

$$\%\,\mathrm{of\ input} = \frac{[\mathrm{DNA}_{\mathrm{sample}}] - [\mathrm{DNA}_{\mathrm{mock}}]}{[\mathrm{DNA}_{\mathrm{input}}]} \times 100 \qquad ([2])$$

where DNA concentrations were computed from equation [1]. $\mathrm{DNA}_{\mathrm{sample}}$ is the ChIP DNA sample, $\mathrm{DNA}_{\mathrm{mock}}$ is the IgG mock IP control, and $\mathrm{DNA}_{\mathrm{input}}$ is the input DNA used in ChIP. Remember that, if the chromatin amount used in ChIP (**Section** 3.5, Step 1) was adjusted based on measurement of the input samples (**Section** 3.4, Step 14), then $\mathrm{DNA}_{\mathrm{input}}$ in equation [2] should be an average of the input for all the samples.

3.7. Analysis

The enrichment (percent of input) determined using the above calculations is, in itself, not a meaningful number. To determine the significance of the enrichment at a region of interest, this region must be compared to another region where the factor of interest is not expected to bind (the negative control region). The enrichment at the negative control region gives a baseline which is assumed to represent zero binding and the significance of the enrichment at the region of interest depends on the signal at this region being significantly above the baseline.

Another means of determining the significance of enrichment of a factor at a particular locus is to compare that enrichment in cells where the factor is present to those where the factor has been knocked out. The enrichment in the knockout cells represents the zero binding baseline, and enrichment is significant at the region of interest only if it is above this baseline.

4. Notes

1. The cross-linking times and formaldehyde concentrations used here are suggestions and may need to be optimized depending on the cell/tissue type used as well as on the factor being immunoprecipitated. Longer cross-linking times or higher formaldehyde concentrations can improve the immunoprecipitation of some factors by increasing the number of cross-links between the factor and the DNA. Conversely, longer cross-linking times can be detrimental for pull-down of some factors because epitopes in the factor may be masked by the cross-linking. At the upper range of fixing, tissues or cells may become resistant to shearing of the chromatin by sonication.
2. Both PMSF and leupeptin have short half-lives in aqueous solutions at room temperature. It is important to prepare the lysis/sonication buffer fresh and keep it on ice before use.
3. The chromatin preparations can be stored at –80°C for months without loss of pull-down efficiency; however, repeated thawing and freezing can reduce this efficiency. To avoid frequent thawing of chromatin, make aliquots just large enough for each experiment you are planning.
4. The amount of chromatin used here is a suggested starting point. In our experience in some cases, using smaller amounts of chromatin can increase the difference between the IP and mock signals by decreasing the background without significantly decreasing the signal from the specific pull-down.

5. To reduce non-specific binding to the protein A beads, blocking reagents may be used to block the beads both prior to the IP and during the IP. To make blocking buffers, add 5% BSA (fraction V) and 100 μg/mL sheared salmon sperm DNA (ssDNA) to aliquots of lysis/sonication buffer and IP buffer. The chromatin should be diluted in lysis/sonication buffer with BSA and ssDNA before incubating with antibody. Also, the beads should be pre-incubated with 200 μL of IP buffer with BSA and ssDNA while resuspended on a rotating platform for 0.5 h. This buffer should be aspirated off the beads before transferring the chromatin/antibody mix.
6. For some antibodies the amount required may need to be determined empirically; however, 1–2 μg per sample is sufficient for many antibodies. For a mock IP (control for non-specific binding) either the same antibody blocked with saturating amounts of an epitope-specific peptide, a pre-immune IgG, or no antibody can be used.
7. In our experience, polyclonal antibodies are more likely to work in ChIP than monoclonal antibodies.
8. If an ultrasonic bath is not available, samples may need to be incubated for 1–2 h at 4°C depending on the antibody (some antibodies may require longer times up to overnight incubations; this should be determined empirically).
9. Non-specific binding of the chromatin to the protein A beads accounts for the majority of the mock signal. Therefore, reducing the amount of beads used may reduce the mock signal (improving the (IP – mock) difference). The 20 μL suggested here is far above what is necessary to bind the antibodies, and this amount is only used as it is convenient to visualize the pellet while aspirating the washes.
10. Keep the slurry in suspension while pipetting and use a tip with the end cut off to avoid clogging.
11. Tris–HCl (17 m*M*) and EDTA (1.7 m*M*) (final pH 9.0) may be substituted here to improve DNA stability over time. Check to make sure that PCR amplification is not negatively affected by the use of this buffer.

Acknowledgment

We thank members of the KB lab for valuable discussions of the method. This work was supported by NIH DK45978 and GM45134 to K.B.

References

1. Bernstein, E. and Allis, C. D. (2005) RNA meets chromatin. *Genes Dev.* **19**, 1635–1655.
2. Schubeler, D. and Elgin, S. C. (2005) Defining epigenetic states through chromatin and RNA. *Nat. Genet.* **37**, 917–918.
3. Felsenfeld, G. and Groudine, M. (2003) Controlling the double helix. *Nature* **421**, 448–453.
4. Sims, R. J., 3rd, Mandal, S. S. and Reinberg, D. (2004) Recent highlights of RNA-polymerase-II-mediated transcription. *Curr. Opin. Cell Biol.* **16**, 263–271.
5. Thiriet, C. and Hayes, J. J. (2005) Chromatin in need of a fix: phosphorylation of H2AX connects chromatin to DNA repair. *Mol. Cell* **18**, 617–622.
6. Kuo, M. H. and Allis, C. D. (1999) In vivo cross-linking and immunoprecipitation for studying dynamic protein:DNA associations in a chromatin environment. *Methods* **19**, 425–433.
7. Orlando, V., Strutt, H. and Paro, R. (1997) Analysis of chromatin structure by in vivo formaldehyde cross-linking. *Methods* **11**, 205–214.
8. Solomon, M. J. and Varshavsky, A. (1985) Formaldehyde-mediated DNA–protein crosslinking: a probe for in vivo chromatin structures. *Proc. Natl. Acad. Sci. U.S.A.* **82**, 6470–6474.
9. Thorne, A. W., Myers, F. A. and Hebbes, T. R. (2004) Native chromatin immunoprecipitation. *Methods Mol. Biol.* **287**, 21–44.
10. Solomon, M. J., Larsen, P. L. and Varshavsky, A. (1988) Mapping protein–DNA interactions in vivo with formaldehyde: evidence that histone H4 is retained on a highly transcribed gene. *Cell* **53**, 937–947.
11. Nelson, J. D., Denisenko, O., Sova, P. and Bomsztyk, K. (2006) Fast chromatin immunoprecipitation assay. *Nucleic Acids Res.* **34**, e2.
12. Huebert, D. J., Kamal, M., O'Donovan, A. and Bernstein, B. E. (2006) Genome-wide analysis of histone modifications by ChIP-on-chip. *Methods* **40**, 365–369.
13. Johnson, D. S., Mortazavi, A., Myers, R. M. and Wold, B. (2007) Genome-wide mapping of in vivo protein–DNA interactions. *Science* **316**, 1497–1502.
14. Chen, R., Weng, L., Sizto, N. C., Osorio, B., Hsu, C. J., Rodgers, R. and Litman, D. J. (1984) Ultrasound-accelerated immunoassay, as exemplified by enzyme immunoassay of choriogonadotropin. *Clin. Chem.* **30**, 1446–1451.
15. Nelson, J. D., Flanagin, S., Kawata, Y., Denisenko, O. and Bomsztyk, K. (2008) Transcription of laminin {gamma}1 chain gene in rat mesangial cells: constitutive and inducible RNA polymerase II recruitment and chromatin states. *Am. J. Physiol. Renal. Physiol.* **294**, F525–533.
16. Zager, R. A., Johnson, A. C., Naito, M. and Bomsztyk, K. (2008) Maleate nephrotoxicity: mechanisms of injury and correlates with ischemic/hypoxic tubular cell death. *Am. J. Physiol. Renal. Physiol.* **294**, F187–197.
17. Denisenko, O. and Bomsztyk, K. (2008) Epistatic interaction between the K-homology domain protein HEK2 and SIR1 at HMR and telomeres in yeast. *J. Mol. Biol.* **375**, 1178–1187.

Chapter 4

μChIP: Chromatin Immunoprecipitation for Small Cell Numbers

John Arne Dahl and Philippe Collas

Abstract

Chromatin immunoprecipitation (ChIP) is a technique of choice for studying protein–DNA interactions. ChIP has been used for mapping the location of modified histones on DNA, often in relation to transcription or differentiation. Conventional ChIP protocols, however, require large number of cells, which limits the applicability of ChIP to rare cell samples. ChIP assays for small cell numbers (in the range of 10,000–100,000) have been recently reported; however, these remain lengthy. Our laboratory has elaborated fast ChIP assays suitable for small cell numbers (100–100,000) and for the immunoprecipitation of histone proteins or transcription factors under cross-linking conditions. We describe here a rapid micro (μ)ChIP assay suited for multiple parallel ChIPs from a single chromatin batch from 1,000 cells. The assay is also applicable to a single immunoprecipitation from 100 cells.

Key Words: Chromatin immunoprecipitation, ChIP, histone, acetylation, methylation, epigenetics.

1. Introduction

Interactions between proteins and DNA are essential for many cellular functions such as genomic stability, DNA replication and repair, chromosome segregation, transcription, and epigenetic silencing of gene expression. ChIP has become a technique of choice in the study of protein–DNA interactions and for unraveling transcriptional regulatory circuits within the cell (1). ChIP has been used for mapping the location of post-translationally modified histones, transcription factors, chromatin modifiers, and other non-histone DNA-associated proteins. This mapping may be restricted to specific genomic sites (2–8) or expanded to the genome-wide level (9–16).

Philippe Collas (ed.), *Chromatin Immunoprecipitation Assays,* Methods in Molecular Biology 567,
DOI 10.1007/978-1-60327-414-2_4, © Humana Press, a part of Springer Science+Business Media, LLC 2009

In a typical ChIP assay, DNA and proteins are reversibly cross-linked to maintain the association of proteins with target DNA sequences. However, when analyzing histone modifications cross-linking may be omitted (native ChIP) (3, 17). Chromatin is subsequently sheared, usually by sonication, to ~500 bp fragments and cleared for large complexes by centrifugation. The supernatant (chromatin) is used for immunoprecipitation of specific protein–DNA complexes using antibodies coupled to beads. The immunoprecipitated complexes are washed under stringent conditions; the precipitated chromatin is eluted; the cross-link is reversed; the proteins are digested; and the DNA is purified. Genomic sequences associated with the precipitated protein can be identified by polymerase chain reaction (PCR), cloning and sequencing, high-throughput sequencing, or hybridization to microarrays (ChIP-on-chip). Parameters and variations of the ChIP assay and tools implemented to investigate the profiles of DNA–protein interactions have recently been addressed elsewhere (1, 18–25).

In spite of the versatility in the nature of DNA-bound proteins and cell types that can be examined by ChIP, the assay has been hampered by a requirement for large cell numbers (in the range of 10^6–10^7), which has prevented the application of ChIP to rare cell samples. Another drawback has been the length of the procedure which can take up to 4 days. These limitations have prompted the development of variations on the ChIP assay. (i) A carrier ChIP (CChIP) assay (4) relies on a single immunoprecipitation from 100 cells and involves the inclusion of carrier chromatin from *Drosophila* cells to reduce loss and facilitate precipitation. However, the assay is cumbersome and entails radioactive labeling of PCR products for detection. It is also unclear whether it is suitable for precipitation of transcription factors. Furthermore, the use of foreign carrier chromatin predicts that primers used for detection of immunoprecipitated sequences must be highly species specific. (ii) Still with the aim of reducing cell numbers for ChIP, a micro-ChIP protocol for 10,000 cells without carrier chromatin was reported (15). Interestingly, the assay allows the analysis of histone or RNA polymerase II (RNAPII) binding throughout the genome by ChIP-on-chip. The assay takes 4 days. (iii) A fast ChIP assay (6, 26) has shortened two steps of conventional ChIP and reduced the assay to 1 day. An ultrasonic bath has been applied to speed up antibody binding to target proteins, and DNA isolation has been sped up by the use of a resin-based (Chelex-100) DNA isolation (26). Nonetheless, the fast protocol requires large number of cells (in the range of 10^6–10^7). (iv) We have developed a quick and quantitative (Q^2)ChIP assay suitable for up to 1,000 histone ChIPs or 100 transcription factor ChIPs from 100,000 cells (7). Q^2ChIP can be undertaken in 1 day. (v) Recently, a microplate-based ChIP assay (matrix-ChIP) was reported, which increases throughput and simplifies the assay (27). All steps are carried out in microplate wells without sample transfers. Matrix-ChIP enables 96 ChIPs for histones and DNA-bound proteins in 1 day (27). (vi) The lower limit on cell numbers has been further pushed by our recent report of a miniaturized ChIP assay (μChIP) suitable for up to eight parallel ChIPs of histones and/or RNAPII from a single batch of 1,000 cells, or for a single ChIP from 100 cells without carrier chromatin (28) (**Fig. 4.1**). The assay has been validated by assessing

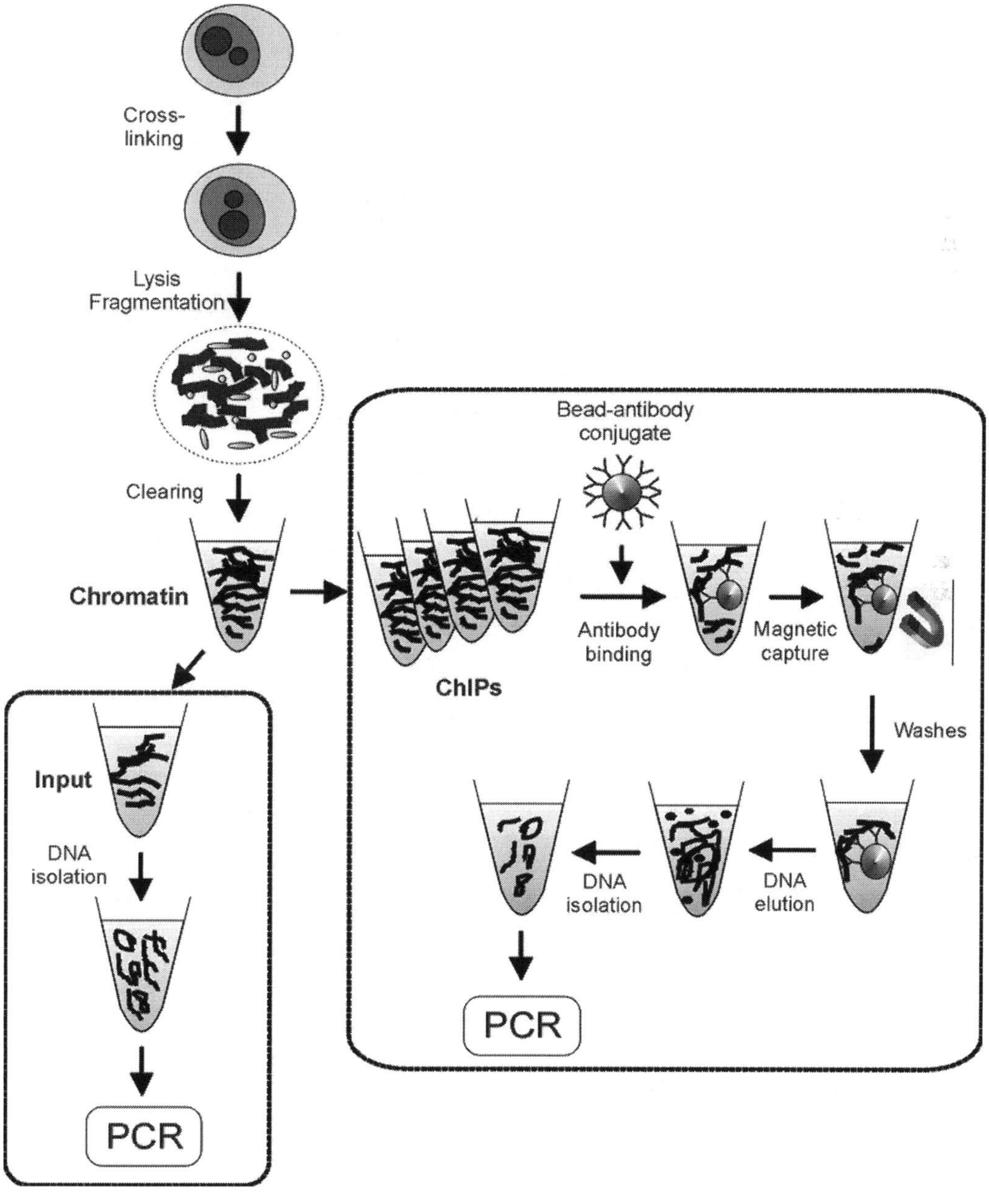

Fig. 4.1. The μChIP assay.

several post-translational modifications of histone H3 and RNAPII binding to developmentally regulated promoters in embryonal carcinoma cells and biopsies (28). The profiles of histone modifications identified from chromatin prepared from 1,000 cells and from starting batches of 100 cells are similar and reflect the expression status of the genes (**Fig. 4.2**). This communication describes the μChIP assay as it is used in our laboratory. Applications of the assay to small tissue biopsies have been reported elsewhere (28).

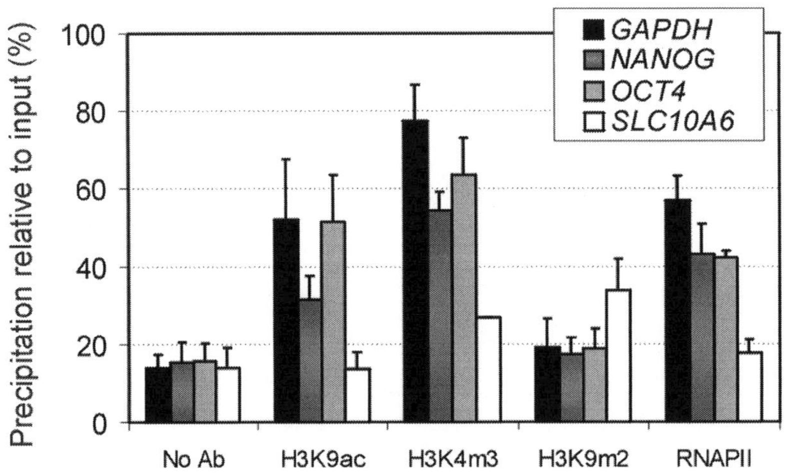

Fig. 4.2. μChIP analysis of histone and RNAPII binding in 100 cells as starting material. The graph shows H3K9ac, H3K4m3, H3K9m2, and RNAPII binding to the *GAPDH*, *NANOG*, *OCT4*, and *SLC10A6* promoters in separate 100 human embryonal carcinoma (NCCIT) cell batches for each antibody, and for a no-antibody (No Ab) control. Data are expressed as mean percent precipitation relative to input chromatin ±SD.

2. Materials

2.1. Laboratory Equipment

1. Siliconized pipette tips.
2. Filtered pipette tips (10 ∝Λ, 200 ∝Λ, 1,000 ∝Λ).
3. Magnetic rack for 200 μL tube strips (Diagenode, cat. no. kch-816-001).
4. 200-μL PCR tubes in eight-tube strip format (Axygen, cat. no. 321-10-051).
5. 0.6- and 1.5-mL centrifuge tubes.
6. Magnetic holder for 1.5 mL tubes.
7. Probe sonicator (Sartorius Labsonic M sonicator with 3 mm diameter probe, or similar).

8. Rotator placed at 4°C.
9. Table-top centrifuge.
10. Minicentrifuge.
11. Vortex.
12. Thermomixer (Eppendorf, model no. 5355–28402 or similar).
13. Heating block.
14. Thermal cycler with real-time capacity.

2.2. Reagents

1. 36.5% formaldehyde.
2. Dynabeads® protein A (Invitrogen, cat. no. 100.02D). The beads should be well suspended before pipetting. Use Dynabeads® protein A beads with rabbit IgGs and Dynabeads® protein G (Invitrogen, cat. no. 100.04D) with mouse IgGs.
3. 5 *M* NaCl.
4. 400 m*M* EGTA.
5. 500 m*M* EDTA.
6. 1 *M* Tris–HCl, pH 7.5.
7. 1 *M* Tris–HCl, pH 8.0.
8. Glycine: 1.25 *M* stock solution in PBS.
9. Chelex-100 (BioRad, cat. no. 142-1253): 10% (wt/vol) Chelex in MilliQ water.
10. Acrylamide carrier (Sigma-Aldrich, cat. no. A9099).
11. Proteinase K: 20 mg/mL solution in MilliQ water.
12. Protease inhibitor mix (Sigma-Aldrich, cat. no. P8340).
13. PMSF: 100 m*M* stock solution in 100% ethanol.
14. Sodium butyrate: 1 *M* stock solution in MilliQ water. Na-butyrate is a histone deacetylase inhibitor and should be used for anti-acetylated epitope ChIPs.
15. Phosphate buffered saline (PBS).
16. PBS/Na-butyrate solution 20 m*M* butyrate in 1X PBS. Make immediately before use.
17. PBS/Na-butyrate/formaldehyde fixative: 20 m*M* butyrate, 1% (vol/vol) formaldehyde, 1 m*M* PMSF, and protease inhibitor mix in 1X PBS. Make immediately before use.
18. Phenol:chloroform:isoamylalcohol (25:24:1).
19. Chloroform:isoamylalcohol (24:1).
20. 3 *M* NaAc.
21. IQ SYBR® Green (BioRad, cat. no. 170-8882).
22. Antibodies of choice. Use ChIP-grade antibodies when available (*see* **Note 1**).

2.3. Buffers

1. Lysis buffer: 50 m*M* Tris–HCl, pH 8.0, 10 m*M* EDTA, 1% (wt/vol) SDS, protease inhibitor mix (1:100 dilution from stock), 1 m*M* PMSF, 20 m*M* Na-butyrate. Protease inhibitor mix, PMSF, and Na-butyrate should be added immediately before use.
2. RIPA buffer: 10 m*M* Tris–HCl, pH 7.5, 140 m*M* NaCl, 1 m*M* EDTA, 0.5 m*M* EGTA, 1% (vol/vol) Triton X-100, 0.1% (wt/vol) SDS, 0.1% (wt/vol) Na-deoxycholate.
3. RIPA ChIP buffer: 10 m*M* Tris–HCl, pH 7.5, 140 m*M* NaCl, 1 m*M* EDTA, 0.5 m*M* EGTA, 1% (vol/vol) Triton X-100, 0.1% (wt/vol) SDS, 0.1% (wt/vol) Na-deoxycholate, protease inhibitor mix (1:100 dilution from stock), 1 m*M* PMSF, 20 m*M* Na-butyrate. Protease inhibitor mix, PMSF, and Na-butyrate should be added immediately before use.
4. TE buffer: 10 m*M* Tris–HCl, pH 8.0, 10 m*M* EDTA.
5. Elution buffer: 20 m*M* Tris–HCl, pH 7.5, 5 m*M* EDTA, 50 m*M* NaCl.
6. Complete elution buffer: 20 m*M* Tris–HCl, pH 7.5, 5 m*M* EDTA, 50 m*M* NaCl, 20 m*M* Na-butyrate, 1% (wt/vol) SDS, 50 μg/mL proteinase K. Na-butyrate, SDS, and proteinase K should be added just before use.

3. Methods

3.1. Preparation of Antibody–Bead Complexes

1. Prepare a slurry of Dynabeads® protein A (if using rabbit IgGs). For 16 ChIPs, including two negative controls, place 180 μL of well-suspended Dynabeads® protein A stock solution into a 1.5 mL tube, place the tube in the magnetic holder, allow beads to be captured, remove the buffer, remove from the magnet, and add 500 μL RIPA buffer. Ensure the stock bead suspension is homogenous before pipetting.
2. Vortex, capture the beads, remove the buffer, add another 500 μL RIPA buffer.
3. Vortex, capture the beads, remove the buffer, add 170 μL RIPA buffer.
4. Vortex the beads and place the tube on ice.

5. Aliquot 90 μL RIPA buffer into 200 μL PCR tubes (one tube per ChIP), place on ice and add 10 μL washed Dynabeads® protein A–bead slurry from Step 4 and 2.4 μg antibody to each tube. To the negative control samples, do not add the antibody, or add a pre-immune antibody preferably of the same isotype as the ChIP antibodies. Place at 40 rpm on a rotator for 2 h at 4°C (*see* **Note 2**).

3.2. Cross-Linking of DNA and Proteins

1. Add 20 m*M* Na-butyrate from the 1 *M* stock to the cell culture and mix gently. Na-butyrate is added immediately before collecting cells and cross-linking to avoid artifactual histone hyperacetylation. Na-butyrate only needs to be included when acetylated epitopes are assessed.
2. Discard the medium to remove dead cells (if cells are growing adherent) and add room temperature (20–25°C) PBS/Na-butyrate (10 mL per 175 cm^2 culture flask).
3. Harvest cells by trypsinization or as per your standard protocol according to cell type. Trypsin or other harvesting solution should contain 20 m*M* Na-butyrate.
4. Count cells and resuspend 1,000 (or 100) cells in 500 μL PBS/Na-butyrate in a 0.6 mL tube at room temperature (*see* **Note 3**).
5. Add 13.5 μL formaldehyde (1% (vol/vol) final concentration), mix by gentle vortexing, and incubate for 8 min at room temperature (*see* **Note 4**).
6. Add 57 μL of the 1.25 *M* glycine stock (125 m*M* final concentration) and incubate for 5 min at room temperature. Pellets of cross-linked cells can be stored at -80°C for at least 1 month.

3.3. Preparation of Chromatin from 1,000 Cells

The procedure described here is for preparing chromatin from 1,000 cells (starting material). It is, however, also suited for up to 50,000 cells with adjustments in sonication conditions. A procedure for assessing chromatin fragmentation by sonication of small cell numbers has recently been published (29).

1. Centrifuge formaldehyde cross-linked cells at 470*g* for 10 min at 4°C in a swing-out rotor with soft deceleration settings. Slowly aspirate and discard the supernatant, leaving ~30 μL of the solution with the cell pellet to ensure that none of the loosely packed cells are aspirated.
2. Resuspend the cells in 500 μL ice-cold PBS/Na-butyrate by gentle vortexing and centrifuge at 470*g* for 10 min at 4°C as in Step 1.
3. Repeat the washing procedure (Step 2) once. Upon aspiration of the last wash, leave 20 μL PBS/Na-butyrate with the cell pellet.

4. Add 120 μL room temperature lysis buffer, vortex for 2 × 5 s, leave on ice for 5 min, and resuspend cells by vortexing. Ensure that no liquid is trapped in the lid.
5. Sonicate on ice for 3 × 30 s, with 30 s pauses on ice between each 30 s session, using the probe sonicator. With the Labsonic M sonicator, use the following pulse settings: cycle 0.5, 30% power (*see* **Note 5**).
6. Add 400 μL RIPA ChIP buffer to the tube (which contains ~140 μL lysate) and mix by vortexing.
7. Centrifuge at 12,000*g* for 10 min at 4°C, aspirate the supernatant (chromatin), and transfer it into a clean 1.5 mL tube chilled on ice (*see* **Note 6**). To avoid aspirating the sedimented material, leave ~50 μL supernatant in the tube after aspiration.
8. Add 410 μL RIPA ChIP buffer to the remaining volume, mix by vortexing, and centrifuge at 12,000*g* for 10 min at 4°C.
9. Aspirate the supernatant, leaving ~20 μL with the (invisible) pellet and pool it with the first supernatant. This yields ~930 μL of chromatin suitable for eight parallel ChIPs and one input reference. Discard the pellets. Diluting the chromatin reduces SDS concentration to 0.1%, which is suitable for immunoprecipitation with most antibodies.
10. Aliquot 100 μL chromatin each into, e.g., eight chilled 0.2 mL tubes (in strip format) containing antibody–bead complexes held to the wall in the magnetic rack (on ice), and from which the RIPA buffer has been pipetted out.
11. Add 100 μL chromatin to a tube chilled on ice. This is used as input chromatin. A 1.5 mL tube is used in this step if DNA is to be purified with phenol:chloroform:isoamylalcohol. For DNA isolation using Chelex-100, a 0.6 mL tube is preferred.

3.4. Preparation of Chromatin from 100 Cells

This procedure is for preparing chromatin when starting with 100 cells, but can also be applied to up to 1,000 cells. When starting with 100 cells, only one immunoprecipitation can be performed per sample. Prepare an additional sample for reference input chromatin.

1. Centrifuge formaldehyde cross-linked cells at 470*g* for 10 min at 4°C in a swing-out rotor with soft deceleration settings. Aspirate the supernatant; leave ~30 μL of the solution with the pellet.
2. Add 500 μL ice-cold PBS/Na-butyrate, resuspend the cells by gentle vortexing, and centrifuge at 470*g* for 10 min at 4°C using a swing-out rotor with soft deceleration settings.

3. Repeat the washing procedure (Step 2) once. Leave ~20 μL of PBS/Na-butyrate with the pellet (invisible) after removing the last wash.
4. Add 120 μL lysis buffer, vortex for twice 5 s, and incubate for 3 min on ice (*see* **Note 7**).
5. Centrifuge the nuclei at 860*g* for 10 min at 4°C using a swing-out rotor with soft deceleration settings and discard the supernatant; leave 20–30 μL of lysis buffer in the tube.
6. Add 120 μL RIPA ChIP buffer and vortex for 10 s.
7. Sonicate each tube on ice for twice 30 s, with 30 s pauses on ice between each 30 s session, using the probe sonicator (cycle 0.5 and 30% power with the Labsonic M). Repeat for each tube while leaving the sonicated samples on ice. Note that when starting with 100 cells, it is impossible to visualize chromatin fragmentation by agarose gel electrophoresis. Instead, we use a PCR-based assay (29).
8. Pipette the lysate several times using a siliconized pipette tip and transfer into a 0.2 mL PCR tube containing antibody-coated beads and from which the RIPA buffer has been removed.

3.5. Immunoprecipitation and Washes

1. Remove the tube strip from the magnetic rack to release the antibody–bead complexes into the chromatin suspension and place the tubes on a rotator at 40 rpm for 2 h at 4°C. This step can be carried out overnight at 4°C if necessary, but prolonged incubation may enhance background.
2. Centrifuge the tubes in a minicentrifuge for 1 s to bring down any solution trapped in the lid during the incubation on the rotator, and capture the immune complexes by placing the tubes in the chilled magnetic rack.
3. Discard the supernatant, add 100 μL ice-cold RIPA buffer, and remove the tubes from the magnetic rack to release the immune complexes into the buffer. Resuspend the complexes by gentle manual agitation and place the tubes on a rotator at 40 rpm for 4 min at 4°C.
4. Repeat Steps 2 and 3 twice. Briefly spin the tubes in a minicentrifuge for 1 s to bring down any liquid trapped in the lid prior to placing the tubes in the magnetic rack.
5. Centrifuge the tubes in a minicentrifuge for 1 s.
6. Remove the supernatant, add 100 μL TE buffer, and incubate on a rotator at 4°C for 4 min at 40 rpm.
7. Centrifuge the tubes in a minicentrifuge for 1 s.

8. Place the tubes on ice (not in the magnetic rack), transfer the content of each tube into separate clean 0.2 mL tubes on ice, capture the complexes in the magnetic rack, and remove the TE buffer.

3.6. DNA Recovery by Phenol:Chloroform: Isoamylalcohol Extraction

We have used two procedures for recovering DNA from the ChIP material and from input chromatin involving (i) a phenol:chloroform:isoamylalcohol extraction and (ii) a resin-mediated DNA isolation (Chelex-100).

3.6.1. DNA Recovery from ChIP Material

Combined DNA Elution, Cross-Link Reversal, Proteinase K Digestion, Followed by DNA Purification by Phenol:Chloroform:Isoamylalcohol Extraction

1. Place the tubes from **Section 3.5**, Step 8 in a rack and add 150 µL complete elution buffer to each tube.
2. Incubate for 2 h on the Thermomixer at 68°C, 1,300 rpm. Meanwhile, prepare the input sample as described in **Section 3.6.2.** DNA elution from immune complexes, cross-link reversal, and protein digestion are combined into one step.
3. Remove tubes from the Thermomixer and centrifuge for 3 s with a minicentrifuge.
4. Capture the beads using the magnetic rack, collect the supernatant, and place it in a clean 1.5 mL tube.
5. Add 150 µL complete elution buffer to the remaining ChIP material and incubate on the Thermomixer for 5 min at 68°C, 1,300 rpm.
6. Remove the tubes from the Thermomixer, capture the beads using the magnetic rack, collect the supernatant, and combine it with the first supernatant.
7. Add 200 µL elution buffer to the eluted ChIP material.
8. Extract DNA once with an equal volume of phenol:chloroform:isoamylalcohol, centrifuge at 15,000*g* for 5 min to separate the phases and transfer 460 µL of the aqueous (top) phase to a clean tube.
9. Extract once with an equal volume of chloroform isoamylalcohol, centrifuge at 15,000*g* for 5 min, and transfer 400 µL of the aqueous phase to a clean tube. Use filtered tips when adding phenol:chloroform:isoamylalcohol and chloroform:isoamylalcohol to prevent dripping during transfer.
10. Add 44 µL of 3 *M* NaAc (pH 7.0), 10 µL of 0.25% (wt/vol) acrylamide carrier, and 1 mL 96% ethanol at –20°C. Mix thoroughly and incubate for at least 1 h at –80°C. DNA can be left at –80°C for several hours or days if more convenient.

11. Thaw the tubes and centrifuge at 20,000*g* for 15 min at 4°C.
12. Remove the supernatant, add 1 ml of 70% ethanol at -20°C, and vortex briefly to wash the DNA pellet. Centrifuge at 20,000*g* for 10 min at 4°C. Repeat this step once more.
13. Remove the supernatant and dissolve the DNA in 30 μL TE for ChIPs from chromatin from 100 cells or 60 μL for a ChIP from chromatin from 1,000 cells. DNA can be immediately used for PCR or stored at –20°C for up to 1 week (*see* **Note 8**).

3.6.2. DNA Recovery from Input Chromatin

1. To input chromatin samples, add 200 μL of elution buffer and 7.5 μL of a 10 × dilution (2 mg/mL) of the proteinase K solution, vortex, and incubate for 2 h on a heating block at 68°C.
2. Remove samples from the heating block and add 200 μL elution buffer.
3. Continue from Step 8 in **Section 3.6.1**, processing the input samples and the ChIP samples in parallel.

3.7. DNA Recovery Using Chelex-100

This DNA recovery procedure describes a Chelex-100-mediated DNA purification reported previously (26), with modifications for small cell number ChIP and to speed up handling.

3.7.1. DNA Recovery from ChIP Samples

1. To the washed ChIP samples, add 40 μL of 10% Chelex-100, release immune complexes, and vortex for 10 s. Make sure the Chelex-100 beads are in suspension while pipetting and that the opening of the pipette tip is large enough not to hinder the beads.
2. Boil ChIP samples and input samples (prepared as described in Step 4, **Section 3.7.2**) for 10 min in a PCR machine and cool to room temperature.
3. Add 1 μL proteinase K solution, vortex, and incubate at 55°C, 30 min, 1,300 rpm in the Thermomixer.
4. Boil for 10 min, centrifuge for 10 s in a minicentrifuge, and keep tubes upright for ~1 min on the bench, with no magnet, to allow beads to settle.
5. Using a siliconized tip, transfer 30 μL of the supernatant into a clean 0.6 mL tube chilled on ice. Take great care to avoid transfer of beads.
6. Add 10 μL MilliQ H_2O to the remaining beads, vortex, and centrifuge for 10 s in a minicentrifuge.
7. After the beads settle, collect 12 μL of the supernatant, pool with the first supernatant, and vortex (*see* **Note 9**).

3.7.2. DNA Recovery from Input Chromatin

1. To input chromatin samples, add 10 μL acrylamide carrier and 250 μL 96% ethanol at –20°C. Vortex thoroughly and place at –80°C for 30 min.
2. Thaw, immediately centrifuge at 20,000*g* for 15 min at 4°C, and wash the pellet in 500 μL of 70% ethanol. Dry the pellet.
3. Add 40 μL of 10% (wt/vol) Chelex-100 to the dried pellet and vortex for 10 s.
4. Continue from Step 2, **Section 3.7.1.**, processing input and ChIP samples in parallel.

3.8. Set-Up of Real-Time PCR and Analysis of Data

1. Prepare a master mix and aliquot for individual 25 μL qPCR reactions (MilliQ water 6.5 μL; SYBR Green Master Mix (2X) 12.5 μL; forward primer (20 μ*M* stock) 0.5 μL; reverse primer (20 μ*M* stock) 0.5 μL; DNA template, 5 μL) for all ChIP and input samples with each primer pair.
2. Prepare a standard curve with genomic DNA. Make sure to include a wide range of DNA concentrations (e.g., 0.005–20 ng/μL) to cover the range of your ChIP DNA samples. Use 5 μL DNA in each PCR. Establish one standard curve for each primer pair and for each PCR plate.
3. Set up a real-time PCR program, using your real-time PCR system, with a 40-cycle program.
4. Acquire the data using your real-time PCR data acquisition program.
5. Calculate the amount of DNA in each sample using the standard curve.
6. Export the data into Excel spreadsheets.
7. Determine the amount of precipitated DNA relative to input as [(Amount of ChIP DNA)/(Amount of input DNA)] × 100. We analyze at least three independent ChIPs, each in duplicate qPCRs and express the data as percent (±SD) precipitated DNA relative to input DNA (**Fig.** 4.2) (*see* **Note 10**).

4. Notes

1. We have used with this protocol the following anti-histone antibodies: anti-H3K9ac (Upstate, cat. no. 06-942), anti-H3K9m2 (Upstate, cat. no. 07-441), anti-H3K9m3 (Upstate, cat. no. 07-442), anti-H3K27m3 (Upstate, cat. no. 05-851), anti-H3K9m3 (Diagenode, cat. no. pAb-056-050), anti-H3K4m2

(Abcam, cat. no. Ab7766), anti-H3K4m3 (Abcam, cat. no. Ab8580). We have also used an anti-RNAPII antibody (Santa Cruz Biotechnology, cat. no. sc-899); the procedure should be tested for other antibodies.

2. This incubation step should be carried out during cross-linking, cell lysis, and chromatin preparation and if necessary can be prolonged until all chromatin samples are ready for immunoprecipitation. We recommend using 0.2 mL PCR tubes in an eight-tube strip format, which fits in the magnetic rack.
3. Up to 50,000 cells can be used using the same protocol. More cells allow the analysis of more genomic loci by PCR. To prevent cell lysis during pipetting, use a 1,000 μL pipette tip or a 200 μL pipette tip with a cut end.
4. Formaldehyde cross-links DNA to proteins located within 2 Å of DNA (30). To simplify the cross-linking step and enhance cell recovery, we consistently cross-link cells in suspension. Time of cross-linking may vary with the protein to be immunoprecipitated, but for most applications, 8–10 min cross-linking is sufficient.
5. Sonication should produce chromatin fragments of ~500 bp (range may be 200–1,200 bp). The sonication regime indicated is suitable for a variety of cultured cell lines but must be optimized for each cell type, particularly for primary cells. Do not allow samples to foam as foaming reduces sonication efficiency. If foaming occurs, ensure that the sonicator probe is placed deep enough, a few millimeters from the bottom of the tube, or reduce sonication intensity.
6. To avoid aspirating the sedimented material, which is invisible, leave ~50 μL supernatant in the tube after aspiration.
7. Keeping cells in lysis buffer for over 3 min prior to centrifugation increases the chance of SDS precipitating. If the SDS precipitates during centrifugation, remove the lysis buffer, add 200 μL RIPA ChIP buffer, dissolve the SDS by vortexing, and centrifuge the nuclei as in Step 5, **Section 3.4.**
8. TE volume depends on the number of cells in the ChIP. Note that low DNA concentrations lead to degradation of the DNA more rapidly than at high concentrations. Thus, we recommend to immediately use DNA for PCR for ChIPs from 1,000 cells or less.
9. The volumes collected must be identical between samples if ChIP results are to be compared. Chelex-100 enhances DNA recovery but yields larger volumes than phenol:chloroform:isoamylalcohol extraction. Final ChIP results are similar with either isolation method (26, 28). The DNA can be immediately used for PCR or stored at –20°C for up to 1 week.

10. If no PCR signal is detected, several factors may be implicated. (1) There is not enough chromatin in the ChIP assay: increase the amount of cells or chromatin (note that it may be difficult to extract all chromatin from certain primary cell types); (2) the ChIP did not work: use ChIP-grade antibodies if possible; do an antibody titration; (3) the PCR did not work: set up a control qPCR with the same primers on genomic DNA and optimize PCR conditions; ensure there is no carryover Chelex-100 with the template. If PCR signals are weaker than expected, there might not be enough DNA template. If variations in PCR signal intensity are detected between ChIP replicates, this may be due to (1) inconsistent chromatin preparations between samples: ensure that insoluble debris are removed by sedimentation after fragmentation; do not to carry over debris when aspirating the chromatin supernatant; (2) inconsistent sonication: practice sonication on larger cell numbers (e.g., 100,000) until fragmentation is reproducible; (3) variable amounts of Dynabeads between samples: ensure magnetic beads are well suspended while pipetting; (4) too little and variable amounts input DNA template (high Ct values): increase the amount of input DNA template in the PCR and ensure consistency between replicates; ensure that ethanol-precipitated DNA is fully dissolved before PCR.

Acknowledgments

Our work is supported by the FUGE, YFF, STAMCELLER, and STORFORSK programs of the Research Council of Norway and by the Norwegian Cancer Society.

References

1. Collas, P. and Dahl, J. A. (2008) Chop it, ChIP it, check it: the current status of chromatin immunoprecipitation. *Front. Biosci.* **13**, 929–943.
2. O'Neill, L. P. and Turner, B. M. (1995) Histone H4 acetylation distinguishes coding regions of the human genome from heterochromatin in a differentiation-dependent but transcription-independent manner. *EMBO J.* **14**, 3946–3957.
3. O'Neill, L. P. and Turner, B. M. (1996) Immunoprecipitation of chromatin. *Methods Enzymol.* **274**, 189–197.
4. O'Neill, L. P., Vermilyea, M. D. and Turner, B. M. (2006) Epigenetic characterization of the early embryo with a chromatin immunoprecipitation protocol applicable to small cell populations. *Nat. Genet.* **38**, 835–841.
5. Azuara, V., Perry, P., Sauer, S., Spivakov, M., Jorgensen, H. F., John, R. M., Gouti, M., Casanova, M., Warnes, G., Merkenschlager, M. and Fisher, A. G. (2006) Chromatin signatures of pluripotent cell lines. *Nat. Cell Biol.* **8**, 532–538.

6. Nelson, J. D., Denisenko, O., Sova, P. and Bomsztyk, K. (2006) Fast chromatin immunoprecipitation assay. *Nucleic Acids Res.* **34**, e2.
7. Dahl, J. A. and Collas, P. (2007) Q^2ChIP, a quick and quantitative chromatin immunoprecipitation assay unravels epigenetic dynamics of developmentally regulated genes in human carcinoma cells. *Stem Cells* **25**, 1037–1046.
8. Attema, J. L., Papathanasiou, P., Forsberg, E. C., Xu, J., Smale, S. T. and Weissman, I. L. (2007) Epigenetic characterization of hematopoietic stem cell differentiation using miniChIP and bisulfite sequencing analysis. *Proc. Natl. Acad. Sci. U.S.A.* **104**, 12371–12376.
9. Bernstein, B. E., Kamal, M., Lindblad-Toh, K., Bekiranov, S., Bailey, D. K., Huebert, D. J., McMahon, S., Karlsson, E. K., Kulbokas, E. J., III, Gingeras, T. R., Schreiber, S. L. and Lander, E. S. (2005) Genomic maps and comparative analysis of histone modifications in human and mouse. *Cell* **120**, 169–181.
10. Boyer, L. A., Lee, T. I., Cole, M. F., Johnstone, S. E., Levine, S. S., Zucker, J. P., Guenther, M. G., Kumar, R. M., Murray, H. L., Jenner, R. G., Gifford, D. K., Melton, D. A., Jaenisch, R. and Young, R. A. (2005) Core transcriptional regulatory circuitry in human embryonic stem cells. *Cell* **122**, 947–956.
11. Bernstein, B. E., Mikkelsen, T. S., Xie, X., Kamal, M., Huebert, D. J., Cuff, J., Fry, B., Meissner, A., Wernig, M., Plath, K., Jaenisch, R., Wagschal, A., Feil, R., Schreiber, S. L. and Lander, E. S. (2006) A bivalent chromatin structure marks key developmental genes in embryonic stem cells. *Cell* **125**, 315–326.
12. Loh, Y. H., Wu, Q., Chew, J. L., Vega, V. B., Zhang, W., Chen, X., Bourque, G., George, J., Leong, B., Liu, J., Wong, K. Y., Sung, K. W., Lee, C. W., Zhao, X. D., Chiu, K. P., Lipovich, L., Kuznetsov, V. A., Robson, P., Stanton, L. W., Wei, C. L., Ruan, Y., Lim, B. and Ng, H. H. (2006) The Oct4 and Nanog transcription network regulates pluripotency in mouse embryonic stem cells. *Nat. Genet.* **38**, 431–440.
13. Lee, T. I., Jenner, R. G., Boyer, L. A., Guenther, M. G., Levine, S. S., Kumar, R. M., Chevalier, B., Johnstone, S. E., Cole, M. F., Isono, K., Koseki, H., Fuchikami, T., Abe, K., Murray, H. L., Zucker, J. P., Yuan, B., Bell, G. W., Herbolsheimer, E., Hannett, N. M., Sun, K., Odom, D. T., Otte, A. P., Volkert, T. L., Bartel, D. P., Melton, D. A., Gifford, D. K., Jaenisch, R. and Young, R. A. (2006) Control of developmental regulators by Polycomb in human embryonic stem cells. *Cell* **125**, 301–313.
14. Guenther, M. G., Levine, S. S., Boyer, L. A., Jaenisch, R. and Young, R. A. (2007) A chromatin landmark and transcription initiation at most promoters in human cells. *Cell* **130**, 77–88.
15. Acevedo, L. G., Iniguez, A. L., Holster, H. L., Zhang, X., Green, R. and Farnham, P. J. (2007) Genome-scale ChIP-chip analysis using 10,000 human cells. *Biotechniques* **43**, 791–797.
16. Zhao, X. D., Han, X., Chew, J. L., Liu, J., Chiu, K. P., Choo, A., Orlov, Y. L., Sung, W. K., Shahab, A., Kuznetsov, V. A., Bourque, G., Oh, S., Ruan, Y., Ng, H. H. and Wei, C. L. (2007) Whole-genome mapping of histone H3 Lys4 and 27 trimethylations reveals distinct genomic compartments in human embryonic stem cells. *Cell Stem Cell* **1**, 286–298.
17. O'Neill, L. P. and Turner, B. M. (2003) Immunoprecipitation of native chromatin: NChIP. *Methods* **31**, 76–82.
18. Hudson, M. E. and Snyder, M. (2006) High-throughput methods of regulatory element discovery. *Biotechniques* **41**, 673, 675, 677.
19. Dunn, J. J., McCorkle, S. R., Everett, L. and Anderson, C. W. (2007) Paired-end genomic signature tags: a method for the functional analysis of genomes and epigenomes. *Genet. Eng. (NY)* **28**, 159–173.
20. Aiba, K., Carter, M. G., Matoba, R. and Ko, M. S. (2006) Genomic approaches to early embryogenesis and stem cell biology. *Semin. Reprod. Med.* **24**, 330–339.
21. Clark, D. J. and Shen, C. H. (2006) Mapping histone modifications by nucleosome immunoprecipitation. *Methods Enzymol.* **410**, 416–430.
22. Negre, N., Lavrov, S., Hennetin, J., Bellis, M. and Cavalli, G. (2006) Mapping the distribution of chromatin proteins by ChIP on chip. *Methods Enzymol.* **410,** 316–341.
23. Wu, J., Smith, L. T., Plass, C. and Huang, T. H. (2006) ChIP-chip comes of age for genome-wide functional analysis. *Cancer Res.* **66**, 6899–6902.
24. Bulyk, M. L. (2006) DNA microarray technologies for measuring protein–DNA interactions. *Curr. Opin. Biotechmol.* **17**, 422–430.
25. O'Geen, H., Nicolet, C. M., Blahnik, K., Green, R. and Farnham, P. J. (2006)

Comparison of sample preparation methods for ChIP-chip assays. *Biotechniques* **41**, 577–580.

26. Nelson, J. D., Denisenko, O. and Bomsztyk, K. (2006) Protocol for the fast chromatin immunoprecipitation (ChIP) method. *Nat. Protoc.* **1**, 179–185.
27. Flanagin, S., Nelson, J. D., Castner, D. G., Denisenko, O. and Bomsztyk, K. (2008) Microplate-based chromatin immunoprecipitation method, Matrix ChIP: a platform to study signaling of complex genomic events. *Nucleic Acids Res.* **36**, e17.
28. Dahl, J. A. and Collas, P. (2008) MicroChIP – A rapid micro chromatin immunoprecipitation assay for small cell samples and biopsies. *Nucleic Acids Res.* **36**, e15.
29. Dahl, J. A. and Collas, P. (2008) A rapid micro chromatin immunoprecipitation assay (μChIP). *Nat. Protoc.* **3**, 1032–1045.
30. Orlando, V. (2000) Mapping chromosomal proteins in vivo by formaldehyde-crosslinked-chromatin immunoprecipitation. *Trends Biochem. Sci.* **25**, 99–104.

Chapter 5

Fish'n ChIPs: Chromatin Immunoprecipitation in the Zebrafish Embryo

Leif C. Lindeman, Linn T. Vogt-Kielland, Peter Aleström, and Philippe Collas

Abstract

Chromatin immunoprecipitation (ChIP) is arguably the assay of choice to determine the genomic localization of DNA- or chromatin-binding proteins, including post-translationally modified histones, in cells. The increasing importance of the zebrafish, *Danio rerio*, as a model organism in functional genomics has recently sparked investigations of ChIP-based genome-scale mapping of modified histones on promoters, and studies on the role of specific transcription factors in developmental processes. ChIP assays used in these studies are cumbersome and conventionally require relatively large number of embryos. To simplify the procedure and to be able to apply the ChIP assay to reduced number of embryos, we re-evaluated the protocol for preparation of embryonic chromatin destined to ChIP. We found that manual homogenization of embryos rather than protease treatment to remove the chorion enhances ChIP efficiency and quickens the assay. We also incorporated key steps from a recently published ChIP assay for small cell numbers. We report here a protocol for immunoprecipitation of modified histones from mid-term blastula zebrafish embryos.

Key words: Chromatin immunoprecipitation, ChIP, embryo, histone modification, zebrafish.

1. Introduction

The importance of zebrafish as a model system for studying vertebrate embryogenesis or even human disease has been strongly established (1–4). Advantages of zebrafish are that several hundreds of synchronized embryos can be produced from a few females, generation interval is short (3–4 months), embryos are transparent, and development is rapid (1,000 cell-stage at 3 h post-fertilization, hpf) and external, so all developmental stages are accessible for manipulation and observation, in contrast to

Philippe Collas (ed.), *Chromatin Immunoprecipitation Assays,* Methods in Molecular Biology 567,
DOI 10.1007/978-1-60327-414-2_5, © Humana Press, a part of Springer Science+Business Media, LLC 2009

most other vertebrate models. Zebrafish are also well suited for functional genomics investigations (4). Large-scale mutagenesis screens can be undertaken and stable transgenic lines are easy to establish. The seventh assembly of the zebrafish genome (Zv7) reports 1, 563, 441, 531 bp with 24,147 protein-coding genes (www.sanger.ac.uk/Projects/D_rerio). Although not finally annotated, access to the genome sequence allows the identification of gene orthologs. Forward genetics has through positional cloning enabled discoveries of over 2,000 zebrafish developmental gene relationships (4). Reverse genetics through antisense Morpholino oligonucleotides (5), TILLING targeted mutagenesis (6), and zinc finger nucleases (7, 8), and the emergence of zebrafish expression arrays with probes from oligonucleotide libraries based on transcription units predicted by improved bioinformatics, places zebrafish functional genomics at a level comparable to that of mouse or human.

Embryo development proceeds from a cascade of gene activation and repression events in response to extracellular signals and local determinants. Resulting changes in gene expression in specific cell types regulate differentiation. The coordinate activation and repression of genes requires intricate regulatory networks (9, 10). These networks are controlled by binding of transcriptional regulators to key gene regulatory sequences. Binding of these factors is itself modulated by modifications of DNA (DNA methylation) or chromatin (such as post-translational modifications of histones). Interactions between proteins and DNA, therefore, are essential to the regulation of gene expression.

To date, the tool of choice for studying protein–DNA interactions and unraveling transcriptional regulatory circuits in cells is chromatin immunoprecipitation (ChIP) [reviewed in (11)]. ChIP has been widely used for mapping the positioning of post-translationally modified histones, transcription factors, or other DNA-binding proteins on specific genomic regions in a variety of cell types and species, including mouse blastocysts (12). In a ChIP assay, DNA and proteins are reversibly cross-linked, chromatin is fragmented, usually by sonication, to ~500 bp fragments and antibodies to the protein of interest (e.g., a modified histone), are used to immunoprecipitate a specific protein–DNA complex. Immune complexes are washed, the chromatin is eluted, cross-links are reversed, and the ChIP DNA is purified. Genomic sequences associated with the precipitated protein can be identified by polymerase chain reaction (PCR), high-throughput sequencing (ChIP-seq), microarray hybridization (ChIP-on-chip), or other methods (11).

Only recently has ChIP been applied to zebrafish embryos. A whole embryo ChIP assay for zebrafish was published in 2006 to establish a proof-of-concept that the procedure was applicable in this species for investigating the enrichment of modified histones (acetylated histone H4) or c-Myc on specific promoters (13). ChIP has also been used for identification of transcriptionally active promoters bearing trimethylated H3 lysine 4 (H3K4m3) in gastrula-stage embryos using a ChIP-on-chip approach (14), and to investigate the role of the transcription factor Trf3 in the initiation of hematopoiesis in the zebrafish embryo (15). These protocols rely on protease (pronase) treatment to remove the chorion prior to preparing nuclei and isolating chromatin. We have found that pronase is detrimental to the efficiency of ChIP and have re-evaluated the procedure for preparation of chromatin. We also take advantage of critical steps in our recently published miniaturized and quick (1 day) ChIP assays (16–18) to produce a revised protocol for efficient immunoprecipitation of modified histones from mid-term blastula (MBT) zebrafish embryos (**Fig. 5.1**).

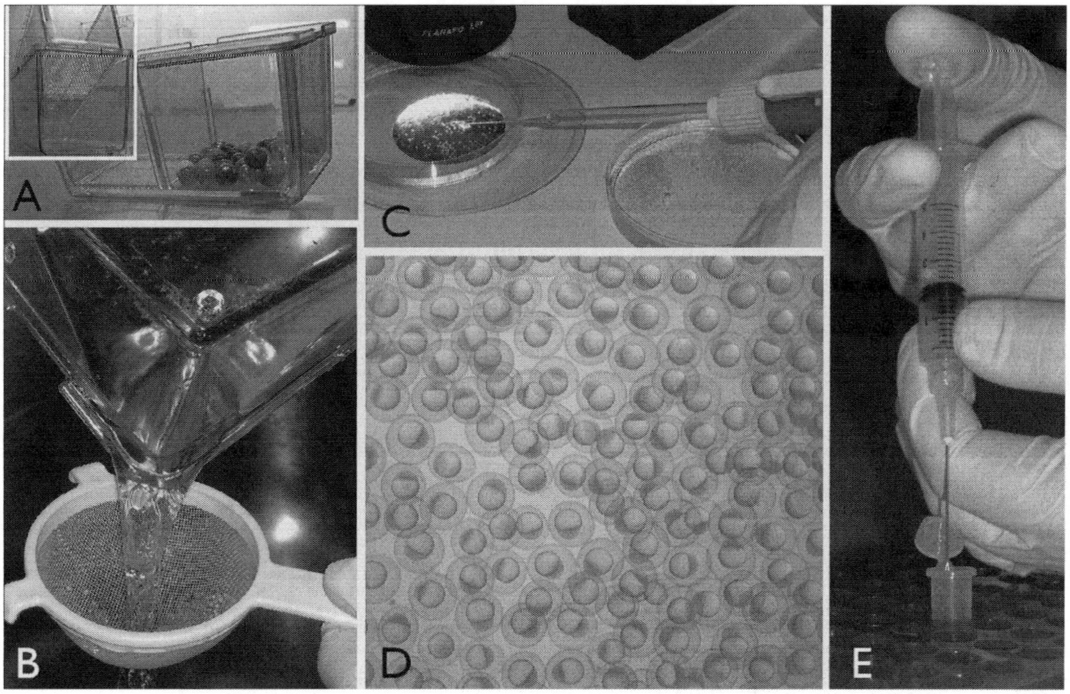

Fig. 5.1. Zebrafish embryo preparation for ChIP assays. (**A**) Breeding tank with a grid in the inner tank; the inner tank is subdivided into two compartments to separate fish of different sex. Marbles are added to the inner tank as enhancement of breeding behavior; marbles are added to both sides (not shown here). (**B**) Harvesting of newly fertilized embryos in a sieve. Embryos can be seen in the sieve. (**C**) Embryos are screened under a dissecting microscope to eliminate unhealthy eggs. (**D**) Selected MBT stage embryos. (**E**) Embryos are homogenized through a 21G needle using a 5 mL syringe.

2. Materials and Reagents

2.1. Materials

2.1.1. Preparation of Zebrafish Embryos

1. Zebrafish, e.g., AB strain (Zebrafish International Resource Center; http://zfin.org/zirc/).
2. Reverse osmosis water production system with filters and UV sterilization (www.zebrafish.no for details).
3. Breeding chambers (2 L) made from autoclavable, FDA-approved, food-grade polycarbonate (Aquatic Habitats, parts no. BTANK2, BINSERT2, BDIVIDER2 and BLID2).
4. Glass marbles (purchased from toy store).
5. Thermo Plate (TOKAI HIT, Model: MATS-U4020WF, or similar).
6. Incubator set to 28°C.
7. Stereo microscope.
8. Digital camera fitted to the microscope.
9. 90 mm plastic Petri dishes.
10. Sieve (purchased from drug store; *see* **Fig.** 5.1B).
11. Glass Pasteur pipettes with glassfirm-pi-pump.

2.1.2. ChIP Assay

1. Filter 10, 200, and 1,000 μL pipette tips.
2. Magnetic rack suited for 200 μL tube strips (Diagenode).
3. 200 μL PCR tubes in eight-tube strip format (Axygen).
4. 0.6 and 1.5 mL centrifuge tubes.
5. Magnetic holder for 1.5 mL tubes.
6. Probe sonicator (e.g., Sartorius Labsonic M sonicator with 3 mm diameter probe at setting 0.5 cycle and 30% power).
7. Rotator (e.g., Science Lab Stuart SB3) placed at 4°C.
8. Tabletop centrifuge.
9. Minicentrifuge.
10. Vortex.
11. Thermomixer (e.g., Eppendorf).
12. Heating block.
13. Real-time thermal cycler.

2.2. Reagents

2.2.1. Preparation of Zebrafish Embryos

1. Instant Ocean (Synthetic sea salt).
2. 1 *M* HCl.

2.2.2. ChIP Assay

1. 36.5% formaldehyde.
2. Dynabeads® Protein A (Invitrogen, cat. no. 100.02D). Beads should be well suspended before pipetting. Use Dynabeads® Protein A beads with rabbit IgGs and Dynabeads® Protein G (Invitrogen, cat. no. 100.04D) with mouse IgGs.
3. 5 *M* NaCl.
4. 400 m*M* EGTA.
5. 500 m*M* EDTA.
6. 1 *M* Tris–HCl, pH 7.5 and 1 *M* Tris–HCl, pH 8.0.
7. Glycine: 1.25 *M* stock solution in PBS.
8. Acrylamide carrier.
9. Proteinase K: 20 mg/mL solution in MilliQ water.
10. Protease inhibitor mix (Sigma-Aldrich, cat. no. P8340).
11. Phenylmethylsulfonyl fluoride (PMSF): 100 m*M* stock solution in 100% ethanol.
12. Na-butyrate: 1 *M* stock solution in MilliQ water.
13. Phosphate buffered saline (PBS).
14. PBS/Na-butyrate solution: 20 m*M* butyrate in 1X PBS. Make immediately before use.
15. PBS/Na-butyrate/formaldehyde fixative: 20 m*M* butyrate, 1 m*M* PMSF, and protease inhibitor mix in 1X PBS. Make up immediately before use.
16. Phenol:chloroform:isoamylalcohol (25:24:1).
17. Chloroform:isoamylalcohol (24:1).
18. 3 *M* NaAc.
19. IQ SYBR® Green (BioRad).
20. Antibodies to the protein to be ChIPed, preferably ChIP-grade.

2.3. Buffers and Solutions

2.3.1. Preparation of Zebrafish Embryos

1. System water for breeding and incubating embryos: purify water by sterile filtration, UV sterilization, and reverse osmosis. Reconditioned by adding, per liter, 0.15 g Instant Ocean (Synthetic sea salt), 0.05 g Na-bicarbonate, and 0.035 g $CaCl_2$. If necessary adjust pH to 7.5 with 1 *M* HCl.
2. Egg water: 60 mg/L Instant Ocean salt in milliQ water. Autoclave.

2.3.2. ChIP Assay

1. Lysis buffer: 50 m*M* Tris–HCl, pH 8.0, 10 m*M* EDTA, 1% (wt/vol) SDS, protease inhibitor mix (1:100 dilution from stock), 1 m*M* PMSF, 20 m*M* Na-butyrate. Protease inhibitor mix, PMSF, and Na-butyrate should be added immediately before use.
2. RIPA buffer: 10 m*M* Tris–HCl, pH 7.5, 140 m*M* NaCl, 1 m*M* EDTA, 0.5 m*M* EGTA, 1% (vol/vol) Triton X-100, 0.1% (wt/vol) SDS, 0.1% (wt/vol) Na-deoxycholate.

3. RIPA ChIP buffer: 10 m*M* Tris–HCl, pH 7.5, 140 m*M* NaCl, 1 m*M* EDTA, 0.5 m*M* EGTA, 1% (vol/vol) Triton X-100, 0.1% (wt/vol) SDS, 0.1% (wt/vol) Na-deoxycholate, protease inhibitor mix (1:100 dilution from stock), 1 m*M* PMSF, 20 m*M* Na-butyrate. Protease inhibitor mix, PMSF, and Na-butyrate should be added immediately before use.
4. TE buffer: 10 m*M* Tris–HCl, pH 8.0, 10 m*M* EDTA.
5. Elution buffer: 20 m*M* Tris–HCl, pH 7.5, 5 m*M* EDTA, 50 m*M* NaCl, 20 m*M* Na-butyrate, 1% (wt/vol) SDS, 50 μg/mL proteinase K. Na-butyrate, SDS, and proteinase K should be added just before use.

3. Methods

3.1. Preparation of Zebrafish Embryos

In this protocol, the ChIP assay is described for embryos at the late MBT stage (>1,000 cells), i.e., between the "high" and "oblong" stages defined on http://www.neuro.uoregon.edu/k12/Table%201.html. At 28°C, this corresponds to 3.5 h post-fertilization (hpf).

1. Set up breeding tanks on the day before you want embryos.
2. Breeding in 2 L tanks with one fish pair. Set up a breeding tank by placing an inner tank with a bottom grid into the 2 L fish tank; the inner tank is divided by a separator into two compartments to separate the fish by sex. Add marbles to both sides of the inner tank and place a lid on top (**Fig. 5.1A**).
3. On the next morning, remove the separator in the 2 L breeding tanks. Avoid stressing the fish and do not feed.
4. After 30–60 min, collect embryos (*see* **Note 1**); pour the embryos from the 2 L tank into an embryo sieve (**Fig. 5.1B**).
5. Thoroughly rinse the embryos in the sieve with system water and transfer them into a 90 mm Petri dish containing room temperature (21–28°C range) system water (*see* **Note 2**).
6. Incubate the embryos for 1 h at 28°C.
7. Using a dissection microscope, select, count, and transfer all healthy embryos to a new 90 mm Petri dish containing system water (**Fig. 5.1C**).
8. To harvest late MBT stage embryos, prolong incubation in the Petri dish for another ~1.5 h at 28°C on a thermoplate or in an incubator (*see* **Note 2**).
9. Document state of embryo development and level of synchronization by a camera fitted to the microscope (**Fig. 5.1D**).

3.2. Cross-Linking of DNA and Proteins

1. Using a transfer pipette, transfer 500 MBT embryos in PBS containing 20 m*M* Na-butyrate, protease inhibitors, and PMSF into a 5 mL syringe fitted with a 21G needle (**Fig. 5.1E**).
2. Let the embryos sink to the bottom of the syringe and remove the PBS with the pipette, leaving ~0.5 mL buffer on top of the embryos.
3. Push the piston and force the embryos through the needle into a 1.5 mL tube. This one-step lysis is usually sufficient to break all the embryos. Wash the needle with a small volume PBS/Na-butyrate, PMSF, and protease inhibitors to collect any leftover in the syringe.
4. Immediately cross-link the cells by adding formaldehyde to 1% vol/vol final concentration, vortexing, and incubating for exactly 8 min at room temperature. Briefly spin (1–2 s) in the minicentrifuge to collect the liquid from the lid.
5. Add glycine to 0.125 *M* to quench the formaldehyde. Vortex, place the tube on ice, and incubate for 5 min. From this step onward, handling of chromatin is carried out on ice.
6. Centrifuge the tube at 470*g* for 10 min at 4°C to sediment cells and fragments from the chorion; carefully remove and discard the supernatant with a 1 mL pipette.
7. Add 500 μL ice-cold PBS/Na-butyrate, PMSF, and protease inhibitors and resuspend the cells by vortexing. Centrifuge at 470*g* for 10 min at 4°C and discard the supernatant.
8. Add another 500 μL PBS/Na-butyrate, PMSF, and protease inhibitors. Transfer to a 0.6 mL tube and centrifuge at 470*g* for 5 min.
9. Remove all the supernatant with a pipette. The cells can be stored as a dry pellet at −80°C for several weeks.

3.3. Preparation of Antibody–Bead Complexes

1. Prepare a slurry of Dynabeads® Protein A or G, depending on the origin of the antibody. For each ChIP to be performed, place 10 μL of well-suspended bead stock solution in a 1.5 mL tube. Place beads in an additional tube for a no-antibody (bead only) control. Work on ice for all steps.
2. Place the tubes in a magnetic holder, capture the beads, remove the supernatant, and add 2.5 volumes of RIPA buffer.
3. Vortex, spin briefly in a minicentrifuge, capture the beads, remove the buffer, and add one volume of RIPA buffer.
4. Repeat Step 3.
5. For each ChIP reaction, add 90 μL RIPA buffer to each 200 μL tube. We find it convenient to use eight-tube PCR strips from Axygen.
6. Add 10 μL of well-dispersed slurry of Dynabeads® Protein.

7. Add a titrated amount of antibody (we routinely use 2.4 μg of anti-modified histone ChIP-grade antibody) (*see* **Note 3**).
8. Incubate on a rotator at 40 rpm at 4°C for 2 h, or overnight if suitable.

3.4. Preparation of Chromatin

1. To a tube containing cells, add lysis buffer to a total volume ~300 μL. Resuspend the pellet with a pipette without making bubbles. We found that starting with a frozen or fresh cross-linked cell pellet has no noticeable influence on ChIP efficiency or results.
2. Cut the end of a 1 mL pipette tip and transfer 120 μL of cell suspension to two 0.6 mL tubes. Incubate on ice for 5–10 min.
3. Sonicate on ice each tube for 8 × 30 s with 30 s pauses on ice between sonication rounds.
4. Centrifuge at 12,000*g* for 10 min at 4°C. Pool 90 μL of the supernatants (chromatin) in a clean 1.5 mL tube.
5. Vortex, spin for 1–2 s in a minicentrifuge, and use 2 μL of chromatin to measure A_{260} with a nanodrop, using lysis buffer with all additives as blank. When starting with 500 embryos, A_{260} should be ~6 U.
6. Dilute the chromatin to 0.2 U A_{260} in RIPA ChIP buffer.
7. Mix well and spin in a minicentrifuge. The diluted chromatin can be stored for several months at –80°C.

3.5. Immunoprecipitation and Washes

1. Spin the tubes with antibody–bead complexes in a minicentrifuge for 1–2 s to bring down any solution trapped in the lid; capture the beads by placing the tubes in a chilled magnetic rack.
2. Remove the RIPA buffer.
3. Remove the tube strips from the magnetic rack and add 100 μL diluted chromatin to each ChIP reaction and to the negative-control ChIP. In addition, place 100 μL input chromatin in a 1.5 mL tube. Put on ice.
4. Place the tubes on the rotator at 40 rpm for 2 h at 4°C. This step can be carried out overnight at 4°C if necessary, but prolonged incubation may enhance background.
5. Centrifuge the tubes in a minicentrifuge for 1 s and capture immune complexes by placing the tubes in the chilled magnetic rack.
6. Discard the supernatant, add 100 μL ice-cold RIPA buffer, and remove tubes from the rack to release immune complexes into the buffer. Resuspend the complexes by gentle manual agitation and place the tubes on rotator at 40 rpm for 4 min at 4°C.
7. Repeat Step 6 twice.

8. Centrifuge the tubes in a minicentrifuge for 1s.
9. Remove the supernatant, add 100 μL TE buffer, and incubate on a rotator at 4°C for 4 min at 40 rpm.
10. Centrifuge the tubes for 1s.
11. Place the tubes on ice (not in the magnetic rack), transfer the content of each tube into separate clean 0.2 mL tubes, capture the complexes in the magnetic rack, and remove the TE buffer.

3.6. DNA Recovery from the Immunoprecipitated Material

1. To each ChIP reaction, add 150 μL ChIP elution buffer. Incubate on thermomixer at 1,300 rpm for 2 h at 68°C.
2. Spin down, capture the beads in the magnetic rack, and transfer the eluate from each tube to clean 1.5 mL tubes.
3. Remove the tube strips from the magnetic rack and add 150 μL ChIP elution buffer. Incubate 15 min on thermomixer as in Step 1.
4. Spin down, capture the beads in the magnetic rack, remove the eluate, and pool it with the first eluate from Step 2.
5. To the pooled eluate (300 μL total volume), add 200 μL ChIP elution buffer.
6. Add proteinase K to 2 mg/mL of the input chromatin sample and incubate at 68°C, 1,300 rpm, on thermomixer for 2 h.
7. Add 500 μL phenol:chloroform:isoamylalcohol, vortex, and centrifuge at 15,000*g* for 5 min. Transfer 450 μL of the upper (aqueous) phase to a new tube.
8. To this aqueous phase, add 450 μL chloroform:isoamyalcohol, vortex, and centrifuge at 15,000*g* for 5 min. Transfer 400 μL of the upper (aqueous) phase to a clean 1.5 mL tube.
9. To this aqueous phase, add 10 μL acrylamide carrier, 40 μL NaAc, and 1 mL 96 or 100% ethanol. Mix by vortexing and inversion and place the tubes at –80°C for 2 h.
10. Centrifuge at 20,000*g* for 10 min at 4°C.
11. Discard the supernatant, wash the pellet with 1 mL 70% ethanol, and let the DNA pellet detach from the tube wall. Centrifuge at 20,000*g* for 10 min, 4°C. Remove the ethanol.
12. Repeat Step 11.
13. Let the DNA pellet dry in open tubes for 1 h.
14. Add 50 μL TE buffer and dissolve the DNA overnight at 4°C.

3.7. Analysis of ChIP DNA by Real-Time PCR

1. Prepare a master mix and aliquot for individual 25 μL qPCR reactions (MilliQ water 6.5 μL; SYBR Green Master Mix (2X) 12.5 μL; forward primer (20 μ*M* stock) 0.5

μL; reverse primer (20 μ*M* stock) 0.5 μL; DNA template, 5 μL) for all ChIP and input samples with each primer pair (*see* **Note 4**).

2. Prepare a standard curve with fragmented genomic DNA, using, e.g., 0.005–20 ng/μL DNA to cover the range of ChIP DNA samples. Use 5 μL DNA in each PCR. Establish one standard curve for each primer pair and for each PCR plate.
3. Set up a real-time PCR 40-cycle program.
4. Acquire the data using your real-time PCR data acquisition program.
5. Calculate the amount of DNA in each sample using the standard curve.
6. Export the data into Excel spreadsheets.
7. Determine the amount of precipitated DNA relative to input as [(Amount of ChIP DNA)/(Amount of input DNA)] × 100 (**Fig. 5.2**).

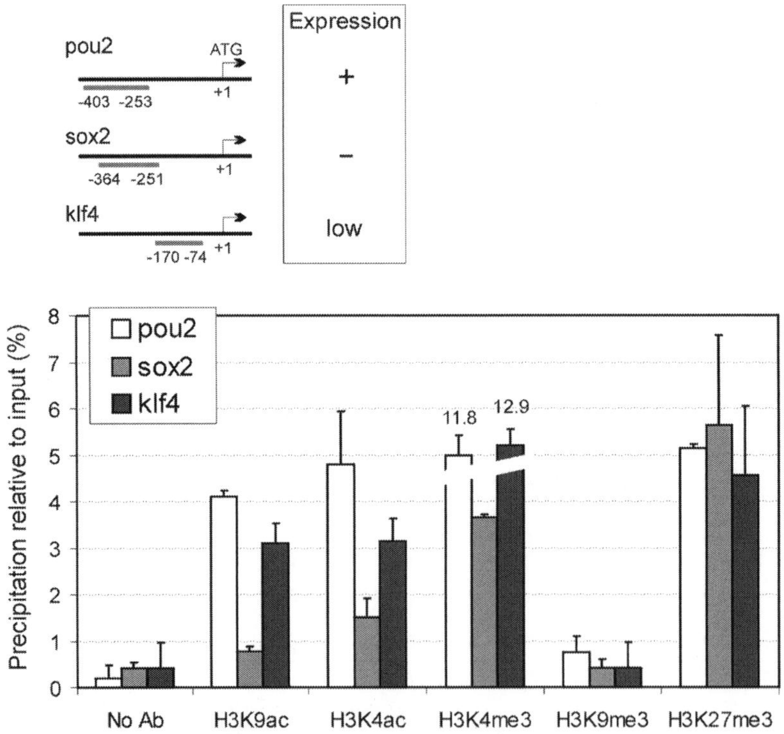

Fig. 5.2. ChIP analysis of post-translationally modified histones in late MBT stage zebrafish embryos. ChIPs were performed using antibodies against indicated histone H3 and H4 modifications as described in this protocol and ChIP DNA was analyzed by quantitative PCR. Promoters of the *pou2*, *sox2,* and *klf4* genes were examined in duplicate ChIPs. Data are expressed as percent precipitated relative to input DNA for each ChIP. Promoter regions relative to the ATG (+1) and expression status of each gene in late MBT stage embryos are shown.

4. Notes

1. It has proven difficult to achieve synchronized breeding when simultaneously breeding many tanks/pairs of fish in order to get sufficient numbers of embryos. For this reason, we often allow 1 h for the breeding/fertilization to take place before collection of embryos (*see* **Section 3.1.**) (**Fig. 5.1B**).
2. For practical reasons, it is difficult to keep a constant temperature of 28°C. If working at lower temperature, time for embryos to reach the late MBT stage is extended by 30–60 min. We always document the state and distribution of embryo stages by taking a picture at the time of harvest. A representative picture of embryo stages ready for ChIP is shown in **Fig. 5.1D**. We have found that a pool of 500 late MBT stage embryos provide enough chromatin for approximately 50 ChIP assays.
3. With zebrafish embryos, we have used the following anti-histone antibodies: anti-H3K9ac (Upstate, cat. no. 06-942), anti-H3K27m3 (Upstate, cat. no. 07-449), anti-H3K9m3 (Diagenode, cat. no. pAb-056-050), anti-H3K4m3 (Abcam, cat. no. Ab8580), and H4Ac (Upstate, cat. no. 06-942).
4. The following primer pairs were used in the data presented here: *pou2* (F) 5′-GATACACCTCGCGTTCCCAAACATGTC-3′ and (R) 5′-TTGCTAATCAATCGGAGTTGGAGGCAG-3′; *sox2* (F) 5′-TGCTGACCGTCCGTAACC-3′ and (R) 5′-ACAACCATTCATAGAGCGACTG-3′; *klf4* (F) 5′-ATCTGATAGGCTACAACTAC-3′ and (R) 5′-TTGGCTGGATGTCTACC-3′. Annealing temperature was 60°C for all primers.

Acknowledgments

This work is supported by a FUGE grant from the Research Council of Norway to PA and PC.

References

1. Ackermann, G. E. and Paw, B. H. (2003) Zebrafish: a genetic model for vertebrate organogenesis and human disorders. *Front. Biosci.* **8**, d1227–d1253.
2. Chen, T., Zhang, Y. L., Jiang, Y., Liu, S. Z., Schatten, H., Chen, D. Y. and Sun, Q. Y. (2004) The DNA methylation events in normal and cloned rabbit embryos. *FEBS Lett.* **578**, 69–72.
3. Berghmans, S., Jette, C., Langenau, D., Hsu, K., Stewart, R., Look, T. and Kanki, J. P. (2005) Making waves in cancer research: new models in the zebrafish. *Biotechniques* **39**, 227–237.
4. Alestrom, P., Holter, J. L. and Nourizadeh-Lillabadi, R. (2006) Zebrafish in functional genomics and aquatic biomedicine. *Trends Biotechnol.* **24**, 15–21.

5. Ekker, S. C. and Larson, J. D. (2001) Morphant technology in model developmental systems. *Genesis* **30**, 89–93.
6. McCallum, C. M., Comai, L., Greene, E. A. and Henikoff, S. (2000) Targeting induced local lesions IN genomes (TILLING) for plant functional genomics. *Plant. Physiol.* **123**, 439–442.
7. Doyon, Y., McCammon, J. M., Miller, J. C., Faraji, F., Ngo, C., Katibah, G. E., Amora, R., Hocking, T. D., Zhang, L., Rebar, E. J., Gregory, P. D., Urnov, F. D. and Amacher, S. L. (2008) Heritable targeted gene disruption in zebrafish using designed zinc-finger nucleases. *Nat. Biotechnol.* **26**, 702–708.
8. Meng, X., Noyes, M. B., Zhu, L. J., Lawson, N. D. and Wolfe, S. A. (2008) Targeted gene inactivation in zebrafish using engineered zinc-finger nucleases. *Nat. Biotechnol.* **26**, 695–701.
9. Boyer, L. A., Lee, T. I., Cole, M. F., Johnstone, S. E., Levine, S. S., Zucker, J. P., Guenther, M. G., Kumar, R. M., Murray, H. L., Jenner, R. G., Gifford, D. K., Melton, D. A., Jaenisch, R. and Young, R. A. (2005) Core transcriptional regulatory circuitry in human embryonic stem cells. *Cell* **122**, 947–956.
10. Loh, Y. H., Wu, Q., Chew, J. L., Vega, V. B., Zhang, W., Chen, X., Bourque, G., George, J., Leong, B., Liu, J., Wong, K. Y., Sung, K. W., Lee, C. W., Zhao, X. D., Chiu, K. P., Lipovich, L., Kuznetsov, V. A., Robson, P., Stanton, L. W., Wei, C. L., Ruan, Y., Lim, B. and Ng, H. H. (2006) The Oct4 and Nanog transcription network regulates pluripotency in mouse embryonic stem cells. *Nat. Genet.* **38**, 431–440.
11. Collas, P. and Dahl, J. A. (2008) Chop it, ChIP it, check it: the current status of chromatin immunoprecipitation. *Front. Biosci.* **13**, 929–943.
12. O'Neill, L. P., Vermilyea, M. D. and Turner, B. M. (2006) Epigenetic characterization of the early embryo with a chromatin immunoprecipitation protocol applicable to small cell populations. *Nat. Genet.* **38**, 835–841.
13. Havis, E., Anselme, I. and Schneider-Maunoury, S. (2006) Whole embryo chromatin immunoprecipitation protocol for the in vivo study of zebrafish development. *Biotechniques* **40**, 34, 36, 38.
14. Wardle, F. C., Odom, D. T., Bell, G. W., Yuan, B., Danford, T. W., Wiellette, E. L., Herbolsheimer, E., Sive, H. L., Young, R. A. and Smith, J. C. (2006) Zebrafish promoter microarrays identify actively transcribed embryonic genes. *Genome Biol.* 7, R71.
15. Hart, D. O., Raha, T., Lawson, N. D. and Green, M. R. (2007) Initiation of zebrafish haematopoiesis by the TATA-box-binding protein-related factor Trf3. *Nature* **450**, 1082–1085.
16. Dahl, J. A. and Collas, P. (2007) Q^2ChIP, a quick and quantitative chromatin immunoprecipitation assay unravels epigenetic dynamics of developmentally regulated genes in human carcinoma cells. *Stem Cells* **25**, 1037–1046.
17. Dahl, J. A. and Collas, P. (2008) MicroChIP – A rapid micro chromatin immunoprecipitation assay for small cell samples and biopsies. *Nucleic Acids Res.* **36**, e15.
18. Dahl, J. A. and Collas, P. (2008) A rapid micro chromatin immunoprecipitation assay (μChIP). *Nat. Protoc.* **3**, 1032–1045.

Chapter 6

Epitope Tagging of Endogenous Proteins for Genome-Wide Chromatin Immunoprecipitation Analysis

Zhenghe Wang

Abstract

The development of chromatin imsmunoprecipitation methods coupled with DNA microarray (ChIP-chip) technology has enabled genome-wide identification of *cis*-DNA regulatory elements to which transcription factors bind. Nonetheless, the ChIP-chip technology requires antibodies with extremely high affinity and specificity for the target transcription factors. Unfortunately, such antibodies are not available for most human transcription factors. In principle, this problem can be circumvented by utilizing ectopically expressed epitope-tagged proteins recognizable by well-characterized antibodies. However, such expression is no longer endogenous. To surmount this problem, we have successfully developed a facile method to knock in a 3xFlag epitope into the endogenous gene loci of transcription factors. The knock-in approach provides a general solution for the study of proteins for which antibodies are sub-standard or not available.

Key words: Epitope tag, ChIP-chip, recombinant adeno-associated virus, knock-in, colorectal cancer.

1. Introduction

The human genome encodes approximately 25,000 proteins. Characterizing all 25,000 depends on the availability of high-quality antibodies that can be used for multiple applications including Western blot, immunofluorescence (IF), and immunoprecipitation (IP). For analysis of transcription factors and other DNA-binding proteins, "ChIP-grade" antibodies capable of immunoprecipitating the protein of interest within the context of chromatin are most often desired (1). Notwithstanding, ChIP-grade antibodies exist for only a small fraction of

Philippe Collas (ed.), *Chromatin Immunoprecipitation Assays,* Methods in Molecular Biology 567,
DOI 10.1007/978-1-60327-414-2_6, © Humana Press, a part of Springer Science+Business Media, LLC 2009

chromatin-associated proteins. This is particularly problematic for ChIP-chip or ChIP-sequencing studies, where the use of more than one antibody is highly recommended. The antibody problem can be circumvented by generating cell lines that stably express epitope-tagged proteins recognizable by available antibodies, but this approach is far from ideal given that expression is no longer endogenous, which may complicate interpretation of results. Moreover, the construction of recombinant plasmids containing both full-length cDNA and epitope sequences can be cumbersome, particularly for proteins encoded by large transcripts.

Epitope tagging by homologous recombination-mediated knock-in (KI) is an effective means for biochemical and cellular studies of proteins in recombination-prone organisms, such as yeast (2). Applying this approach to somatic mammalian cells is not feasible due to low frequency of homologous recombination between exogenous plasmid and specific genomic loci. Recent studies have shown that this problem can be circumvented by delivering constructs with recombinant adeno-associated virus (rAAV), which can increase the frequency of homologous recombination to as much as 2% (3). We have successfully developed a method whereby rAAV is used to "knock in" epitope tag sequences into targeted loci in human somatic cells (4). The tagged proteins, which harbor three Flag epitopes in tandem (3xFlag), can be exploited for Western blot, IP, IF, and ChIP-chip analyses (4). Here, step-by-step protocol is described for the 3xFlag KI approach.

2. Materials

2.1. Targeting Vector Construction

1. pTK-Neo-USER-3xFlag targeting vector.
2. Restriction enzymes: Xba I, Nt.BbvC I (New England Biolabs).
3. Hi-fidelity platinum Taq polymerase (Invitrogen).
4. USER enzyme (New England Biolabs).
5. Subcloning Efficiency™ DH5α™ Competent Cells (Invitrogen).
6. LB agar plates with 100 μg/mL ampicillin.

2.2. rAAV Targeting Virus Generation

1. Dulbecco's Modified Eagle's Medium (DMEM) supplemented with 10% fetal bovine serum (FBS) and 1% Pen/Strep.
2. HEK 293T cells.
3. Phosphate buffered saline.

4. Opti-MEM ® I Reduced Serum Media (Invitrogen, Carlsbad, CA).
5. Lipofectamine™ Transfection Reagent (Invitrogen, Carlsbad, CA).
6. pAAV-RC Plasmid and pHelper Plasmid (Stratagene, La Jolla, CA).
7. Cell scraper.

2.3. Gene Targeting of Human Cells

1. McCoy's 5A Medium (Invitrogen) supplemented with 10% FBS and 1% Pen/Strep.
2. DLD1 colorectal cancer cells (ATCC, Manassas, VA).
3. Trypsin–EDTA.
4. 96-well tissue culture plates.
5. Geneticin.

2.4. Genomic DNA Preparation

1. Lyse-N-go reagent (Pierce, Rockford, IL).
2. Trypsin–EDTA without phenol red.

2.5. Targeted Clone Screening

1. 96-well PCR plates.
2. Platinum Taq polymerase (Invitrogen).

2.6. Excision of the Neomycin Resistance Gene

1. Adeno-Cre recombinase (Adeno-Cre).
2. 6-well and 24-well plates.

3. Method

The 3xFlag tag sequences are inserted before the stop codon of target genes through rAAV-mediated homologous recombination (outlined in **Fig. 6.1**). The entire procedure can be arbitrarily divided into six major steps: (1) Targeting vector construction; (2) rAAV targeting virus generation; (3) Gene targeting of human cells; (4) Genomic DNA preparation; (5) Targeted clone screening; and (6) Excision of the Neomycin resistance gene. It takes ~45 days to generate 3xFlag knock-in clones in DLD1 cells.

We also developed a one-step highly efficient targeting vector construction strategy (**Fig. 6.2**). Recently, the New England Biolabs has developed the USER (uracil-specific excision reagent) cloning technique, which facilitates assembly of multiple DNA fragments in a single reaction by in vitro homologous recombination and single-strand annealing (5). In this system, the vector contains a cassette with two inversely oriented nicking

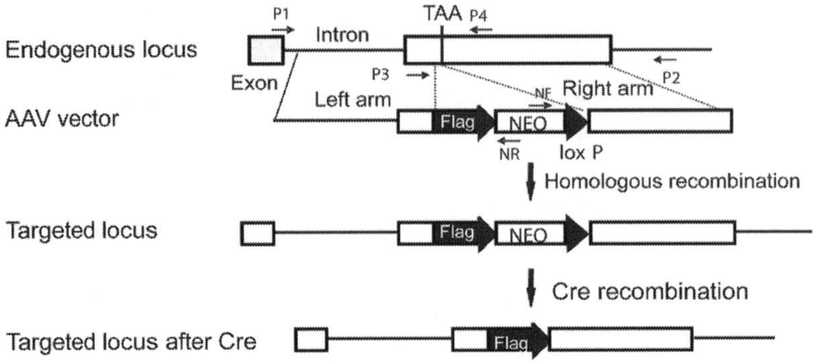

Fig. 6.1. Schematic diagram of tagging endogenous protein with 3xFlag. rAAV targeting vectors contain a left and right arm homologous to sequences in the target gene, flanking a *NEO Lox P-3x Flag* cassette. Clones are then screened by genomic PCR with primers complementary to the neomycin resistance gene and upstream of the left (indicated as P1 and NR) or downstream of the right (indicated as NF and P2) homologous arms. The neomycin gene cassette is excised with Cre-recombinase and genomic PCR using primers P3 and P4 identifies clones with the correct excision.

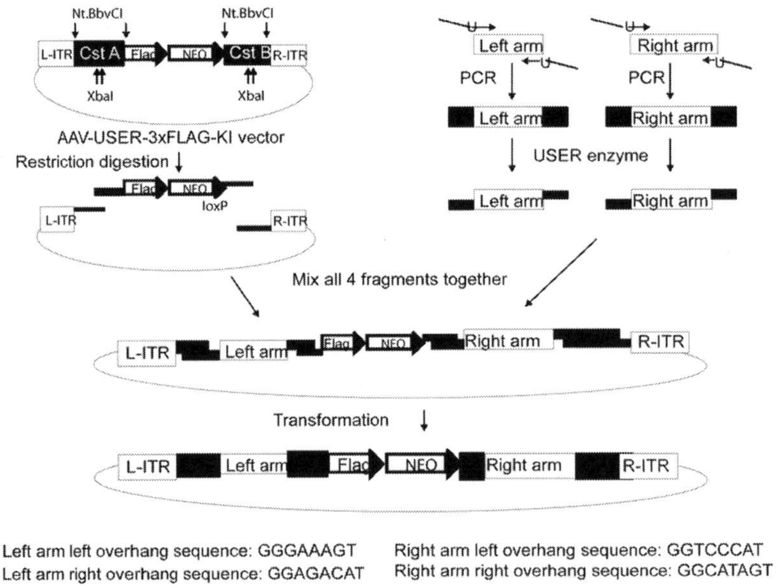

Fig. 6.2. Diagram of targeting vector construction by USER cloning.

endonuclease sites separated by restriction endonuclease site(s). The vector is then digested and nicked with restriction endonucleases, yielding a linearized vector with eight-nucleotide single-stranded, non-complimentary overhangs. To generate target molecules for cloning into this vector, a single deoxyuridine (dU) residue is placed eight nucleotides from the 5′-end of each PCR primer. In addition to the dU, the PCR primers contain sequence that is compatible with each unique overhand on the vector. After amplification, the dU is excised from the PCR products with a uracil DNA glycosylase and an endonuclease (the USER enzyme), generating PCR products flanked by 3′ eight-nucleotide single-stranded extensions that are complementary to the vector overhangs. When mixed together, the linearized vector and PCR products directionally assemble into a recombinant molecule through complementary single-stranded extensions. To make the rAAV-mediated targeting vector compatible with the USER cloning system, we inserted cassette A (Cst A) between L-ITR and 3xFlag sequences, and cassette B (Cst B) between the right lox P site and R-ITR of the AAV-3xFlag knock-in vector to generate the AAV-USER-3xFlag-KI vector (**Fig. 6.2**). These cassettes contain two inversely oriented nicking endonuclease sites (Nt. BbvCI) separated by restriction endonuclease sites (Xba I). After treatment with Nt.BbvC I and Xba I restriction enzymes, the AAV-USER-3xFlag-KI vector is digested into a 3xFlag-lox P-Neo-lox P fragment flanked by two 5′ single-stranded overhangs (**Fig. 6.2**) and a vector backbone flanked by two 5′ overhangs (**Fig. 6.2**). PCR is then used to amplify left and right homologous arms from genomic DNA. The sequence GGGAAAGdU is added to the 5′ of the forward left-arm primers, and GGAGACAdU is added to the reverse left-arm primers. GGTCCCAdU is added to the forward right-arm primers and GGCATAGdU to the reverse left-arm primers. The PCR products are then treated with the USER enzymes to generate single-stranded overhangs. Finally, the left and right arms are mixed with the two vector fragments followed by bacterial transformation (**Fig. 6.2**).

3.1. Targeting Vector Construction

3.1.1. Design of PCR Primers

Using Primer 3 (http://frodo.wi.mit.edu/cgi-bin/primer3/primer3_www.cgi), design primers as follows (*see* **Note 1**):

For the left arm:

Forward primer: add GGGAAAGdU to the 5′ end of the designed PCR primer.

Reverse primer: add GGAGACAdUnn to the 5′ end of the reverse sequences of the upstream of stop codon (the first n could be A, T, G, or C; the second n could be any nucleotides but A so that the 3xFlag is in frame fused with the targeted gene, and avoid to introduce a stop codon before the 3xFlag).

For the right arm:

Forward primer: add GGTCCCAdU to the downstream sequencesof stop codon.

Reverse primer: add GGCATAGdU to the 5′ end of the designed PCR primer.

3.1.2. Amplification of Left and Right Arms

1. Use DLD1 genomic DNA (or genomic DNA from the cell that you intend to target) as the templates. The left and right arms are generated by PCR in two separate reactions (20 μL each) according to the following receipt and cycling conditions:

10 μL reaction	1 reaction	15 reactions
H_2O	5.58	83.7
10X HiFi buffer	1	15
10 m*M* dNTPs	0.2	3
50 m*M* $MgSO_4$	0.4	6
DMSO	0.6	9
Primer 1 (50 μ*M*)	0.06	0.9
Primer 2 (50 μ*M*)	0.06	0.9
DNA	2	30
HifiTaq (5 U/μL)	0.1	1.5

PCR cycling conditions:
94°C for 2 min; 1 cycle
94°C for 10 s, 64°C for 30 s, 68°C for 1 min; 4 cycles
94°C for 10 s, 61°C for 30 s, 68°C for 1 min; 4 cycles
94°C for 10 s, 58°C for 30 s, 68°C for 1 min; 4 cycles
94°C for 10 s, 55°C for 30 s, 68°C for 1 min; 20 cycles
68°C for 5 min; 1 cycle

Extension time should be set according to the length of the arm, at 1 kb per min.

2. Run the PCR products on a 1% agarose gel and purify both fragments.

3.1.3. Vector Digestion

1. Digest 5 μg of pTK-Neo-USER-3xFlag vectors DNA with 40 U of Xba I overnight at 37°C in a total volume of 100 μL.
2. Add 20 U of XbaI the next morning together with 20 U of Nt.BbvCI to the digestion mixture, and incubate for 2 h at 37°C.
3. Run the digestion mixture on 1% agarose gel and excise both fragments. The large fragment is named as B and the small fragment is named as S.
4. Extract both the B and S fragments with a gel extraction kit.

3.1.4. Insertion of PCR Fragments into the USER Vectors

1. Mix B (~30 ng), S, left arm and right arm together in a 1:10:10:10 molar ratio.
2. Add 1 μL of 10S TE buffer, pH 8.0, and 1 μL of USER™ enzyme mixture (1 U/μL) to 8 μL of the mixture prepared in Step 1 above (*see* **Note 2**).
3. Incubate the reaction mixture for 20 min at 37°C, followed by 20 min at 25°C.

3.1.5. Transformation

1. Mix the entire USER-treated reaction mixture (10 μL) with 50 μL of chemically competent *E. coli* cells and transfect by heat shock. Do not use electroporation for transfection.
2. Plate them on LB agar plates supplemented with ampicillin (100 μg/mL).

3.2. rAAV Targeting Virus Generation

1. Plate HEK293T cells in a T75 flask one day prior to transfection to achieve a 40–80% confluence at the time of transfection.
2. Prepare two wells of a 24-well tissue culture plate and add 750 μL of OptiMEM into each well.
3. In one well, add 3 μg each of the targeting vector, pAAV-RC, and pHelper plasmids and mix well. In the second well, add 54 μL of lipofectamine transfection reagent.
4. Drip the DNA mixture into the lipofectamine mixture and let it sit for 10–30 min while preparing the HEK293T cells to be transfected.
5. Rinse the cells once with sterile PBS and once with OptiMEM, then add 7.5 mL of OptiMEM and keep the cells in incubator.
6. Add the lipofectamine/DNA mixture into the HEK293T cells, rock gently, and return the cells to the incubator.
7. After 3–4 h, remove the OptiMEM medium and replace with complete medium (DMEM supplemented with 10% FBS and 1% Pen/Strep).
8. Grow the cells for 72 h prior to harvesting virus.
9. Scrap the transfected cells and pool them with the culture medium in a 15 mL conical tube. The floating cells contain a lot of viruses.
10. Spin cells down at 800*g* for 3 min and aspirate medium.
11. Suspend the cells into 1 mL of sterile PBS.
12. Freeze and thaw the pellet three cycles. Each cycle consists of 10 min freezing in a dry ice–ethanol bath, and 10 min thawing in a 37°C water bath, vortex after each thawing.

13. Spin the lysate at 10,000*g* for 5 min in a micro-centrifuge to remove cell debris.
14. Divide the supernatant containing rAAV into three aliquots (~330 μL each) and freeze them at –80°C. In general, one-third of the virus generated from one T75 cm^2 flask is sufficient for infection of one 25 cm^2 flask containing the cells to be targeted.

3.3. Gene Targeting of Human Cells

1. Grow DLD1 cells or cells of your interest to be targeted in a T25 flask at 60–80% confluence.
2. Wash cells once with PBS.
3. Add ~330 μL of rAAV and then 1.5 mL of the appropriate growth media (McCoy's 5A for DLD1 cells) to the flask.
4. Incubate at 37°C for 2–5 h.
5. Add 5 mL of growth media into the flask and grow for 48 h.
6. Harvest cells by trypsinization and resuspend cells in 100 mL of medium containing 1 mg/mL geneticin.
7. Distribute 50 mL of cell suspension into two 96-well plates (250 μL/well).
8. Add 50 mL of geneticin-containing medium to the remaining 50 mL of cell suspension.
9. Repeat Steps 7 and 8 until you have a stack of 10–20 96-well plates. The purpose of this step is to serially dilute cells so that you will get one geneticin-resistant clone/well.
10. Wrap the plates with Saran Wrap to minimize evaporation and incubate them at 37°C for 10–14 days prior to consolidating single clones.
11. Check the plates on day 10 and mark the single clones under the microscope.
12. Consolidate the single clones, once they grow to 1/3–1/2 of the wells.
13. Dump the medium from the 96-well plates, add 50 μL of trypsin into each of the marked wells, and incubate the plates at 37°C for >20 min.
14. Prepare a set of 96-well plates with 200 μL medium added into each well.
15. Transfer all of single clones into the new 96-well plates and grow cells to confluence. If you cannot get enough single clones, you can screen multiple clones.

3.4. Genomic DNA Preparation

1. To a monolayer or a large colony in a 96-well tissue culture plate, add 25–30 μL trypsin–EDTA without phenol red. This should be roughly 2,000–5,000 cells/ μL. Incubate at 37°C for 10 min.

2. Using a multi-channel pipette, aliquot 5 μL of Lyse-N-Go reagent to each well of a 96-well PCR plate (*see* **Note 3**).
3. Shake the tissue culture plate gently to dislodge cells. Pipette 2 μL of cell suspension from each well to the PCR plate containing Lyse-N-Go reagent.
4. Add 200 μL of fresh medium back to the plate with the trypsinized cells and keep growing them.
5. Cycle as per manufacturer's recommendations: 65°C, 30 s, 8°C, 30 s 65°C, 1.5 min, 97°C, 3 min, 8°C, 1 min, 65°C, 3 min, 97°C, 1 min, 65°C, 1 min, 80°C, 5 min.
6. Spin down the reactions to get it at the bottom of the tube.
7. Add 20 μL of ddH20 (PCR grade) to each well, spin down, and use 2 μL for the PCR.

3.5. Targeted Clone Screening

1. Design forward PCR primers upstream of the left arm (close to 5′ end of left arm and avoid repetitive sequences). Those primers are designated as left-arm screening primers.
2. Design reverse PCR primers downstream of the right arm (close to 3′ end of left arm and avoid repetitive sequences). Those primers are designated as right-arm screening primers.
3. Pair the left-arm screening primers with NR (GTTGTGCCCAGTCATAGCCG) or pair the right-arm screening primers with NF (TCTGGATTCATC-GACTGTGG) to perform PCRs for screening targeted clones.
4. Perform all PCR reactions with platinum Taq DNA polymerase using the conditions specified by the manufacturer. The reaction volume is 10 μL in 96-well plates using the following receipt and cycling conditions (*see* **Note 4**):

10 μL reaction	1	18	26	50	75	102
H_2O	5.7	73.8	147.7	284	426	579.4
10X buffer	1	13	26	50	75	102
10 m*M* dNTPs	0.2	2.6	5.2	10	15	20.4
50 m*M* $MgCl_2$	0.3	3.9	7.8	15	22.5	30.6
DMSO	0.6	7.8	15.6	30	45	61.2
Primer 1 (50 μ*M*)	0.06	0.78	1.56	3	4.5	6.12
Primer 2 (50 μ*M*)	0.06	0.78	1.56	3	4.5	6.12
DNA	2	36	52	100	150	204
Taq (5U/μL)	0.1	1.3	2.6	5	7.5	10.2

PCR cycling conditions:
94°C for 2 min; one cycle

94°C for 10 s, 64°C for 30 s, 68°C for 1–3 min; four cycles

94°C for 10 s, 61°C for 30 s, 68°C for 1–3 min; four cycles

94°C for 10 s, 58°C for 30 s, 68°C for 1–3 min; four cycles

94°C for 10 s, 55°C for 30 s, 68°C for 1–3 min; 35 cycles.

Extension time should be set according to the length of the arm at 1 kb per min.

3.6. Excision of the Neomycin Resistance Gene

1. Design a pair of primers surrounding the stop codon to amply a fragment ~200 bp (Cre screening primers).
2. Transfer the positive clones to 24-well plate to expend them (From now on, do not add geneticin into medium). Pick at least two of the targeted clones for excision of the neomycin resistance gene.
3. Once confluence, split two-thirds of the cells to a six-well plate to grow as a stock, and transfer the remaining one-third of the cells to a new 24-well plate for adeno-Cre virus infection.
4. Add adeno-Cre virus to the 24-well and grow for 24 h.
5. Dilute the cells and plate into 96-well plates so that you will have single clones. Incubate the plates for 2 weeks. On day 10, mark single clones.
6. Consolidate 24 clones for each of the Cre-ed clones. Prepare genomic DNA as describes in **Section 3.4**.
7. Perform PCR with the Cre screening primers. The clones with neomycin resistance gene being excised should give two bands (as shown in **Fig. 6.3**) (*see* **Notes 5 and 6**).

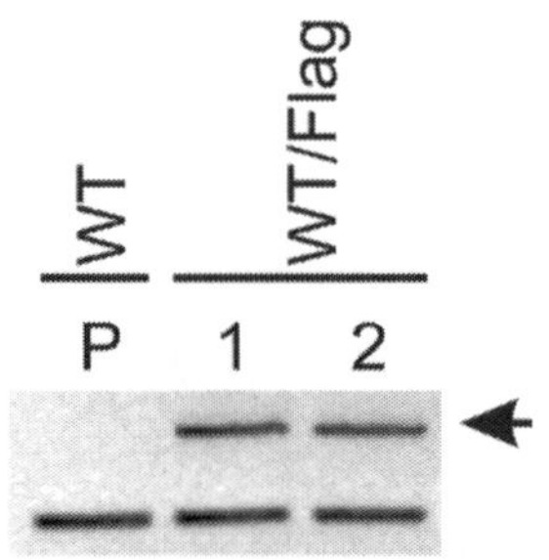

Fig. 6.3. Genomic PCR 3xFlag knock-in clones. Parental (P) and 3xFlag knock-in cells (clone 1 and clone 2). Arrow points to the targeted allele.

4. Notes

1. For left-arm reverse and right-arm forward primers, you do not have many choices. Just use the sequences around stop codon. Sometimes, it is hard to find good pairs of PCR primers. In this case, amplify a big fragment using left forward primer (P1) and the reverse Cre screening primer P4 (**Fig. 6.1**) first, and then perform nest-PCR to amplify the left arm. You can use the same strategy to amply the right arm.
2. The USER cloning system is rapid and highly efficient (>80% cloning efficiency). If you have trouble with this system, we also have a targeting vector for the traditional restriction and ligation cloning method. We are happy to send it to you per request.
3. Lyse-N-Go is a reagent from Pierce that is useful for the rapid, inexpensive production of template DNA from cells. Such templates have been used successfully for a number of PCR reactions in which products of up to 5 kb have been amplified robustly. However, Qiagen genomic DNA prep kit is an expensive alternative to produce better quality DNA.
4. After getting the positive clones, make new genome DNA using QIAamp DNA Blood Mini Kit and confirm with two pairs of screening primers across both arms (i.e., left-arm screen primer + NR and right-arm screening primer + NF, **Fig. 6.1**).
5. It is imperative to confirm expression of Flag tagged proteins by Western blot.
6. We have successfully targeted DLD1, RKO, LOVO, and HCT116 colorectal cancer cells so far. Other cell lines should be targetable as well.

Acknowledgments

The author would like to thank Dr. Chao Wang for proof reading. This work was supported by RO1 CA127590 and HG004722.

References

1. Bitinaite, J., Rubino, M., Varma, K. H., Schildkraut, I., Vaisvila, R. and Vaiskunaite, R. (2007) USER friendly DNA engineering and cloning method by uracil excision. *Nucleic Acids Res.* **35**, 1992–2002.
2. Kim, T. H. and Ren, B. (2006) Genome-wide analysis of Protein–DNA interactions. *Annu. Rev. Genomics Hum. Genet.* **7**, 81–102.
3. Kohli, M., Rago, C., Lengauer, C., Kinzler, K. W. and Vogelstein, B. (2004) Facile

methods for generating human somatic cell gene knockouts using recombinant adeno-associated viruses. *Nucleic Acids Res.***32**, e3.

4. Lee, T. I., Rinaldi, N. J., Robert, F., Odom, D. T., Bar-Joseph, Z., Gerber, G. K., Hannett, N. M., Harbison, C. T., Thompson, C. M., Simon, I., Zeitlinger, J., Jennings, E. G., Murray, H. L., Gordon, D. B., Ren, B., Wyrick, J. J., Tagne, J. B., Volkert, T. L., Fraenkel, E., Gifford, D. K. and Young, R. A. (2002) Transcriptional regulatory networks in *Saccharomyces cerevisiae. Science* **298**, 799–804.
5. Zhang, X., Guo, C., Chen, Y., Shulha, H. P., Schnetz, M. P., LaFramboise, T., Bartels, C. F., Markowitz, S., Weng, Z., Scacheri, P. C. and Wang, Z. (2008) Epitope tagging of endogenous proteins for genome-wide ChIP-chip studies. *Nat. Methods* **5**, 163–165.

Chapter 7

Flow Cytometric and Laser Scanning Microscopic Approaches in Epigenetics Research

Lorant Szekvolgyi, Laszlo Imre, Doan Xuan Quang Minh, Eva Hegedus, Zsolt Bacso, and Gabor Szabo

Abstract

Our understanding of epigenetics has been transformed in recent years by the advance of technological possibilities based primarily on a powerful tool, chromatin immunoprecipitation (ChIP). However, in many cases, the detection of epigenetic changes requires methods providing a high-throughput (HTP) platform. Cytometry has opened a novel approach for the quantitative measurement of molecules, including PCR products, anchored to appropriately addressed microbeads (Pataki et al. 2005. *Cytometry* **68**, 45–52). Here we show selected examples for the utility of two different cytometry-based platforms of epigenetic analysis: ChIP-on-beads, a flow-cytometric test of local histone modifications (Szekvolgyi et al. 2006. *Cytometry* **69**, 1086–1091), and the laser scanning cytometry-based measurement of global epigenetic modifications that might help predict clinical behavior in different pathological conditions. We anticipate that such alternative tools may shortly become indispensable in clinical practice, translating the systematic screening of epigenetic tags from basic research into routine diagnostics of HTP demand.

Key words: Chromatin immunoprecipitation (ChIP), flow cytometry, ChIP-on-beads, laser scanning cytometry (LSC).

1. Introduction

Epigenetic changes associated with gene regulation play a major role in the establishment of altered differentiation states. Specific modifications often correlate with gene activation or repression; for instance H3K4ac and H3K4me3 are permissive for gene activation whereas H3K9me2, H3K27me3, and methylation of CpG islands in promoter regions correlate with transcriptional silencing. Often, activating and repressive marks co-exist at gene start sites, reflecting perhaps epigenetic heterogeneity among otherwise similar cells, establishing a fine balance that could determine the gene expression patterns in the tissue.

Philippe Collas (ed.), *Chromatin Immunoprecipitation Assays,* Methods in Molecular Biology 567,
DOI 10.1007/978-1-60327-414-2_7, © Humana Press, a part of Springer Science+Business Media, LLC 2009

The 'epigenetic code' has become an indispensable concept in basic research, and its principles are also utilized to develop drugs and diagnostic tools (1–3) several genes being epigenetically misregulated have been shown to associate with different kinds of cancer, highlighting the role of the 'language' of covalent modifications in tumorigenesis (4, 5). For instance, based on the patterns of modifications, two disease subtypes with different risks of tumor recurrence have been characterized in prostate cancer patients, independently from tumor stage, preoperative prostate-specific antigen levels, and capsule invasion (6).

The chromatin of cancer cells often exhibits both an overall (global) DNA hypomethylation and hypermethylation of specific regions, leading to 'DNA methylation imbalance' (7). The recurrence of global DNA hypomethylation in many types of human cancer is suggestive of its significant role in carcinogenesis, perhaps by inducing genomic instability and/or activating oncogenes (8, 9). However, global hypomethylation is subject to a high degree of variability, unaccounted for by our current level of understanding (10, 11). In addition to neoplastic transformation, problems of epigenetic regulation, including CpG methylation disorders are also involved in a wide range of pathological phenomena (12, 13). In most eukaryotes, methylation of DNA occurs at the cytosine residues of cytosine-phospho-guanine (CpG) dinucleotides. The enzymes responsible for the production of 5-methylcytosine (5-mc) involving the fifth carbon atom of cytosine in CpG dinucleotides are the DNA methyltransferases DNMT1, DNMT3a, and DNMT3b, of which the first is involved in the maintenance of methylation during DNA replication, while all appear to be important in the establishment of methylation patterns in most physiological and pathological settings (14–16).

1.1. Flow- and Laser Scanning Cytometry in Epigenetics Research

Our understanding of epigenetics has been transformed in recent years by a succession of technological innovations. Approaches involving microarrays and, most recently ultra-high throughput (deep) sequencing technology have been applied to map cytosine methylation, chromatin modifications, and ncRNAs across entire genomes. Genome-scale studies of histone modifications and other aspects of chromatin structure typically rely on an immunological procedure, chromatin immunoprecipitation (ChIP) (17), in which specific antibodies are used to enrich chromatin. ChIP is a powerful tool in epigenetics; however, in many cases the detection of epigenetic changes or transcription factor binding associated with the regulation of certain genes would require ChIP-based methods that provide high-throughput (HTP) potential. Monitoring local as well as global changes of epigenetic markers could be extremely useful in diagnostics as well as in basic research.

Flow-cytometric analysis provides a novel means for the quantitative measurement of molecules also in cell-free solutions, anchoring them to appropriately addressed microbeads. The utility and power of this approach has been demonstrated in the case of various assays of molecular diagnostic value: immunoassays, sensitive measurement of protease or nuclease activity, detection of deletion/insertion of sequences by heteroduplex analysis, etc., that could all be adapted to a 'lab-on-beads' platform, i.e., the flow-cytometric analysis of microbead-captured macromolecules (1, 18, 19). Many samples can be simultaneously analyzed in a FACSarray instrument using fluorescent dyes matching its optical channels.

Beyond lending a HTP platform for the analysis of gene-specific epigenetic markers, cytometry also makes global analysis of epigenetic changes possible, most conveniently in its on-slide format, by microscope-based cytometers. Laser scanning cytometry (LSC) provides a robust method for analyzing single-cell events on slides (20, 21). It generates quantitative fluorescence data similar to flow cytometry, but the analyzed cells are attached to the surfaces of microscopic slides or culture chambers. The main advantages of LSC are that (i) the possible correlation between the simultaneously measured parameters is detected at the individual cell resolution, i.e., with a sensitivity surpassing that of flow cytometry; (ii) the instrument is able to relocate each cell for additional measurements, thus the analysis of functional features of live cells can be combined with measurements that require fixed cells; and (iii) measurements can be performed in an automated fashion, preprogrammed for several slides.

Examples highlighted in this review demonstrate the value of two different HTP platforms for epigenetic analysis, namely ChIP-on-beads and assessment of global epigenetic traits by LSC. These methods might help introduce systematic screening of different epigenetic tags into clinical practice, especially of those that correlate with therapeutic success. It will be shown that sequence-specific capture of PCR-amplified ChIP-fragments on microbeads allows a robust detection of histone-tail modifications in the promoter region of a well-characterized gene, tissue transglutaminase type 2 (*TGM2*). We also assess the prospects of laser scanning cytometry for the analysis of epigenetic changes involving the whole genome via the example of a global DNA methylation test.

1.2. High-Throughput Screening of Local Epigenetic Changes by ChIP-on-Beads

We have investigated the cellular levels of H4K acetylation and H3K4 methylation of the histone tails at the promoter of the *TGM2* gene, to test whether these covalent modifications can be detected using a flow-cytometric platform. As shown earlier (2) and briefly recapped herein, the flow-ChIP method, nick-named ChIP-on-beads, can be easily implemented in a routine flow-cytometric clinical laboratory without relying on real-time QPCR. In the ChIP-on-beads assay, a standard ChIP is performed and then this DNA is used as template in an end-point PCR reaction. The sense and anti-sense primers are tagged at their 5′ ends with fluorescent dyes (e.g., Fam, Cy3) and biotin, respectively. Small aliquots of the *Fam/biotin*-ended PCR products are then bound to streptavidin-conjugated microbeads and quantified by flow cytometry. Of note, PCRs must be stopped in the linear phase to ensure reliable quantification; this should be initially determined in pilot QPCR experiments. The similarity of data obtained by QPCR and by flow cytometry has been shown (2).

As shown in **Fig.7.1A**, the fluorescence intensity of the microbeads increases linearly with the quantity of the fluoresceinated PCR products added, allowing the expression of ChIP-PCR

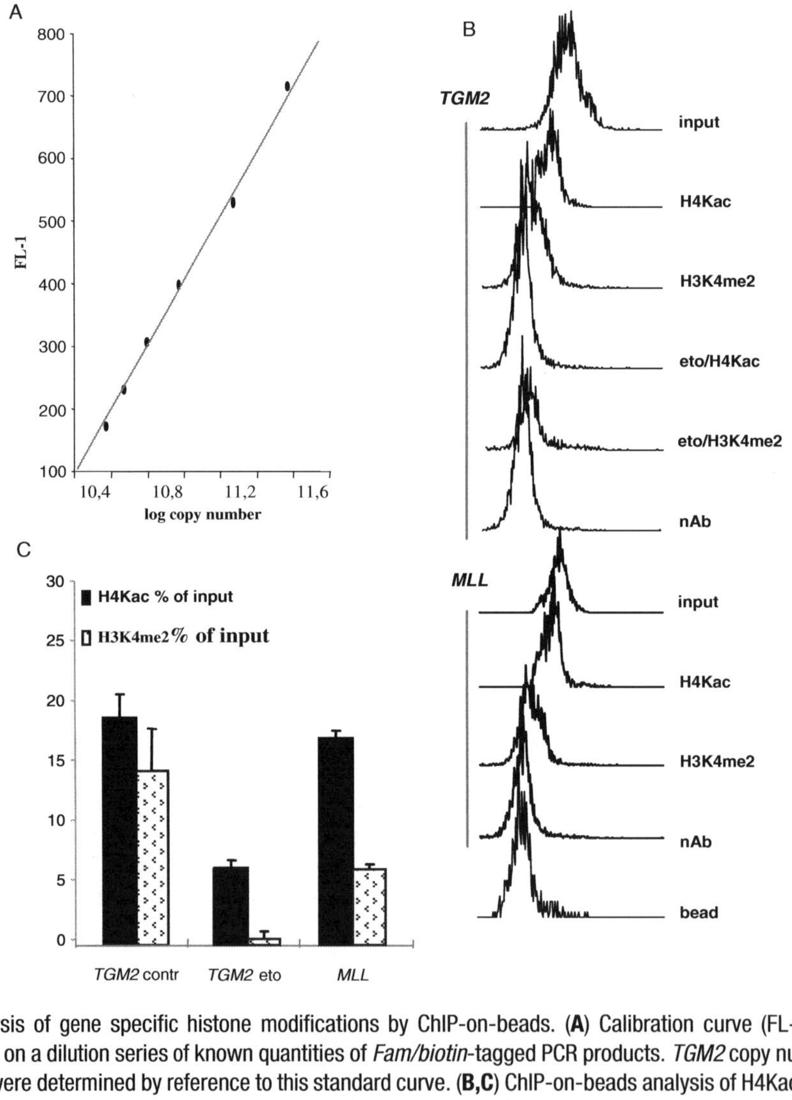

Fig. 7.1. Analysis of gene specific histone modifications by ChIP-on-beads. (**A**) Calibration curve (FL-1 vs. log copy number) based on a dilution series of known quantities of *Fam/biotin*-tagged PCR products. *TGM2* copy numbers of ChIP-PCR samples were determined by reference to this standard curve. (**B,C**) ChIP-on-beads analysis of H4Kac and H3K4me2 histone modifications at the *TGM2* g ene promoter and at exon 9 of the *MLL* gene, in Jurkat cells. Apoptosis was induced by etoposide treatment (Eto). (**B**) Flow-cytometric fluorescence distribution histograms of *Fam/biotin*-labeled ChIP-PCR samples captured on streptavidin-conjugated microbeads. (**C**) The level of modified histones within the *TGM2* and *MLL* genes are expressed as percent of input values (Y axis), based on the means of fluorescence distribution and after subtracting the background (i.e., no-antibody % of input values). Panels (**B**) and (**C**) were reproduced from (2).

yields as absolute copy numbers. The flow-cytometric fluorescence distribution means are used to calculate the fraction of DNA copy numbers in the ChIP samples relative to the input DNA (**Fig.7.1B**). Comparing control and early-apoptotic Jurkat cells for changes in the level of H4Kac and H3K4me within the promoter of *TGM2*, we observed a significant decrease in both histone modifications (**Fig.7.1C**), suggestive of the closure of chromatin structure early upon apoptosis. In comparison, the observed histone modifications at exon 9 of the *MLL* gene, used as positive control, were in accordance with its known histone-code profile (22); in contrast, the β-globin gene, used as negative control, gave <0.1% Ab/input ratios (not shown).

1.3. Testing Global Epigenetic Changes by Laser Scanning Microscopy: Studies on DNA Methylation

It is often important to consider in what global context local epigenetic changes occur (23). Moreover, global changes of certain epigenetic modifications may have their independent diagnostic value, especially when analyzed in correlation with other phenotypic markers, an opportunity offered by up-to-date laser scanning microscopic systems (20, 21). Development of antibodies and chimeric methyl CpG-binding antibody-like proteins (24–27), both recognizing CpG with high specificity, has opened novel perspectives for the diagnostic analysis of global methylation states. Anti-5mC antibodies are commercially available through various sources (e.g., Abcam and Biocarta US).

In experiments using the recombinant mCpG-binding antibody-like MBD-Fc protein (26–28), the overall level of CpG methylation has been quantified in the HCT116 cell line (**Fig.7.2**). As shown by confocal laser scanning microscopy

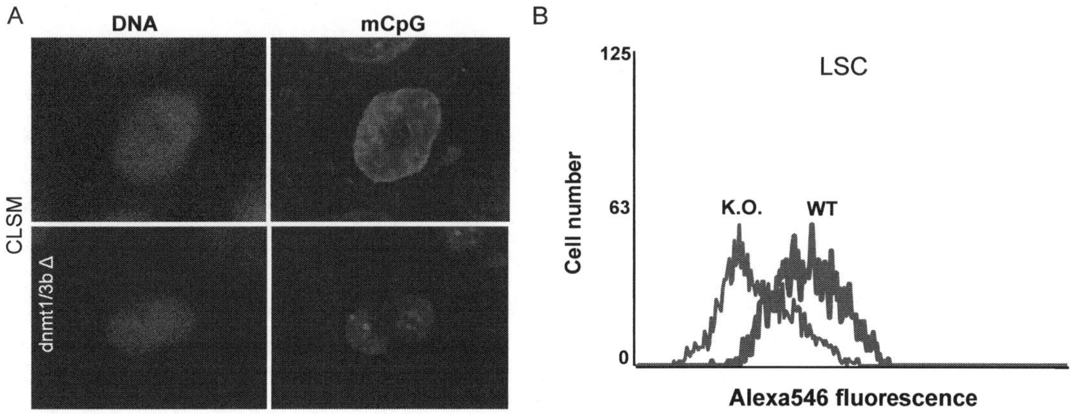

Fig. 7.2. Global DNA methylation analyzed by confocal laser scanning microscopy and laser scanning cytometry. *WT:* wild-type *DNMT1/DNMT3b* HCT116 cells immunolabeled with the MBD-Fc fusion protein. *K.O.: dnmt1/dnmt3b* knock-out HCT116 cells immunolabeled with the MBD-Fc fusion protein. *Left slides*: DNA stained by Hoechst. *Right slides*: methylated DNA (mCpGs). (**A**) Methylated CpG dinucleotides visualized by confocal laser scanning microscopy (CLSM). (**B**) Sample analyzed by laser scanning cytometry (LSC). MCpG (*red*) fluorescence was quantified in the slide-attached cells (n>400) and presented (in arbitrary units) as fluorescence distribution histograms.

(CLSM), mCpGs have been efficiently labeled by indirect immunofluorescence in *DNMT1/3b* wild-type and, to a lesser extent, *dnmt1/3b* double knock-out cells. The level of mCpGs has been quantified in a sizable population of cells by an iCys laser scanning cytometer and iCyte 2.6 software (CompuCyte, USA). As shown in **Fig.7.2**, the fluorescence distributions of the Alexa546-labeled mCpGs are significantly different in the *DNMT1/3b*$^{+/-}$ cells; this result demonstrates the utility of LSC for the fine assessment of global methylation states in different cell types (e.g., differentiated vs. stem cells) or in a specific cell type (e.g., in human peripheral lymphocytes isolated from blood samples) before and after drug treatment or chemotherapy. Since LSC can be performed in an automated fashion, such studies could be made on large sets of biopsy material so as to establish the exact role of global DNA methylation in human pathological diagnosis of various diseases.

Data presented herein have demonstrated that if combined, flow cytometry and conventional PCR offer a powerful tool in the quantitative analysis of ChIP results. We have found high levels of H4Kac and H3K4me at the *TGM2* gene core promoter (**Fig.7.1**). These levels significantly decreased upon apoptosis and this was accompanied by the down-regulation of TGM2 mRNA expression (2), suggesting that this enzyme does not contribute to the early manifestations of apoptosis in Jurkat cells. Differences in the global level of DNA methylation in HCT116 wild-type and methylation defective cells have been revealed by LSC, the on-slide version of flow cytometry (**Fig.7.2**). Both assays can be easily implemented, and readily applied in a HTP format. We envisage the utility of these platforms primarily in clinical screening efforts addressing one, or a few, epigenetic markers in many samples simultaneously, depending on cost/time considerations and availability of instrumentation/expertise.

Although the epigenetic changes are heritable, they appear to be readily reversed by specific drug treatments as opposed to gene mutations. We expect that the epigenetic silencing of, e.g., tumor suppressor genes will soon become a frequent target of HTP screening studies because these mechanisms may be as important in carcinogenesis as the inactivating mutations. Drugs targeting the enzymes that remove or add these chemical tags are at the forefront of research: diseases to be targeted include cancer, imprinting disorders, autoimmune diseases, certain neurological disorders, diabetes, cardiopulmonary diseases, in which mis-steps in epigenetic programming have been directly implicated. Pharmaceutical companies have set up programs on histone decacetylases (HDACs) and DNA methyltransferases (DNMTs) and their inhibitors, as they have the potential to re-activate specific tumor suppressor genes; clinical trials being on the way are promising the prospect of eliciting tumor regression by modulation of epigenetic regulation.

Based on the above, we anticipate that epigenetic analysis will enter routine diagnostic practice whenever monitoring epigenetic markers can help predict clinical behavior. When large sets of samples are to be assessed, high-throughput platforms for the accurate evaluation of the ChIP results are of general interest. In view of the fact that most routine techniques can be adapted to flow cytometry which exceeds more conventional methods in sensitivity and reproducibility, the approaches shown can provide a universal platform for almost any kind of lab purposes. Whether ChIP-QPCR, ChIP-on-beads, or LSC-based assays of global epigenetic changes will be selected as the approach of choice for such screening projects will be determined by the particular task undertaken, and the capabilities of the clinical laboratories. We believe that these alternative ChIP platforms can help bring epigenetic analysis within reach for routine laboratories, especially for those involved in clinical diagnostics.

2. Materials

2.1. Cell Culture

1. McCoy's medium (Sigma-Aldrich).
2. Solution of trypsin: stock solution at 0.5%, working solution at 0.05% in 1X phosphate buffered saline (PBS); store at –20°C.
3. Glutamine: stock solution at 200 m*M*, final concentration at 2 m*M* in ddH_2O; store at –20°C.
4. Etoposide (Sigma-Aldrich): stock solution at 40 m*M*, working concentration at 40 μ*M*.

2.2. Detection of Methylated CpGs by Immunofluorescence

1. 1X PBS: 1.37 *M*NaCl, 27 m*M*KCl, 100 m*M*Na_2HPO_4, 18 m*M*KH_2PO_4; adjust to pH 7.4 with HCl if necessary.
2. Labeling solution: 1X PBS/10%BSA; store at –20°C.
3. Primary antibody (1.9 mg/mL): MBD-Fc, a recombinant antibody which was made of human MBD domain (methyl binding domain) fused with an Fc fragment of a human IgG1 and expressed in *Drosophila* S2 cells (26–28); store at 4°C.
4. Secondary antibody (2 mg/mL): Alexa546-conjugated anti-human IgG (Invitrogen); store at 4°C.
5. Hoechst 33342 (Invitrogen): stock solution: 1 m*M*, working solution: 4 μ*M*, final concentration: 2 μ*M*, diluted in 1X PBS; store at –20°C.
6. Prolong Gold (Invitrogen).

2.3. ChIP-on-Beads

1. Nucleus isolation buffer: 5 m*M* pipes, pH 8.0, 85 m*M* KCl, 0.5% NP-40, protease inhibitors (Sigma-Aldrich, cat no. P8340).
2. Sonication buffer: 1% SDS, 10 m*M* EDTA, 50 m*M* Tris-HCl, pH 8.0, protease inhibitors.
3. IP buffer: 0.01% SDS, 1.1% Triton X-100, 1.2 m*M* EDTA, 20 m*M* Tris-HCl pH 8.0, 167 m*M* NaCl, protease inhibitors.
4. Blocked protein A/G Sepharose (Upstate, cat. no. 16-157).
5. Antibodies (Upstate): anti-H4Kac, 2 μg/IP (cat. no. 06-866), anti-H3K4me2, 5 μg/IP (cat. no. 07-030).
6. Wash buffer (WB) A: 0.1% SDS, 1% Triton X-100, 2 m*M* EDTA, 20 m*M* Tris-HCl, pH 8.0, 150 m*M* NaCl, protease inhibitors.
7. WB B: 0.1% SDS, 1% Triton X-100, 2 m*M* EDTA, 20 m*M* Tris-HCl, pH 8.0, 500 m*M* NaCl, protease inhibitors.
8. WB C: 0.25 *M* LiCl, 1% NP-40, 1% Na-deoxycholate, 1 m*M* EDTA, 10 m*M* Tris-HCl, pH 8.0, protease inhibitors.
9. 1X TE: 10 m*M* Tris-HCl, pH 7.5, 1 m*M* EDTA.
10. QIAquick PCR Purification Kit (Qiagen).
11. Primers: forward 5′-*Fam*-GAGACCCTCCAAGTGCGAC-3′, reverse 5′-*Biotin*-CCAAAGCGGGCTATAAGTTA GC-3′.
12. Streptavidin-coated microbeads (6 μm, Polyscience).

3. Methods

3.1. ChIP-on-Beads

1. Treat exponentially growing Jurkat cells with 40 μ*M* etoposide (eto) for 3 h at 37°C to induce apoptosis.
2. Fix cells with 1% formaldehyde for 10 min at room temperature. Stop fixation by adding 2.5 *M* glycine to a final concentration of 0.67 *M*, for 5 min at room temperature. Wash cells twice in ice-cold PBS.
3. Resuspend cells in 1 mL of nucleus isolation buffer and incubate them for 10 min on ice. Vortex tubes in every 2–3 min.
4. Centrifuge isolated nuclei at 500*g* for 3 min, at 4°C. Resuspend pellet in 500 μL sonication buffer.
5. Sonicate chromatin to an average fragment size of 500 bp using a Bioruptor (Diagenode); 0.5 min ON/0.5 min OFF pulses for 2 × 12 min usually produces the desired size distribution.

6. Centrifuge sheared chromatin samples at maximum speed for 20 min. Keep supernatants (leave 50 μL on the bottom of the tubes). Freeze in liquid nitrogen and store samples at –80°C (or proceed immediately).
7. Thaw samples on ice and centrifuge them at maximum speed for 10 min at 4°C. Transfer supernatants into clean tubes (do not disturb pellet on the bottom of the tubes).
8. Dilute chromatin samples 1:10 in IP buffer as follows: 100 μL chromatin 900 μL IP buffer.
9. Pre-clear samples by incubating them on a rotating wheel with 30 μL of blocked protein A/G Sepharose for 30 min at 4°C. Spin samples at 500*g* for 3 min at 4°C. Keep supernatants.
10. Perform immunoselection for >12 h on a rotating wheel by adding the following antibodies to the samples: anti-H4Kac and anti-H3K4me2; as negative control, omit specific Ab but add a specific IgG protein from the same isotype to one of the pre-cleared samples.
11. Preserve 10 μL from the 'negative control' as 'input' DNA and store it at –20°C. Collect immune complexes by adding 40 μL of blocked protein A/G Sepharose to each sample and incubate them for 45 min on a rotator. Spin samples at 500*g* for 3 min.
12. Wash the pelleted immune complexes as follows: 2 × WB A, 2 × WB B, 2 × WB C, 1 × TE. Resuspend pellets in 500 μL TE. At this point thaw input DNA and dilute it to 500 μL; process it together with the IP samples.
13. Reverse cross-links by incubating the samples at 98°C for 10 min. Put samples on ice.
14. Digest residual RNAs with 200 μg/mL RNase A for 30 min at 37°C.
15. Digest proteins by 0.5 mg/mL proteinase K for at least 2 h at 55°C.
16. Purify DNA on PCR clean-up columns (Qiagen). Immunoprecipitated DNA samples (input, negative control, H4Kac/H3K4me2, respectively) are ready to be tagged by *Fam/biotin* PCR.
17. In the *Fam/biotin* PCR, use primers listed in **Section**2.3. Perform PCRs under standard conditions and stop after 15–20 cycles, i.e., in the linear phase. Validate by QPCR (2). Purify the 5′-*Fam/biotin* labeled ChIP-PCR products on PCR clean-up columns.
18. Carry out flow cytometry on a Becton-Dickinson FACScan flow cytometer as follows: 5 μL of the *Fam/biotin*-tagged ChIP-DNA was added to 10,000 streptavidin-coated, plain

beads in 50 μL PBS. Incubate samples for 15 min at room temperature, wash in 1 mL PBS, and run at high speed. Set laser power to 15 mW and detect fluorescence signals through the 530/30 interference filter of the FL1 channel in logarithmic mode. Evaluate results using the BDIS CELLQUEST 3.3 (Becton-Dickinson) software. *TGM2* copy numbers are determined by reference to a standard curve obtained from a dilution series of known quantities of *Fam/biotin*-tagged PCR products (**Fig.**7.1A). Express ChIP yields as percentage of input after subtracting background (no antibody (nAb) % of input).

3.2. Immunofluorescence and Laser Scanning Cytometry

1. Grow HCT116 *DNMT1/3b* wt and *DNMT1/3b* knock-out cells on coverslips overnight.
2. Wash cells in 200 μL 1X PBS, 3 × 3 min.
3. Fix cells in a series of diluted methylalcohol (MetOH) (as shown below); wash cells with 200 μL of diluted MetOH once for 3 min, for each dilution. Start with the 10 × dilution. After washes, incubate cells in concentrated MetOH overnight at –20°C.

	1X PBS (μL)	MetOH (μL)
10 × MetOH	900	100
8 × MetOH	875	125
6 × MetOH	833	167
4 × MetOH	750	250
2 × MetOH	500	500

4. Rehydrate cells in a series of diluted 1X PBS as shown below; wash cells in 200 μL diluted MetOH for 3 min in each dilution. Start with the 10 × dilution. After the final rehydration step, wash with 200 μL 1X PBSs

	MetOH (μL)	1X PBS (μL)
10 × (1X PBS)	900	100
8 × (1X PBS)	875	125
6 × (1X PBS)	833	167
4 × (1X PBS)	750	250
2 × (1X PBS)	500	500

5. In order to relax DNA, place samples into Petri dishes (without the cover) in PBS/1% BSA and irradiate them with UV light for 30 min.
6. Immunolabel samples using the mCpG-specific MBD-Fc fusion protein or a commercially available Anti-5mC as primary antibody for 30 min at room temperature. Wash cells in 200 μL of 1% BSA/PBS, 3 × for3 min.
7. Label samples with an Alexa546-conjugated anti-human IgG secondary antibody, for 30 min at room temperature. Wash cells in 200 μL 1% BSA/PBS 3 × for 3 min.
8. Stain DNA with 50 μL Hoechst 33342 (2 μ*M*) and cover with Prolong Gold antifade.
9. Scan slides (*see* **Note 1**).

4. Notes

1. MCpGs have been visualized using a Zeiss LSM 510 confocal laser-scanning microscope using excitation wavelengths of 543 and 351/364 nm. Fluorescence emission was detected through 560–615 and 385–470 nm band-pass filters. Images were taken in multitrack mode to prevent cross-talk between the channels. Pixel image (512 × 512) stacks of 2–2.5 μm thick optical sections were obtained with a 63 × Plan-Apochromat oil immersion objective (NA 1.4).

The same samples were also analyzed using an iCys laser scanning cytometer (CompuCyte). The instrument used in our studies is equipped with a violet-blue diode, an argon-ion, and a HeNe laser (wavelengths 405, 488, and 633 nm, respectively). The violet and Ar-ion laser lines were used for excitation of Hoechst and Alexa 546 dyes. To identify single nuclei, contouring was based on Hoechst fluorescence detected in the blue channel (460–485 nm). Fluorescence of Alexa 546 (MCpGs) was detected in the orange channel (565–585 nm) based on the contour gained in the blue channel. In single nuclei identified by contouring on fluorescence of the nuclear stain, the integral fluorescence related to the MCpGs divided by the area of the contour was used to describe the methylation level. This corrects for differences in nuclear size. Data evaluation and hardware control were performed using the iCys 2.6 software for Windows XP. Using the 4 × objective to scan an indicated area on a slide, 400–1000 cells were scanned in about 10 min (21). LSC can screen relatively large number of cells on a slide. The cells are distinguished based on their fluorescence properties like in flow cytometry. However, as the position of each cell is fixed on the slide and the instrument saves the positional information, any correlation between the different parameters measured can be detected in a very sensitive manner. In addition, the cells can be relocated and visually analyzed or re-scanned after re-staining with conventional stains or fluorescent markers.

Acknowledgments

The authors thank Drs. Rolf Ohlsson and Anita Göndör (Uppsala, Sweden) for the DNMT-KO and control HCT116 cells and Dr. Michael Rehli (Regensburg, Germany) for the stably transfected *Drosophila Schneider 2*(S2) cell line producing the MBD-Fc fusion protein. This publication was sponsored by OTKA fundings TO48742, OTKA 72762, and the research grant of the Ministry of Public Health ETT 067/2006.

References

1. Pataki, J., Szabo, M., Lantos, E., Szekvolgyi, L., Molnar, M., Hegedus, E., Bacso, Z., Kappelmayer, J., Lustyik, G. and Szabo, G. (2005) Biological microbeads for flow-cytometric immunoassays, enzyme titrations, and quantitative PCR. *Cytometry* **68**, 45–52.
2. Szekvolgyi, L., Balint, B. L., Imre, L., Goda, K., Szabo, M., Nagy, L. and Szabo, G. (2006) Chip-on-beads: flow-cytometric evaluation of chromatin immunoprecipitation. *Cytometry* **69**, 1086–1091.
3. Balint, B. L., Szanto, A., Madi, A., Bauer, U. M., Gabor, P., Benko, S., Puskas, L. G., Davies, P. J. and Nagy, L. (2005) Arginine methylation provides epigenetic transcription memory for retinoid-induced differentiation in myeloid cells. *Mol. Cell Biol.* **25**, 5648–5663.
4. Downs, J. A. and Jackson, S. P. (2003) Cancer: protective packaging for DNA. *Nature***424**, 732–734.
5. Hake, S. B., Xiao, A. and Allis, C. D. (2004) Linking the epigenetic 'language' of covalent histone modifications to cancer. *Br. J. Cancer* **90**, 761–769.
6. Seligson, D. B., Horvath, S., Shi, T., Yu, H., Tze, S., Grunstein, M. and Kurdistani, S. K. (2005) Global histone modification patterns predict risk of prostate cancer recurrence. *Nature* **435**, 1262–1266.
7. Lafon-Hughes, L., Di Tomaso, M. V., Mendez-Acuna, L. and Martinez-Lopez, W. (2008) Chromatin-remodelling mechanisms in cancer. *Mutat. Res.* **658**, 191–214.
8. Fanelli, M., Caprodossi, S., Ricci-Vitiani, L., Porcellini, A., Tomassoni-Ardori, F., Amatori, S., Andreoni, F., Magnani, M., De Maria, R., Santoni, A., Minucci, S. and Pelicci, P. G. (2008) Loss of pericentromeric DNA methylation pattern in human glioblastoma is associated with altered DNA methyltransferases expression and involves the stem cell compartment. *Oncogene* **27**, 358–365.
9. Piyathilake, C. J., Frost, A. R., Bell, W. C., Oelschlager, D., Weiss, H., Johanning, G. L., Niveleau, A., Heimburger, D. C. and Grizzle, W. E. (2001) Altered global methylation of DNA: an epigenetic difference in susceptibility for lung cancer is associated with its progression. *Hum. Pathol.* **32**, 856–862.
10. Estecio, M. R., Gharibyan, V., Shen, L., Ibrahim, A. E., Doshi, K., He, R., Jelinek, J., Yang, A. S., Yan, P. S., Huang, T. H., Tajara, E. H. and Issa, J. P. (2007) LINE-1

hypomethylation in cancer is highly variable and inversely correlated with microsatellite instability. *PLoS ONE* **2**, e399.

11. Ogino, S., Kawasaki, T., Nosho, K., Ohnishi, M., Suemoto, Y., Kirkner, G. J. and Fuchs, C. S. (2008) LINE-1 hypomethylation is inversely associated with microsatellite instability and CpG island methylator phenotype in colorectal cancer. *Int. J. Cancer* **122**, 2767–2773.
12. Shimabukuro, M., Sasaki, T., Imamura, A., Tsujita, T., Fuke, C., Umekage, T., Tochigi, M., Hiramatsu, K., Miyazaki, T., Oda, T., Sugimoto, J., Jinno, Y. and Okazaki, Y. (2007) Global hypomethylation of peripheral leukocyte DNA in male patients with schizophrenia: a potential link between epigenetics and schizophrenia. *J. Psychiatr. Res.* **41**, 1042–1046.
13. Matarazzo, M. R., Boyle, S., D'Esposito, M. and Bickmore, W. A. (2007) Chromosome territory reorganization in a human disease with altered DNA methylation. *Proc. Natl. Acad. Sci. U.S.A.* **104**, 16546–16551.
14. Miranda, T. B. and Jones, P. A. (2007) DNA methylation: the nuts and bolts of repression. *J. Cell Physiol.* **213**, 384–390.
15. Rhee, I., Bachman, K. E., Park, B. H., Jair, K. W., Yen, R. W., Schuebel, K. E., Cui, H., Feinberg, A. P., Lengauer, C., Kinzler, K. W., Baylin, S. B. and Vogelstein, B. (2002) DNMT1 and DNMT3b cooperate to silence genes in human cancer cells. *Nature* **416**, 552–556.
16. Sun, L., Zhao, H., Xu, Z., Liu, Q., Liang, Y., Wang, L., Cai, X., Zhang, L., Hu, L., Wang, G. and Zha, X. (2007) Phosphatidylinositol 3-kinase/protein kinase B pathway stabilizes DNA methyltransferase I protein and maintains DNA methylation. *Cell Signal* **19**, 2255–2263.
17. Kuo, M. H. and Allis, C. D. (1999) In vivo cross-linking and immunoprecipitation for studying dynamic protein:DNA associations in a chromatin environment. *Methods* **19**, 425–433.
18. Taylor, J. D., Briley, D., Nguyen, Q., Long, K., Iannone, M. A., Li, M. S., Ye, F., Afshari, A., Lai, E., Wagner, M., Chen, J. and Weiner, M. P. (2001) Flow cytometric platform for high-throughput single nucleotide polymorphism analysis. *Biotechniques* **30**, 661–666, 668–669.
19. Spiro, A. and Lowe, M. (2002) Quantitation of DNA sequences in environmental PCR products by a multiplexed, bead-based method. *Appl. Environ. Microbiol.* **68**, 1010–1013.
20. Bacso, Z., Everson, R. B. and Eliason, J. F. (2000) The DNA of annexin V-binding apoptotic cells is highly fragmented. *Cancer Res.* **60**, 4623–4628.
21. Bacso, Z. and Eliason, J. F. (2001) Measurement of DNA damage associated with apoptosis by laser scanning cytometry. *Cytometry* **45**, 180–186.
22. Khobta, A., Carlo-Stella, C. and Capranico, G. (2004) Specific histone patterns and acetylase/deacetylase activity at the breakpoint-cluster region of the human MLL gene. *Cancer Res.* **64**, 2656–2662.
23. Beck, S. and Rakyan, V. K. (2008) The methylome: approaches for global DNA methylation profiling. *Trends Genet.* **24**, 231–237.
24. Habib, M., Fares, F., Bourgeois, C. A., Bella, C., Bernardino, J., Hernandez-Blazquez, F., de Capoa, A. and Niveleau, A. (1999) DNA global hypomethylation in EBV-transformed interphase nuclei. *Exp. Cell Res.* **249**, 46–53.
25. Adouard, V., Dante, R., Niveleau, A., Delain, E., Revet, B. and Ehrlich, M. (1985) The accessibility of 5-methylcytosine to specific antibodies in double-stranded DNA of Xanthomonas phage XP12. *Eur. J. Biochem.* **152**, 115–121.
26. Gebhard, C., Schwarzfischer, L., Pham, T. H., Andreesen, R., Mackensen, A. and Rehli, M. (2006) Rapid and sensitive detection of CpG-methylation using methyl-binding (MB)-PCR. *Nucleic Acids Res.* **34**, e82.
27. Gebhard, C., Schwarzfischer, L., Pham, T. H., Schilling, E., Klug, M., Andreesen, R. and Rehli, M. (2006) Genome-wide profiling of CpG methylation identifies novel targets of aberrant hypermethylation in myeloid leukemia. *Cancer Res.* **66**, 6118–6128.
28. Schilling, E. and Rehli, M. (2007) Global, comparative analysis of tissue-specific promoter CpG methylation. *Genomics* **90**, 314–323.

Chapter 8

Serial Analysis of Binding Elements for Transcription Factors

Jiguo Chen

Abstract

The ability to determine genome-wide location of transcription factor binding sites (TFBS) is crucial for elucidating gene regulatory networks in human cells during normal development and disease such as tumorigenesis. To achieve this goal, we developed a method called serial analysis of binding elements for transcription factors (SABE) for globally identifying TFBS in human or other mammalian genomes. In this method, a specific antibody targeting a DNA-binding transcription factor of interest is used to pull down the transcription factor and its bound DNA elements through chromatin immunoprecipitation (ChIP). ChIP DNA fragments are further enriched by subtractive hybridization against non-enriched DNA and analyzed through generation of sequence tags similar to serial analysis of gene expression (SAGE). The SABE method circumvents the need for microarrays and is able to identify immunoprecipitated loci in an unbiased manner. The combination of ChIP, subtractive hybridization, and SAGE-type methods is advantageous over other similar strategies to reduce the level of intrinsic noise sequences that is typically present in ChIP samples from human or other mammalian cells.

Key words: Serial analysis of binding elements (SABE), transcription factor binding sites (TFBS), chromatin immunoprecipitation (ChIP), subtractive hybridization, serial analysis of gene expression (SAGE), functional genomics, protein–DNA interaction, transcriptional regulation, gene expression, human genome.

1. Introduction

A major challenge in the post-genome era is to elucidate global gene transcriptional regulatory networks in human normal and cancer cells (1). Transcription factors control gene expression through binding-specific regulatory sequences and recruiting chromatin-modifying complexes and the general transcription machinery to initiate RNA synthesis (2). Alterations in gene expression required to co-ordinate various biological processes

Philippe Collas (ed.), *Chromatin Immunoprecipitation Assays,* Methods in Molecular Biology 567,
DOI 10.1007/978-1-60327-414-2_8, © Humana Press, a part of Springer Science+Business Media, LLC 2009

such as the cell cycle and normal development and pathological states such as tumorigenesis are in part a consequence of changes in the DNA binding status of various transcription factors. Consequently, sensitive technologies, to accurately and efficiently identify bona fide regulatory elements for specific transcription factors in vivo on a genome-wide scale, will be needed to elucidate human gene regulatory networks.

Global localization analysis of binding sites for sequence-specific transcription factors in vivo can be performed using chromatin immunoprecipitation (ChIP) and determining the genomic location of the ChIP-enriched DNA by microarray hybridization (ChIP-on-chip) (3, 4). This method circumvents the limitations of traditional methods. When coupled with gene expression and other relevant information, ChIP-on-chip assays can be extremely useful in analyzing yeast transcriptional regulatory networks, in which the promoters are well characterized (1, 5). This technique has been broadly used to identify the genomic sites bound by regulators of transcription in yeast and other eukaryotic cells (6, 7). Limited analysis of human transcription factor binding sites using ChIP-on-chip strategies have also been performed with selected promoters of genes of interest (8, 9), with CpG microarrays (10) or with selected chromosomes (11). However, comparable strategy for globally analyzing binding sites of transcription factors to the human genome is currently impracticable due to the enormous size and complexity, and also because regulatory elements are often found at vast distances either upstream or downstream from the core promoter. In fact, only 20–30% of the transcription factor binding sites localize to known promoter regions (11, 12). A solution to this limitation is to use microarrays that interrogate the entire genome. Problems with such "whole-genome tiling" microarrays are cost, reproducibility, and statistical analysis (13).

To overcome these limitations and allow interrogation of entire mammalian genome in an unbiased manner, we developed a novel approach to study genome-wide location analysis of transcription factors in human genome in vivo. This technology, called serial analysis of binding elements (SABE) (12, 14), involves specific ChIP (15), enrichment of ChIP DNA by subtractive hybridization (16), and generation of sequence tags similar to serial analysis of gene expression (SAGE) (17). Similar approaches were developed independently by different groups, attesting to the utility of this approach (12, 18–22). Termed SACO (for serial analysis of chromatin occupancy) (18), STAGE (sequence analysis of genomic enrichments) (19), GMAT (genome-wide mapping technique) (21), or ChIP-PET (22), these techniques including SABE circumvent the need for microarrays to identify immunoprecipitated loci. Compared with tiling genomic microarrays, these methods are considerably more affordable. Although whole-genome tiling arrays will undoubtedly become less expensive, this cost differential is likely to continue for the foreseeable future.

Our approach for generating sequence tags using SABE is different from those similar techniques (SACO, STAGE, or GMAT) in that SABE tags are generated as random 18-mers produced from ChIP DNA fragments. The advantage of this is that the "tag resolution" is not limited by the presence of a four-cutter restriction enzyme site in the ChIP DNA that is used to "anchor" the tags, which makes this technology truly unbiased. Moreover, SABE does not require cloning, re-cloning, and library construction steps of ChIP DNA described in ChIP-PET method (22), which are labor- and time-consuming and cause potential bias. In addition, our technique recruits a subtractive hybridization step, which is essential to reduce the intrinsic noise resulting from isolation of repetitive sequences during ChIP in mammalian cells (23).

2. Materials

2.1. Plasmids

1. Plasmids pTet-Off and pTRE2hyg are used for the construction of tetracycline-inducible cell line expressing transcription factor of interest. Both plasmids are available from Clontech (cat. No. 631017 and 631014, respectively).
2. Plasmid p3FLAG is a mammalian vector for stable expression of fusion protein with a triple FLAG epitope on the N-terminal. p3FLAG was constructed by inserting a triple FLAG epitope (5′-CTAGACC <u>ATG</u> GAC TAC AAA GAC CAT GAC GGT GAT TAT AAA GAT CAT GAC ATC GAT TAC AAG GAT GAC GAT GAC AAG-3′) (start code underlined) into *Nhe*I site of pcDNA3.1/*myc*-His(-)B (Invitrogen cat. No. V855-20). p3FLAG also has c-*Myc* and 6-His epitopes on its C-terminal to meet different purposes. Two similar plasmids to p3FLAG are commercially available (p3xFLAG-CMV^{TM}-10 for N-terminal Met-3xFLAG expression and p3xFLAG-*myc*-CMV^{TM}-26 for N-terminal Met-3xFLAG, C-terminal c-*Myc* (dual tagged) expression, Sigma-Aldrich E4401 and E6401, respectively).
3. Plasmid pZERO-2a is a modified version of cloning vector pZERO-2 (Invitrogen cat. No. K2600-01) specific for SABE library construction. pZERO-2a was made by creating a unique *Aat*II site (GACGTC) between *Spe*I and *EcoR*I of the multiple cloning site of pZERO-2 through site-directed mutagenesis (i.e., G<u>C</u>CG<u>C</u>C to GACGTC). Like pZERO-2, plasmid

pZERO-2a allows direct selection of positive recombinants via disruption of the lethal gene, *ccd*B. Expression of *ccd*B results in the death of cells containing non-recombinant vector.

2.2. Cell Culture and Medium

1. Inducible cell line expressing transcription factor of interest tagged with 3xFLAG epitope.
2. RPMI 1640 or Dulbecco Modified Eagle's Medium (DMEM) supplemented with 10% fetal bovine serum (FBS).
3. Doxycycline (Sigma, St. Louis, MO) is dissolved in water at 50 mg/mL, stored in aliquots at 4°C, and used in cell culture at a concentration of 1 μg/mL.

2.3. Reagents

1. Anti-FLAG M2 affinity gel (Sigma-Aldrich, cat. No. A2220).
2. Normal mouse IgG-agarose (Sigma-Aldrich, cat. No. A0919).
3. Yeast tRNA (1 μg/μL) (Invitrogen, cat. No. # 15401-029).
4. Protease inhibitor cocktail (PIC, 100X, Sigma-Aldrich, cat. No. P8340).
5. RNase A (20 μg/μL, Invitrogen, cat. No. 12091-021).
6. Proteinase K (20 μg/μL, Invitrogen, cat. No. 25530-049).
7. Phenol:chloroform:isoamyl alcohol mixture (25:24:1).
8. Chloroform:isoamyl alcohol mixture (24:1).
9. QIAquick PCR purification kit (Qiagen, cat. No. 28106).
10. Micro Bio-Spin Chromatography Column (Bio-Rad, cat. No. 732-6204).
11. DNA polymerase I, Klenow fragment (NEB, cat. No. M0210L).
12. T4 DNA ligase (NEB, cat. No. #M0202L).
13. Taq DNA polymerase (NEB, cat. No. #M0267L).
14. MmeI (NEB, cat. No. #R0637L).
15. TaiI (Fermentas, cat. No. #ER1142).
16. AatII (NEB, cat. No. #R0117L).
17. 30% acrylamide (29:1) (Bio-Rad, cat. No. 161-0121).
18. 10 bp DNA ladder (Invitrogen, cat. No. 10821-015).
19. SYBR green I nuclear acid gel stain (Invitrogen, cat. No. S7567).
20. Dynabeads M-280 streptavidin (Invitrogen, cat. No. 112-05D).

2.4. Buffers

1. 10X PBS: 80 g NaCl, 2 g KCl, 14.4 g $Na_2HPO_4 \cdot 7H_2O$, 2.4 g KH_2PO_4, H_2O to 1 L. Adjust pH to 7.2, autoclave, and store at RT.
2. Hypotonic buffer: 10 m*M* HEPES, pH 7.4, 10 m*M* KCl, 1.5 m*M* $MgCl_2$, 1X PIC. Add PIC fresh before use.

3. ChIP lysis buffer: 50 m*M* HEPES, pH 7.4, 150 m*M* NaCl, 1 m*M* EDTA, 1% Triton X-100, 1X PIC. Add PIC fresh before use.
4. ChIP high salt buffer: 50 m*M* HEPES, pH 7.4, 500 m*M* NaCl, 1 m*M* EDTA, 1% Triton X-100, 1X PIC. Add PIC fresh before use.
5. ChIP wash buffer: 50 m*M* HEPES, pH 7.4, 250 m*M* LiCl, 1 m*M* EDTA, 1X PIC. Add PIC fresh before use.
6. TE buffer: 10 m*M* Tris-HCl, pH 8.0, 1 m*M* EDTA, 1X PIC. Add PIC fresh before use.
7. Elution buffer: 50 m*M* Tris-HCl, pH 8.0, 10 m*M* EDTA, 1% SDS.
8. 5X Hybridization buffer: 2.5 *M* NaCl, 250 m*M* HEPES, pH 8.3, 1 m*M* EDTA.
9. 10X TBE buffer: 890 m*M* Tris-HCl, pH 8.3, 890 m*M* boric acid, 20 m*M* EDTA.
10. PAGE gel diffusion buffer: 0.5 *M* ammonium acetate, 10 m*M* magnesium acetate, 1 m*M* EDTA, pH 8.0, 0.1% SDS.
11. 2X wash/binding buffer: 2 *M* NaCl, 10 m*M* Tris-HCl, pH 7.5, 1 m*M* EDTA.

2.5. Linkers and Primers

1. Linker LK-A: sense, 5′-AGCACTCTCCAGCATAT-CACTCCAACGT-3′; Anti-sense, 5′- ACGTTGGAGTGA-TATGCTGGAGAGTGCT amino-3′.
2. Linker LK-B: sense, 5′-ACCTGCCGACTATCCAAT-CATCCAACGT-3′; Anti-sense, 5′-ACGTTGGATGATTG-GATAGTCGGCAGGT amino-3′ (*see* **Note 1**).
3. Primer-A: 5′-Biotin-AGCACTCTCCAGCATATCAC-3′.
4. Primer-B: 5′-Biotin-ACCTGCCGACTATCCAATCA-3′ (*see* **Note 2**).

3. Methods

The SABE method involves serial enzymatic reactions and DNA manipulations; therefore, a good practice is to monitor the accuracy and efficiency of each step. Overall, there are several key factors to consider when performing SABE. First, for any selected transcription factor of interest, information of at least one well-defined target gene and binding site for that particular transcription factor is needed. This information of a known target gene is used to design PCR primers to monitor the efficiency of ChIP and subtractive hybridization. Without this information, it is hard to know whether the final ChIP DNA is really enriched or not because ChIP with mammalian cells will definitely bring down a lot of background DNA. The information from in vitro binding assays (EMSA, DNA foot printing, SELEX, etc.) may not necessarily reflect in vivo binding of transcription factors and, therefore, cannot be used for this purpose. Second, it is estimated that at least 43% of the human genome is occupied by repetitive elements (24, 25). ChIP provides only a partial enrichment of specific DNAs, and consequently, the signal-to-noise ratio is too low to make direct analysis of target genes practical. To address this problem, SABE employs a subtractive hybridization step modified from representational difference analysis (16) that enables selective amplification of ChIP-enriched DNA over reference (non-enriched) DNA. This step is essential to reduce the intrinsic noise resulting from isolation of repetitive sequences during ChIP in mammalian cells. Third, the quality of antibody used for immunoprecipitation is very important. Transcription factors generally express at low level in living cells and have a weaker affinity for DNA than histone proteins; therefore, ChIP application of transcription factors is particularly demanding because the antibody must be capable of recognizing the native protein as part of a cross-linked protein–DNA complex. Many antibodies, even those that work well for Western blots, fail this more rigorous test. Different antibodies may also produce significantly different data sets (11). Triple FLAG epitope and the corresponding anti-FLAG M2 antibody provide the most sensitive antigen–antibody detection system to date. Detection of fusion proteins containing 3xFLAG is 20–200 times more sensitive than other tags such as c-myc, 6xHis, GST, or HA and is ideal for ChIP assays of low-level expression transcription factors in mammalian cells (http://www.sigmaaldrich.com/). There are several advantages in using a universal antibody–IP system with transcription factor of interest tagged with 3xFLAG. First is that many transcription factors show poor antigenicity and do not have good antibodies for efficient IP. Second is that some target genes show much less binding capacity than the others to the same transcription factor (26). To get enrichment of these weaker binding sites by ChIP, the transcription factor of interest has to be over-expressed to enhance the binding to these sites. Although over-expression of an epitope-tagged protein may cause artifactual interactions, this concern can be addressed by a subsequent verification step. Third is that using a universal antibody–IP system will produce a unique background related to IP process, which can be easily distinguished from bona fide IP products when applying to different transcription factors.

SABE method is shown in **Fig. 8.1**. An inducible human cell line expressing a transcription factor of interest tagged with 3xFLAG epitope is established. Cells are cross-linked in vivo using formaldehyde and lysed; DNA is sheared by sonication to produce fragments of 200–1,000 bp. Protein–DNA complexes are

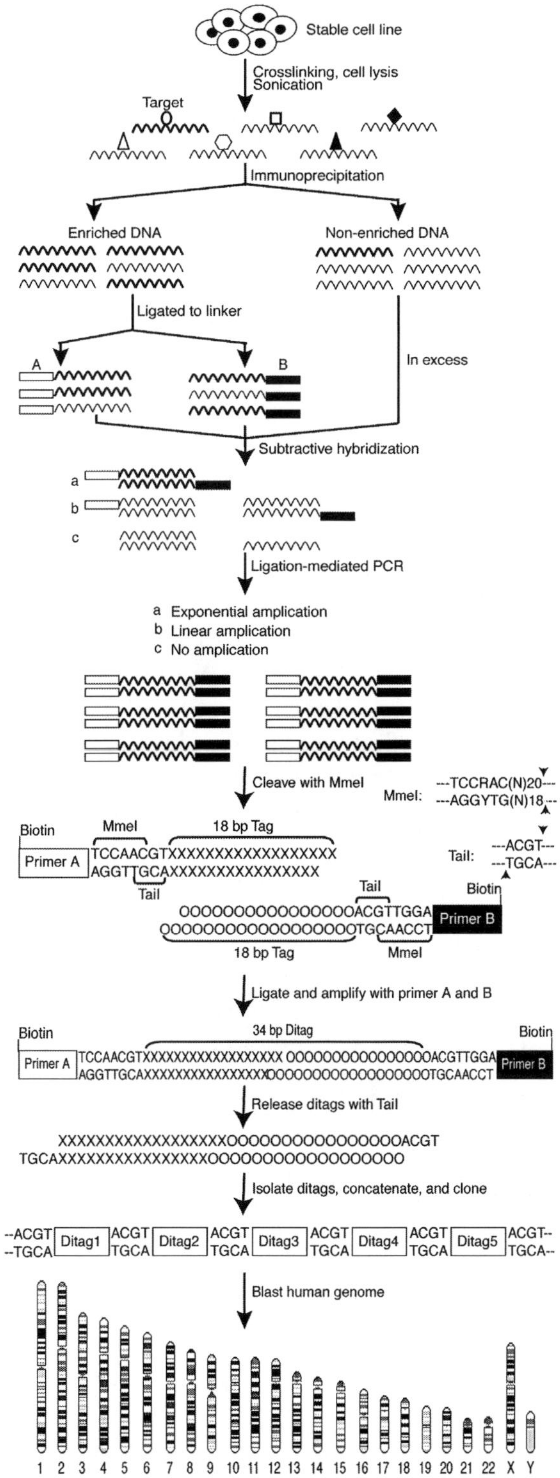

Fig. 8.1. Schematic representation of SABE method (12, 14). Inducible human cell line expressing transcription factor (TF) of interest are cross-linked, lysed, and sonicated. Protein–DNA complexes are immunoprecipitated using specific antibody. ChIP-enriched DNA is ligated to linkers and specific DNA selectively amplified by subtractive hybridization

then immunoprecipitated by using anti-FLAG M2 affinity beads. ChIP-enriched DNA is divided into two and ligated to either linker A or B, then hybridized to excessive amount of non-enriched DNA control (subtractive hybridization), followed by ligation-mediated PCR. After amplification, non-specific DNA sequences will be under-represented in the product mixture relative to specific DNA fragments. To analyze enriched DNA fragments, a strategy modified from SAGE is performed (17). The linkers A and B are designed with overlapping recognition sites for the type III endonuclease, *Mme*I and a 4 bp cutter, *Tai*I (**Fig. 8.1**). Additionally, to facilitate separation of the linkers from the final tag DNAs, the primers contain a 5′ biotin moiety. DNA fragments from subtractive hybridization and PCR amplification are digested with *Mme*I, and the 46 bp fragments, including 28 bp of the linker plus 18 bp of flanking tag sequence, are purified on 8% acrylamide gels. Because *Mme*I leaves a 2 bp 3′ overhang, to maximize information content of the tags, the digested fragments are ligated directly to form ditags, rather than trimming to create blunt ends (**Fig. 8.1**). The ligated ditags are amplified with primers A and B and then released by digestion with *Tai*I. *Tai*I was selected because it maximally overlaps with the *Mme*I site and is more efficient than *Nla*III, the anchoring enzyme used in SAGE (27). After digestion, the ditags can be separated from the biotin-tagged primer fragments by using streptavidin Dynabeads, further purified by electrophoresis, ligated to form concatemers, and directly cloned into pZERO-2a vector containing an *Aat*II site (GACGTC). Clones containing concatemers of 200–2,000 bp are analyzed by sequencing. Ditags can be identified in the sequencing data because each is 34 bp long separated by a *Tai*I sequence (ACGT). The final tag generated by SABE method is 18 bp long, including a 2 bp overlap generated by the *Mme*I digestion (**Fig. 8.1**). Tag sequences are used to blast the human genome database to identify its genomic location. Putative binding sites for the factor of interest can then be identified by analyzing flanking DNA on genes of particular interest for consensus sequences, with the rationale that the SABE tag must reside within a segment no greater than the length of the original sheared immunoprecipitated DNA fragments.

Fig. 8.1. (continued) and ligation-mediated PCR. Sequence tags are released by digestion with *Mme*I and ditags are produced by ligation, which are released by digestion with *Tai*I and separated from biotinylated linkers by using streptavidin magnetic beads. Ditags are concatemerized, cloned, and sequenced. Ditag sequences are 34 bp long and are separated by the *Tai*I recognition sequence (ACGT). Each tag sequence is 18 bp long and can be used to blast human genome database to decide its unique location.

3.1. Cell Culture, Cross-Linking, and Sonication

1. Grow 5 × 10^8 cells expressing the transcription factor of interest tagged with 3xFLAG epitope. The cells should be 80–90% confluent.
2. Collect cells. For adherent cells, aspirate the growth medium from the cells, scrape the cells into 50 mL conical centrifuge tubes using fresh medium, centrifuge for 5 min at 450*g* at room temperature (25°C), and then discard the supernatant. For suspension cells, collect the cells in 50 mL conical centrifuge tubes, centrifuge for 5 min at 450*g* at room temperature, and then discard the supernatant.
3. Re-suspend the cells in 45 mL pre-warmed culture media and collect all cells into one 50 mL conical centrifuge tube. Add 1.25 mL 37% formaldehyde solution to the cell suspension (final concentration: 1% formaldehyde). Incubate at room temperature for 10 min, with occasional inversion, to cross-link the protein of interest with DNA (*see* **Note 3**).
4. Add 5 mL 1.25 *M* glycine to the fixed culture and incubate at room temperature for 5 min, with occasional inversion.
5. Centrifuge cells for 5 min at 420*g* at 4°C and discard supernatant. Wash cells twice with 40 mL ice-cold 1X PBS, spin down cells for 5 min at 420*g* at 4°C after each wash, and discard supernatant. Place cell pellet on ice.
6. Re-suspend cell pellet in 5 mL ice-cold hypotonic buffer. Pass the cells through 27 1/2 gauge needle 10 times on ice to extract the nuclei. Collect the nuclei by centrifuging for 10 min at 10,000*g* at 4°C (*see* **Note 4**).
7. Discard the supernatant and re-suspend the nuclei in 6 mL lysis buffer. Incubate on ice for 30 min (*see* **Note 5**).
8. Shear chromatin by sonicating cell lysate for 10 min with cycles of 10 s of sonication followed by 50 s of pause with a sonicator. Keep cell lysate on ice during sonication. The final size of sheared DNA should be around 200–1,000 bp with average ~500 bp (*see* **Note 6**).
9. Centrifuge the suspension at 12,000*g* for 10 min at 4°C. Transfer supernatant (soluble cell lysate) into a new 15 mL tube. Place the tube on ice.

3.2. Pre-cleaning and Immunoprecipitation

1. Thoroughly suspend the ANTI-FLAG M2 affinity agarose gel and normal control mouse IgG-agarose gel in the vial, in order to make a uniform suspension of the resin. Immediately transfer 400 μL (for 6 mL of cell lysate) of the resin from each agarose gel in its suspension buffer to a separate new 1.5 mL tube to allow a homogenous dispersion of the resin. For resin transfer, use a clean, plastic pipette tip with the end enlarged to allow the resin to be transferred.

2. Centrifuge the resins for 30 s at 6,000*g* using a fixed angle rotor. In order to let the resin settle in the tube flatly, wait for 1–2 min before handling the samples. Aspirate the supernatant with a 27G1/2 needle.
3. Wash the packed gel with 1 mL lysis buffer. Repeat the wash twice. Be sure that the wash buffer is removed and no resin is discarded.
4. Add the normal control mouse IgG-agarose gel to the 6 mL soluble cell lysate to pre-clean the cell lysate. Add ANTI-FLAG M2 affinity agarose gel to 1 mL lysis buffer with 0.1% BSA and 1 μg/μL yeast tRNA to block the gel. Incubate both tubes on the rotating platform at 4°C for at least 1 h.
5. Collect pre-cleaned cell lysate and ANTI-FLAG M2 affinity agarose gel separately by centrifugation for 30 s at 6,000*g* at 4°C. Note pre-cleaned cell lysate is the supernatant in one tube and ANTI-FLAG M2 affinity agarose gel is the pellet in another tube.
6. Transfer ANTI-FLAG M2 affinity agarose gel to pre-cleaned cell lysate. Dilute the cell lysate with 1 volume (6 mL) of lysis buffer. Incubate at 4°C on a rotating platform overnight. Immunoprecipitation may be carried out for a longer time for convenience.

3.3. Washing, Elution, and Reversal of Cross-Link

1. Centrifuge the cell lysate with resin for 5 min at 3,000*g* at 4°C. Transfer the supernatants to a new 15 mL tube and keep as the non-enriched control.
2. Transfer the resin to a new 1.5 mL tube with fresh ChIP lysis buffer. Wash the resin three times sequentially with 1 mL each of the following pre-cooled buffers, all containing 1X PIC:ChIP lysis buffer; ChIP high salt buffer; ChIP wash buffer; and TE buffer. Pellet the resin during each wash by centrifugation for 30 s at 6,000*g* at 4°C and carefully aspirate the supernatant with a 27G1/2 needle.
3. Add 400 μL of elution buffer to the washed resin. As a control, transfer 360 μL of non-enriched control into a 1.5 mL tube and add 40 μL of 10% SDS. Incubate overnight at 65°C in a hybridization oven with rotation to revert the cross-link. This step may be carried out for a longer time for convenience.

3.4. Purification of ChIP DNA

1. Pellet the resin by centrifugation for 30 s at 6,000*g*. Transfer the supernatant to a new tube.
2. Add 3 μL of RNase A (20 μg/μL) to each tube. Incubate samples for 1 h at 37°C. Add 20 μL of proteinase K (20 μg/μL) to each tube. Incubate at 50°C for another hour.

3. Extract twice with 1 volume of phenol:chloroform:isoamyl alcohol mixture (25:24:1). Centrifuge for 3 min at 16,000*g* (13,000 rpm in an Eppendorf centrifuge with a 24-place fixed angle rotor) at 4°C. Transfer the DNA solution (upper aqueous phase) into a new tube after each extraction. Extract once with 1 volume of chloroform:isoamyl alcohol mixture. Centrifuge again for 3 min at 16,000 *g* at 4°C. Transfer the DNA solution into a new tube.
4. Add 1/10 volume of 3 *M* NaAc (pH 5.3). Add 3 volumes of cold 95–100% ethanol and mix briefly. Incubate at –20°C for at least 2 h.
5. Centrifuge at 16,000*g* for 20 min at 4°C. Pour off the supernatant, add 1 mL cold 70% ethanol, vortex briefly, and centrifuge again at the same speed for 3 min at 4°C. Carefully remove the supernatant with a pipette.
6. Let the pellet dry for a couple of minutes and re-suspend the pellet in 50 μL of TE; incubate at 65°C for 10 min.
7. Measure the DNA yield and purity using a spectrophotometer. The yield using anti-FLAG M2 affinity gel generally is 50–100 μg. Adjust both ChIP-enriched DNA and non-enriched DNA concentration to 1 μg/μL. The DNA can be stored for several months at –20°C.
8. Test specific enrichment of ChIP DNA over non-enriched DNA using known target and binding sites information for the transcription factor of interest. This information will be used in the subtractive hybridization step (*see* **Note** 7).

3.5. Blunting of ChIP-Enriched DNA

1. To blunt ChIP-enriched DNA, set up the following reaction mix: 50 μL of ChIP DNA, 1 μg/μl, 30 μL of 10X EcoPol buffer, 1 μL dNTP mix (10 m*M* each), 10 μL Klenow fragment (5 U/μL), and 209 μL water.
2. Mix by pipetting and incubate at RT for 15 min. Stop the reaction by adding 6 μL of 0.5 *M* EDTA and heating at 75°C for 20 min.
3. Extract once with phenol:chloroform:isoamyl alcohol mixture. Centrifuge for 3 min at 16,000*g* at 4°C. Transfer the DNA solution to a new tube. Extract once again with chloroform:isoamyl alcohol mixture and centrifuge for 3 min at 16,000*g* at 4°C. Transfer the DNA solution to a new tube.
4. Add 1/10 volume of 3 *M* NaAc. Add 3 volumes of cold 95–100% ethanol and centrifuge at 16,000*g* for 20 min at 4°C. Wash with 1 mL cold 70% ethanol. Dry the pellet and re-suspend the DNA pellet in 30 μL TE. Incubate at 65°C for 10 min.
5. Measure the DNA yield and adjust the DNA concentration to 1 μg/μL. The DNA can be stored for several months at –20°C.

3.6. Ligation of ChIP-Enriched DNA and Subtractive Hybridization

1. Separate ChIP-enriched DNA into two equal amounts and set up two ligation reactions, one with linker A and another with linker B. Set up the ligation mix as follows, also include mock ligation (without linker) as negative control: 6 μL of Blunt ChIP DNA (1 μg/μL), 6 μL of 10X DNA ligase buffer, 1 μL of LK-A (45 μ*M*) or LK-B, 3 μL of T4 DNA ligase, and 44 μL of water. Mix by pipetting and incubate for at least 2 h at 16°C. Longer ligation may be optimal. The ligation reaction can be left overnight at 16°C.
2. Recover the DNA using a QIAquick PCR purification kit according to the manufacturer's direction. Briefly, add 5 volumes (300 μL) of buffer PB1 to each of the ligation reaction (60 μL) and mix. To bind DNA, apply the samples to the QIAquick columns and centrifuge for 60 s at 10,000*g*. Wash the columns with 0.75 mL buffer PE. Place each column in a clean 1.5 mL tube and elute DNA in 30 μL buffer EB. Measure DNA concentration and purity. Adjust DNA concentration to 0.1 μg/μL. Test ligation efficiency by PCR using primer A or B. Efficient ligation with linkers will produce a significant amount of PCR products compared with control ligation without linker. The DNA can be stored for several months at –20°C.
3. Set up two hybridization solutions with either LK-A or LK-B ligated DNA as follows: 4 μL of 5X hybridization buffer, 12 μL of LK-A or LK-B DNA (0.1 μg/μL), and 4 μL of non-enriched DNA (1 μg/μL). Overlay with mineral oil, denature at 98°C for 1.5 min, and then hybridize at 65°C for 1.5 h.
4. Mix the two hybridization solutions (LK-A DNA and LK-B DNA), add 8 μL more heat-denatured non-enriched DNA and 2 μL of 5X hybridization buffer. Hybridize overnight at 65°C (*see* **Note 8**).
5. In the final 30 μL hybridization reaction, add the following: 20 μL of 10X PCR reaction buffer (NEB), 6 μL dNTP (10 m*M*), and 142 μL water. Incubate at 85°C for 3 min, and then bring down to 72°C before adding 2 μL of Taq DNA polymerase. Incubate at 72°C for another 10 min.
6. Purify the DNA using a QIAquick PCR purification kit. Briefly, add 5 volumes (1000 μL) of buffer PB1 to the DNA solution (200 μL) and mix. To bind DNA, apply the mixed solution to two QIAquick columns, each with 600 μL and centrifuge for 60 s at 10,000 *g*. Wash the columns with 0.75 mL buffer PE. Place each column in a clean 1.5 mL tube. Elute DNA in 50 μL elution buffer. These are linker-ligated DNAs (LK-DNAs). The final eluted DNA can be stored for several months at –20°C.

3.7. Optimizing PCR Condition and Linker-Mediated PCR

1. Make a serial two-fold dilution of LK-DNA template for a total of 20 dilutions. Set up the PCR reactions as follows: 4 μL (with various concentration) of LK-DNA template, 10 μL of 10X PCR buffer, 3 μL dNTP (10 m*M*), 2 μL primer A, 2 μL primer B, 1 μL Taq DNA polymerase, 78 μL water. Run the PCR as follows: 95°C 3 min; then 95°C 1 min, 58°C 1 min, and 72°C 2 min for 30 cycles; 72°C 10 min, and hold at 4°C.
2. Run 15 μL of each PCR product on a 2% agarose gel. The PCR product should be a smear ranging from 100 to 2,000 bp with an average size of 500 bp. Determine the minimal amount of template DNA required to yield maximum amount of PCR products. Set this amount of template DNA as optimal concentration for the following PCR reactions. Generally the optimal amount of template is 0.1–1 μL.
3. Set up large-scale PCR reactions using optimal template concentration determined at the last step: total 20 PCR reactions are needed for this step; each PCR reaction contains: 4 μL of ChIP-enriched DNA template (optimal concentration), 10 μL of 10X PCR buffer, 3 μL dNTP (10 m*M*), 2 μL primer A, 2 μL primer B, 1 μL Taq DNA polymerase, and 78 μL water.
4. Run PCR as follows: 95°C 3 min; then 95°C 1 min, 58°C 1 min, and 72°C 2 min for 30 cycles; 72°C 10 min; and then hold at 4°C.
5. Collect the PCR products (total 2,000 μL) in a 15 mL tube, and purify using a QIAquick PCR purification kit according to the handbook. Briefly, add 5 volumes (10 mL) of buffer PB1 to the PCR solution (2,000 μL) and mix. To bind DNA, add the mixed solution to six QIAquick columns, each with 600 μL and centrifuge for 60 s at 10,000*g*. Add the remaining solutions to the columns until all solutions have been added to the columns. Repeat the centrifuge step after each loading. Wash the columns with 0.75 mL buffer PE. Place each column in a clean 1.5 mL tube. Elute DNA with 50 μL of buffer EB. Collect the elution from all columns. Measure PCR yield and purity with a spectrophotometer. Adjust DNA concentration to 0.1 μg/μL. Generally the DNA yield will be 20–30 μg. The DNA can be stored for several months at –20°C.

3.8. MmeI Digestion, Isolation of Sequence Tag, and Ditag Formation

1. Set up a MmeI digest reaction as follows: 200 μL of PCR product (at 0.1 μg/μL), 40 μl of NEB buffer, 440 μL of 10X SAM, and 110 μL water.
2. Mix the reaction before adding MmeI enzyme. Then add 10 μL MmeI (2 U/μL) and mix very gently by pipetting six to eight times. Incubate at 37°C for 2 h. The reaction can be left overnight at 37°C (*see* **Note 9**).

3. Set up an 8% PAGE gel (16 × 20 cm) with a 15-well comb (well width: 6.5 mm, thickness: 1.0 mm) in a Bio-Rad PROTEAN II xi cell as follows: 10.7 mL of 30% acrylamide (29:1), 2 mL of 10X TBE, 200 μL APS (10%), 40 μL TEMED, and 27.06 mL water.
4. Add 50 μL of 50% glycerol (do not use loading dye) to the MmeI digestion (400 μL) and mix. Load the reaction directly to the gel, each lane with 60 μL. Total eight lanes are needed. Also include one lane with un-cut control, one with loading dye (bromophenol blue) only and one with 1 μg of 10 bp DNA ladder.
5. Run the gel for 2–4 h at 200 V with water-cooling until bromophenol blue is three-fourths down the gel.
6. Stain the gel with SYBR Green I at a dilution of 1:10,000 in 1X TBE buffer for 30 min with gentle agitation. Visualize the bands under a standard UV trans-illuminator and take a photo as a record. A strong 46 bp band should be seen.
7. Make a hole through the bottom of a 0.5 mL Eppendorf tube using an 18-gauge needle.
8. Using a new razor blade, excise the 46 bp band from the gel. Collect the gel slices from two lanes into one 0.5 mL tube with a hole and place the tube on a 2 mL screwed tube. Total four tubes are needed. Centrifuge for 1 min at 16,000*g*. The excised bands will be broken into small pieces and collected in the 2 mL tube.
9. Add 1 mL of gel diffusion buffer to the 2 mL tube containing the gel pieces. Incubate at 65°C for 2 h to elute the DNA from the gel with agitation.
10. Pass the gel solution through a Micro Bio-Spin Chromatography Column by centrifuging at 3 min at 16,000*g* to remove any residual polyacrylamide. Collect the DNA solution in 1.5 mL tubes.
11. Fill up the tubes with 1-butanol and mix. Centrifuge 1 min at 16,000*g*. Discard the upper phase containing 1-Butanol. Repeat this step until the volume in each tube is reduced to 200 μL. Transfer all the DNA solutions from four tubes to a new 1.5 mL tube and reduce the volume the DNA solution to 400 μL with 1-Butanol.
12. Extract the DNA solution twice with 1 volume of phenol:-chloroform:isoamyl alcohol mixture. Centrifuge for 3 min at 16,000 *g*at 4°C. Transfer the DNA solution to a new tube. Extract once again with 1 volume of chloroform:isoamyl alcohol mixture and centrifuge for 3 min at 16,000 *g* at 4°C. Transfer the DNA solution to a new tube.
13. Add 1/10 volume of 3 *M* NaAc, add 3 volumes of cold 95–100% ethanol, and vortex. Incubate at –20°C for 2 h.

14. Centrifuge at 16,000*g* at 4°C for 30 min. Carefully remove and discard the supernatant. Wash the DNA pellet twice, each with 800 μL of cold 70% ethanol. Air-dry and re-suspend the DNA pellet in 20 μL of water. These are the 46 bp long MmeI sequence tags. The DNA can be stored for several months at –20°C.
15. Set up a ligation reaction as follows: 17 μL of purified tags, 2 μL of 10X ligation buffer, and 1 μL T4 DNA ligase. Mix gently and incubate overnight at 16°C.
16. Add 180 μL of TE to the ligation reaction. Heat at 65°C for 10 min to inactivate the DNA ligase.

3.9. Optimizing PCR Condition and PCR Amplification of Ditags

1. Make a serial two-fold dilution of ligated ditags for a total of 20 dilutions. Set up the PCR reactions as follows: 10 μL (various concentration) of ligated ditag template, 10 μL of 10X PCR buffer, 3 μL dNTP (10 m*M*), 2 μL primer A, 2 μL primer B, 1 μL Taq DNA polymerase, and 72 μL water.
2. Run the PCR as follows: 95°C 3 min; then 95°C 30 s, 58°C 30 s and 72°C 10 s for 30 cycles; 72°C 10 min; and finally hold at 4°C.
3. After PCR, set up an 8% PAGE gel as indicated before and analyze the PCR products. A clear 90 bp ditag band should be seen. Determine the minimal amount of template DNA (ditags) required to yield a significant 90 bp band. Set this amount of template DNA as optimal concentration and proceed to scale-up PCR. Generally the optimal template amount is 1 μL.
4. Set up the PCR reaction as follows using optimal ditag template concentration determined in last step. Total 20 reactions are needed. One reaction contains: 1 μL of ligated ditag template at optimal concentration, 10 μL of 10X PCR buffer, 3 μL dNTP (10 m*M*), 2 μL primer A, 2 μL primer B, 1 μL Taq DNA polymerase, and 81 μL water.
5. Run the PCR as follows: 95°C 3 min; then 95°C 30 s, 58°C 30 s and 72°C 10 s for 30 cycles; 72°C 10 min, and hold at 4°C.
6. Collect the PCR products into five 1.5 mL tubes, each containing 400 μl. Extract once with 1 volume of phenol:chloroform:isoamyl alcohol mixture and once with chloroform:isoamyl alcohol mixture. Spin for 3 min at 4°C at 16,000*g* after each extraction. Transfer the supernatant to new tubes.
7. Add 1/10 volume of 3 *M* NaAc, add 3 volumes of cold 95–100% ethanol, and vortex. Incubate at –20°C for 2 h.
8. Centrifuge at 16,000 *g* at 4°C for 30 min. Carefully remove and discard the supernatant. Wash the DNA pellet with 70% ethanol. Air-dry and re-suspend the pellet in 20 μL of water. Collect all DNA solutions into one tube. Measure the yield and purity. Adjust DNA concentration to 0.1 μg/μL. The DNA can be stored at –20°C for several months.

3.10. Tail Digestion and Purification of Ditags

1. Set up TaiI digest reaction as follows: 100 μL PCR product (0.1 μg/μL), 20 μl of buffer R⁺10 μL TaiI (10 units/μL), and 70 μL water.
2. Mix gently. Incubate at 65°C for 2 h or overnight at 65°C.
3. Aliquot 100 μL (10 μg/μL) of Dynabeads M-280 streptavidin into a clean 1.5 mL tube. Add 200 μL of 1X wash/binding buffer and vortex to suspend beads. Apply a magnet field to the side of the tube for 1–2 min. Remove and discard the supernatant. Repeat wash once.
4. To 200 μL of TaiI-digested ditags, add an equal volume of 2X wash/binding buffer and mix. Then transfer the solution to the tube containing magnetic beads. Vortex to suspend the particles and incubate at room temperature for 10 min with agitation.
5. Apply a magnet field. Transfer the supernatant into a new tube.
6. Set up a 12% PAGE gel (16 × 20 cm) using a Bio-Rad PROTEAN II xi cell system as follows: 16 mL of 30% acrylamide (29:1), 2 mL of 10X TBE, 200 μL APS (10%), 40 μL TEMED, and 21.76 mL water.
7. Add 50 μL of 50% glycerol (do not use loading dye) to the ditag solution (400 μL) and mix. Load the solution directly to the gel, each lane with 60 μL. Total of eight lanes are needed. Also include one lane with un-cut control, one with loading dye (bromophenol blue) only and one with 1 μg of 10 bp DNA ladder.
8. Run the gel at 200 V for 2 h. Purify the 34 bp ditag band as described in **Section 3.8.** Dissolve the final ditag in 20 μL of water. The precipitated DNA can be stored at –20°C for several months.

3.11. Ligation of Ditags to Form Concatemers and Isolation

1. Set up a ligation reaction as follows: 17 μL of purified ditags, 2 μL of 10X ligation buffer, and 1 μL T4 DNA ligase. Mix gently and incubate overnight at 16°C.
2. Load the ligation solution onto a 1% agarose gel and run the gel.
3. Excise the concatemers of 200–2,000 bp from the gel. Collect the gel slices into 1.5 mL tube.
4. Purify the concatemers using a QIAquick gel purification kit according to the manufacturer's directions. Briefly, weigh the gel slices and add 3 volumes of buffer QG to 1 volume of gel (100 mg ∼100 μL). Incubate at 50°C for 10 min. Add 1 gel volume of isopropanol to the sample and mix. To bind DNA, add the samples to QIAquick columns and centrifuge for 60 s at 10,000*g*. Add 0.5 mL of buffer QG to the column and centrifuge again for 60 s. Wash the columns with 0.75 mL buffer PE. Place each column in a clean 1.5 mL tube. Elute DNA with 50 μL of buffer EB. Measure the DNA concentration and adjust to 10 ng/μL. The DNA can be stored at –20°C for several months.

3.12. Cloning Concatemers and Colony PCR Analysis

1. Transform plasmid pZERO-2a into an F′ *E. coli*strain (e.g., JM109) and spread to LB-Kanamycin plate. Select one single colony and grow in 500 mL of LB medium containing 50 μg/mL Kanamycin and purify plasmid DNA using CsCl gradient ultracentrifugation (*see* **Note 10**).
2. Digest 1 μg of CsCl-purified pZERO-2a plasmid with AatII. Extract the DNA with phenol:chloroform:isoamyl alcohol mixture and chloroform:isoamyl alcohol mixture. Precipitate the DNA with ethanol and dissolve it in 100 μL of TE (10 ng/μL).
3. Set up the ligation reaction as follows: 5 μL of digested vector (10 ng/μL), 5 μL of purified concatemers (10 ng/μL), 1 μL of 10X ligation buffer, 8 μL water, and 1 μL T4 DNA ligase.
4. Incubate at 16°C for 30 min. Longer ligation may be optimal. Transform 10 μL of ligation solution into 100 μL of DH5α competent cells. Plate all transformation mix on LB-Kanamycin plates.
5. Analyze Kanamycin-resistant colonies by colony PCR using M13 forward and reverse primers. Pick up clones bearing an insert between 200 and 2,000 bp.

3.13. Sequencing and Sequence Analysis

1. Grow selected clones and sequence these clones using T7 primer.
2. Analyze the sequencing data. Typically, each clone contains 10–30 ditags. Ditags are 34 bp long and separated by the TaiI recognition site, ACGT. The final tag generated by the SABE method is 18 bp long, including a 2 bp overlap generated by MmeI digestion. Tag sequences are used to blast the human genome database to identify its genomic location. Putative binding sites for the transcription factor of interest can be identified by analyzing flanking sequences for consensus sequences (*see* **Note 11**).

4. Notes

1. Two linkers are used to prevent the formation of pan-like structure during subtractive hybridization and LM-PCR. Both linkers are modified with an amino group at the 3′ end to prevent self-ligation. Linkers should be obtained PAGE-purified after synthesis from oligo company (i.e., Integrated DNA Technologies for linker syntheses).
2. Both primers are biotinylated at the 5′ end to facilitate isolation of ditags. Primers should be obtained PAGE purified.

3. The aim of cross-linking is to fix the transcription factor of interest to its chromatin binding sites. Cross-linking is a time-critical procedure and the optimal length of cross-linking depends on the cell type and transcription factor of choice. Too much cross-linking may mask epitopes for efficient immunoprecipitation and too little cross-linking may lead to incomplete fixation. If uncertain, perform a time-course experiment and run a conventional ChIP assay to optimize cross-linking conditions.
4. Cell lysis can be observed by the addition of the Trypan blue solution to an aliquot of cells. The dye is excluded from the intact cells, but stains the nuclei of lysed cells. Lysis should be 80–90%. If the lysis is not sufficient, perform several more strokes until lysis is complete. If nuclear lysis or clumps of nuclei are visualized, the cell disruption was too vigorous or too many strokes were performed.
5. Foaming during the sonication step can result in insufficient shearing of chromatin DNA. To avoid this, use 6 mL total volume in a 15 mL conical tube and keep sonicator tip 0.5–1 cm deep in cell lysate sample during sonication.
6. Sonication efficiency will vary depending on sonicator, cell type, and extent of cross-linking and will have to be optimized to yield the desired final average length of DNA for each specific cell type. Ideally, the average DNA size of sheared sample should be confirmed by 2% agarose gel electrophoresis stained with ethidium bromide.
7. For all DNA enriched by ChIP experiments, the efficiency of immunoprecipitation must be determined by quantitative real-time PCR analyses (i.e., ratio of the amount of enriched (immunoprecipitated) DNA over that of the non-enriched (left-over) DNA). For this purpose, the knowledge of at least one well-defined binding site for the transcription factor of interest is needed. This knowledge of a known target gene is used to design the primers for real-time PCR and optimize immunoprecipitation conditions. The ratio of enrichment should also be normalized to the level observed at a control region, which is defined as 1.0. In general, if more than 10 fold of enrichment can be achieved by immunoprecipitation step, the following subtractive hybridization step can further increase the signal-to-noise ratio. However, please note that the transcription factor of interest may have different affinity to its individual binding sites. For example, ChIP recovers several 100-fold more p21 and MDM2 promoter DNA while recovers substantially weaker or background p53 binding elements for Bax, AIP1, and PIG3 (26). For this reason, if there are more than one known target genes for the

transcription factor of interest available, choose the one that can achieve higher signal-to-noise ratio. Always test the quality of antibody and optimize ChIP conditions (cross-linking, sonication, immunoprecipitation, etc.) using this knowledge of known target gene(s) for the specific transcription factor and cell type of your choice. If a satisfactory ratio of ChIP enrichment cannot be achieved using its own antibody, consider making a construct of transcription factor tagged with 3xFLAG epitope and using anti-FLAG M2 affinity agarose beads for ChIP, which has been proven to have the highest affinity compared with other epitope tags.

8. The ratio of non-enriched over ChIP-enriched DNA in subtractive hybridization solution is dependent on the ratio of enrichment obtained from the immunoprecipitation step. Typically, 10-fold of ChIP enrichment will be needed to begin the subtractive hybridization step.
9. (a) Make 10 times *S*-adenosylmethionine (10X SAM) solution (500 μ*M*) from its original concentration (32 m*M*) freshly before use. (b) Reaction using MmeI should be done at or near stoichiometric concentration as indicated (1 μg DNA/1 μL MmeI). Excessive amounts of MmeI block cleavage.
10. (a) Plasmid pZERO-2a cannot grow in *E. coli* strains without a *lac*I^q gene (e.g., DH5α); (b) plasmid DNA purified by other methods (i.e., Qiagen plasmid purification kit) contains small amount of *E. coli* genomic DNA, which may be cloned and mistakenly selected for sequencing.
11. Due to the quality of performance for each SABE step, final clones may contain primer dimers and linker sequences. Final clones may also contain *E. coli* genomic sequences if using plasmid DNA purified by methods other than CsCl ultracentrifugation. It is worth noting that there are about 30–40% of the final SABE tags that cannot be assigned unique locations to the human genome due to multiple hits. This is probably because of the repetitive elements in the human genome, whose lengths range from several hundreds to several thousands of base pairs (24, 25).

References

1. Wyrick, J. J. and Young, R. A. (2002) Deciphering gene expression regulatory networks. *Curr. Opin. Genet. Dev.* **12**, 130–136.
2. Ptashne, M. and Gann, A. (1997) Transcriptional activation by recruitment. *Nature* **386**, 569–577.
3. Iyer, V. R., Horak, C. E., Scafe, C. S., Botstein, D., Snyder, M. and Brown, P. O. (2001) Genomic binding sites of the yeast cell-cycle transcription factors SBF and MBF. *Nature* **409**, 533–538.
4. Ren, B., Robert, F., Wyrick, J. J., et al. (2000) Genome-wide location and function

of DNA binding proteins. *Science* **290**, 2306–2309.

5. Lee, T. I., Rinaldi, N. J., Robert, F., et al. (2002) Transcriptional regulatory networks in *Saccharomyces cerevisiae*. *Science* **298**, 799–804.
6. Boyd, K. E. and Farnham, P. J. (1997) Myc versus USF: discrimination at the cad gene is determined by core promoter elements *Mol. Cell. Biol.* **17**, 2529–2537.
7. Zeller, K. I., Haggerty, T. J., Barrett, J. F., Guo, Q., Wonsey, D. R., Dang, C. V. (2001) Characterization of nucleophosmin (B23) as a Myc target by scanning chromatin immunoprecipitation. *J. Biol. Chem.* **276**, 48285–48291.
8. Ren, B., Cam, H., Takahashi, Y., et al. (2002) E2F integrates cell cycle progression with DNA repair, replication, and G(2)/M checkpoints. *Genes Dev.* **16**, 245–256.
9. Odom, D. T., Zizlsperger, N., Gordon, D. B., et al. (2004) Control of pancreas and liver gene expression by HNF transcription factors. *Science* **303**, 1378–1381.
10. Weinmann, A. S., Yan, P. S., Oberley, M. J., Huang, T. H. and Farnham, P. J. (2002) Isolating human transcription factor targets by coupling chromatin immunoprecipitation and CpG island microarray analysis. *Genes Dev.* **16**, 235–244.
11. Cawley, S., Bekiranov, S., Ng, H. H., et al. (2004) Unbiased mapping of transcription factor binding sites along human chromosomes 21 and 22 points to widespread regulation of noncoding RNAs. *Cell* **116**, 499–509.
12. Chen, J. and Sadowski, I. (2005) Identification of the mismatch repair genes PMS2 and MLH1 as p53 target genes by using serial analysis of binding elements. *Proc. Natl. Acad. Sci. U.S.A.* **102**, 4813–4818.
13. Colvis, C. M., Pollock, J. D., Goodman, R. H., et al. (2005) Epigenetic mechanisms and gene networks in the nervous system. *J. Neurosci.* **25**, 10379–10389.
14. Chen, J. (2006) Serial analysis of binding elements for human transcription factors. *Nat. Protoc.* **1**, 1481–1493.
15. Orlando, V. (2000) Mapping chromosomal proteins in vivo by formaldehyde-crosslinked-chromatin immunoprecipitation. *Trends Biochem. Sci.* **25**, 99–104.
16. Lisitsyn, N. and Wigler, M. (1993) Cloning the differences between two complex genomes. *Science* **259**, 946–951.
17. Velculescu, V. E., Zhang, L., Vogelstein, B. and Kinzler, K. W. (1995) Serial analysis of gene expression. *Science* **270**, 484–487.
18. Impey, S., McCorkle, S. R., Cha-Molstad, H., et al. (2004) Defining the CREB regulon: a genome-wide analysis of transcription factor regulatory regions. *Cell* **119**, 1041–1054.
19. Kim, J., Bhinge, A. A., Morgan, X. C. and Iyer, V. R. (2005) Mapping DNA–protein interactions in large genomes by sequence tag analysis of genomic enrichment. *Nat. Methods* **2**, 47–53.
20. Labhart, P., Karmakar, S., Salicru, E. M., et al. (2005) Identification of target genes in breast cancer cells directly regulated by the SRC-3/AIB1 coactivator. *Proc. Natl. Acad. Sci. U.S.A.* **102**, 1339–1344.
21. Roh, T. Y., Ngau, W. C., Cui, K., Landsman, D. and Zhao, K. (2004) High-resolution genome-wide mapping of histone modifications. *Nat. Biotechnol.* **22**, 1013–1016.
22. Wei, C. L., Wu, Q., Vega, V. B., et al. (2006) A global map of p53 transcription-factor binding sites in the human genome. *Cell* **124**, 207–219.
23. Blais, A. and Dynlacht, B. D. (2005) Constructing transcriptional regulatory networks. *Genes Dev.* **19**, 1499–1511.
24. Lander, E. S., Linton, L. M., Birren, B., et al. (2001) Initial sequencing and analysis of the human genome. *Nature* **409**, 860–921.
25. Venter, J. C., Adams, M. D., Myers, E. W., et al. (2001) The sequence of the human genome. *Science* **291**, 1304–1351.
26. Kaeser, M. D. and Iggo, R. D. (2002) Chromatin immunoprecipitation analysis fails to support the latency model for regulation of p53 DNA binding activity in vivo. *Proc. Natl. Acad. Sci. U.S.A.* **99**, 95–100.
27. Saha, S., Sparks, A. B., Rago, C., et al. (2002) Using the transcriptome to annotate the genome. *Nat. Biotechnol.* **20**, 508–512.

Chapter 9

Modeling and Analysis of ChIP-Chip Experiments

Raphael Gottardo

Abstract

Chromatin immunoprecipitation on microarrays, also known as ChIP-chip, is a popular technique for genome-wide localization of DNA-binding proteins. However, the high density (several million genomic sequences for small eukaryote genomes) and the high noise-to-signal ratio of microarrays make the analysis of ChIP-chip data very challenging. In this chapter, we review some of the issues involved in the analysis of ChIP-chip data and present a few statistical methods that can be used to overcome these issues and improve the detection of DNA–protein binding sites.

Key words: Bayesian analysis, binding sites, multiple testing, normalization, statistics.

1. Introduction

Chromatin immunoprecipitation on microarrays, ChIP-chip, is the most widely used method for identifying in vivo DNA–protein bound regions in a high-throughput manner (1). Recently, Affymetrix (Santa Clara, CA), NimbleGen Systems (Madison, WI), and Agilent Technologies (Palo Alto, CA) have developed oligonucleotide arrays that tile all of the non-repetitive genomic sequences of human and other eukaryotes. These tiling arrays, coupled with ChIP, permit the unbiased mapping of DNA–protein binding sites. Annotation of the transcription factor binding sites in a given genome is essential for building genome-wide regulatory networks, which can then be used in health research to better understand diseases and identify new targets for drugs, etc. However, the large amount of data (several million measurements) and the small number of replicates available are very challenging for any statistical analysis.

Philippe Collas (ed.), *Chromatin Immunoprecipitation Assays,* Methods in Molecular Biology 567,
DOI 10.1007/978-1-60327-414-2_9, © Humana Press, a part of Springer Science+Business Media, LLC 2009

Similar to gene expression arrays (2, 3), tiling arrays query each sequence of interest with a short oligonucleotide, referred to as an oligo or probe. The difference is that the probes used do not necessarily represent genes, but short sequences of DNA in a given genome. The ChIP protocol generates an IP-enriched DNA fragment population and measures the enrichment of each probe in this population. In general, a control sample is also generated to calibrate the IP sample, and there are various ways of obtaining control populations (1). In terms of tiling resolution and coverage, this can vary greatly from one manufacturer to the other. For example, Affymetrix tiling arrays contain oligonucleotides of 25 base pairs (bps) in length, spanning the non-repetitive regions of a genome at an average resolution of 35 bps in humans and higher in smaller genomes. Because the original genomic DNA is sheared into segments of an average length of 500–1,000 base pairs (bps) or less, one would expect a DNA–protein bound region to be of an approximate length of 0.5–1 kbps containing a fixed number of probes (the actual number depends on the tiling resolution) with intensities that form a peak-like structure whose center corresponds to probes closest to the actual binding site. In practice, empirical studies suggest that the length of bound regions can be extremely variable (4–6).

The fluorescent intensity values obtained from an oligonucleotide microarray hybridization are not directly comparable because of systematic probe biases due to non-specific binding. If not accounted for, such biases can severely deteriorate any subsequent analysis. It turns out that this problem is closely related to the base composition of the nucleic acid molecules. For example, sequences with a high G/C content tend to induce stronger hybridization, because each G-C pair forms three hydrogen bonds, whereas an A-T pair forms two. The statistical method of normalization aims at making the probe measurements more comparable by reducing these biases. Johnson et al. (7) introduced the first normalization model for ChIP-chip based on probe sequence composition. This model was motivated by sequence-specific probe behavior models for gene expression microarrays (8–10). Other normalization techniques borrowed from gene expression include Lowess (11, 12) and quantile–quantile (13, 14) normalization. However, these techniques do not use the probe sequence information, and as shown by Royce et al. (15), will typically be inferior.

Once the data have been properly normalized, one can proceed with the detection of bound regions. Several approaches are available for analyzing ChIP-chip data. A common approach is to test a hypothesis for each probe using a sliding window statistic and then to try to correct for multiple testing (5, 16). Keles et al. (5) used a scan statistic, which is an average of *t*-statistics across a certain number of probes while Cawley et al. (4) used Wilcoxon's rank sum test within a certain genomic distance sliding window. In each of these situations, two types of error can occur: a false positive (type I error) or a false negative (type II error). When many hypotheses are tested at the same time, the probability of making a type I error increases. One approach to overcoming this problem is to try to control the total number of type I errors or false positives. This can be done using multiple testing procedures to control some measure of the overall type I error. The most common measure in the area of microarrays is the false discovery rate (FDR), which is the proportion of false positives among the total number of discoveries reported (17). A difficulty with sliding window approaches is that the resulting *p*-values (or statistics) are not independent as each test uses information from neighboring probes, and it is challenging to devise powerful multiple adjustment procedures. In addition, the window size is fixed and has to be determined in advance. Alternatives to sliding window approaches include hidden Markov models (18, 19) and Bayesian approaches (6, 14, 20). Bayesian approaches can make the best use of available prior information while borrowing strength from the data when estimating the quantities of interest. Using such Bayesian techniques, inference is usually based on the posterior distribution of the parameters. In this chapter, we review and illustrate two methods that can be used to analyze ChIP-chip data, namely MAT (7) and BAC (6).

2. Materials

2.1. Data

We use two publicly available datasets that have already been analyzed by several research groups.

2.1.1. ER Data

Carroll et al. (21) mapped the association of the estrogen receptor (ER) on chromosomes 21–22. These data contain two conditions (genomic DNA control and IP enriched) with three replicates each. Several binding sites have already been identified and experimentally validated, and we will use this information to compare the different methods presented. In total, we have a set of 83 verified bound regions we can use for validation.

2.1.2. Spike-In Data

The second dataset we use is a spike-in data that was generated as part of the Encode consortium project (22) covering 1% of the human genome using the Affymetrix technology (1.0R arrays). In this experiment, 96 clones approximately 500 bps in length were spiked into sample at $(2n + 1)$-fold enrichment for $n = 1,\ldots, 8$ and compared to genomic DNA. Some of these clones mapped to overlapping locations on the genome and a few of the clones mapped to locations that were not on one or both of the arrays. Control samples consisted of sonicated DNA that were labeled and hybridized on the array. There were 67 unique spike-in regions and the number of probes in each region ranged from 3 to 94 probes, with a median of 21. The size of the regions covered on the array ranged from 65 to 2044 bps, with a median of 470. The probes on the array are 25 bps long and the midpoints of consecutive probes are spaced at an average of 35 bps. The spike-in data set includes five replicate arrays for both the treatment and control samples.

2.2. Software

All results presented in this chapter were obtained using open source implementations of MAT and BAC.

2.2.1. MAT

MAT is written as a python package and can be downloaded at http://chip.dfci.harvard.edu/~wli/MAT/. The webpage contains all instructions for the installation and use of the package.

2.2.2. BAC

BAC is written in the R statistical language with a few functions implemented in C for efficiency. The BAC package is distributed as part of BioConductor (23), an open source and open development software project for the analysis and comprehension of genomic data. The package can be downloaded at http://www.bioconductor.org/packages/bioc/html/BAC.html. The package contains a vignette with detailed instructions on how to use it.

3. Methods

3.1. Normalization

Normalization plays an important role in the analysis of tiling arrays and thus ChIP-chip. Its aim is to remove systematic biases and ease the separation of the true signal due to DNA–protein bound regions from the background noise. MAT (7) was the first normalization model for ChIP-chip based on probe sequence information. In MAT, the normalization is done in two stages: (i) a prediction model for the probe intensities is derived from their sequence compositions; and (ii) each probe is normalized by subtracting its predicted intensity (representing the bias) from the observed intensity. The rationale behind MAT is that any correlation between the observed and predicted intensities would provide evidence of probe sequence specific biases. In MAT, the normalization is performed by fitting, to all probes on a given array, the following linear model,

$$y_p = \alpha + \sum_{j=1}^{25} \sum_{k \in \{A,C,G,T\}} \beta_{jk} I_{pjk} + \sum_{k \in \{A,C,G,T\}} \gamma_k n_{pk}^2 + \delta \log(c_p) + \epsilon_p \quad [1]$$

where y_p is the log transformed intensity from probe p, n_{pk} is the nucleotide count of type k in the sequence of probe p, α is the overall baseline intensity, I_{pjk} is an indicator function equal to one if the nucleotide at position j is k in probe sequence p and 0 otherwise, β_{jk} is the effect of nucleotide k at each position j, γ_k is the effect of the nucleotide count squared, c_p is the number of times the sequence of probe p appears in the genome (copy number), δ is the effect of log probe copy number, and ε_p is an error term. In other words, the first term on the right hand side of equation [1] is the mean log intensity of all probes, the second term accounts for nucleotide positional effect, the third term accounts for overall nucleotide composition, while the last term accounts for the fact that if a probe maps to multiple locations in the genome its intensity will typically be greater. The 81 resulting parameters can be easily estimated via least squares (*see* **Note 1**). When applied to both the ER and the spike-in data, the correlations between observed and predicted intensities ranged from 0.62 to 0.86, suggesting that a significant part of the signal measured by the probes is due to non-specific hybridization. The effect of the MAT normalization applied to both datasets is shown in **Fig. 9.1**. Before normalization the GC content has a strong effect on the log intensities; the greater the GC content, the greater the intensity. After normalization, the effect of the GC content on the log intensities is significantly decreased. **Figure 9.2** shows the effect of each single nucleotide (A, G, C, T) as a function of its position on the probe. One can see that G/C's have the maximum effect particularly if they are towards the middle of a probe. In the next section, we will see that if the probe measurements are not properly normalized, it can severely affect the detection of bound regions.

3.2. Detection of Bound Regions

In addition to normalization, MAT can also detect bound regions with a sliding window approach based on a trimmed mean statistic combined to an FDR estimation procedure (7). The trimmed mean removes the top and bottom 10% of the normalized intensities and averages the remaining 80%. It thus provides robustness against outliers. Assuming that the null distribution of the trimmed mean based statistic is symmetric about the median, for each cutoff value above the median (positive cutoff), a negative cutoff is defined as the value symmetric to the positive cutoff about the median. After merging nearby probes beyond both cutoffs, the region FDR can be estimated as the ratio of negative regions over the total number of regions. MAT can automatically select the proper cutoff so that the region FDR is less than or equal to the user-specified FDR value.

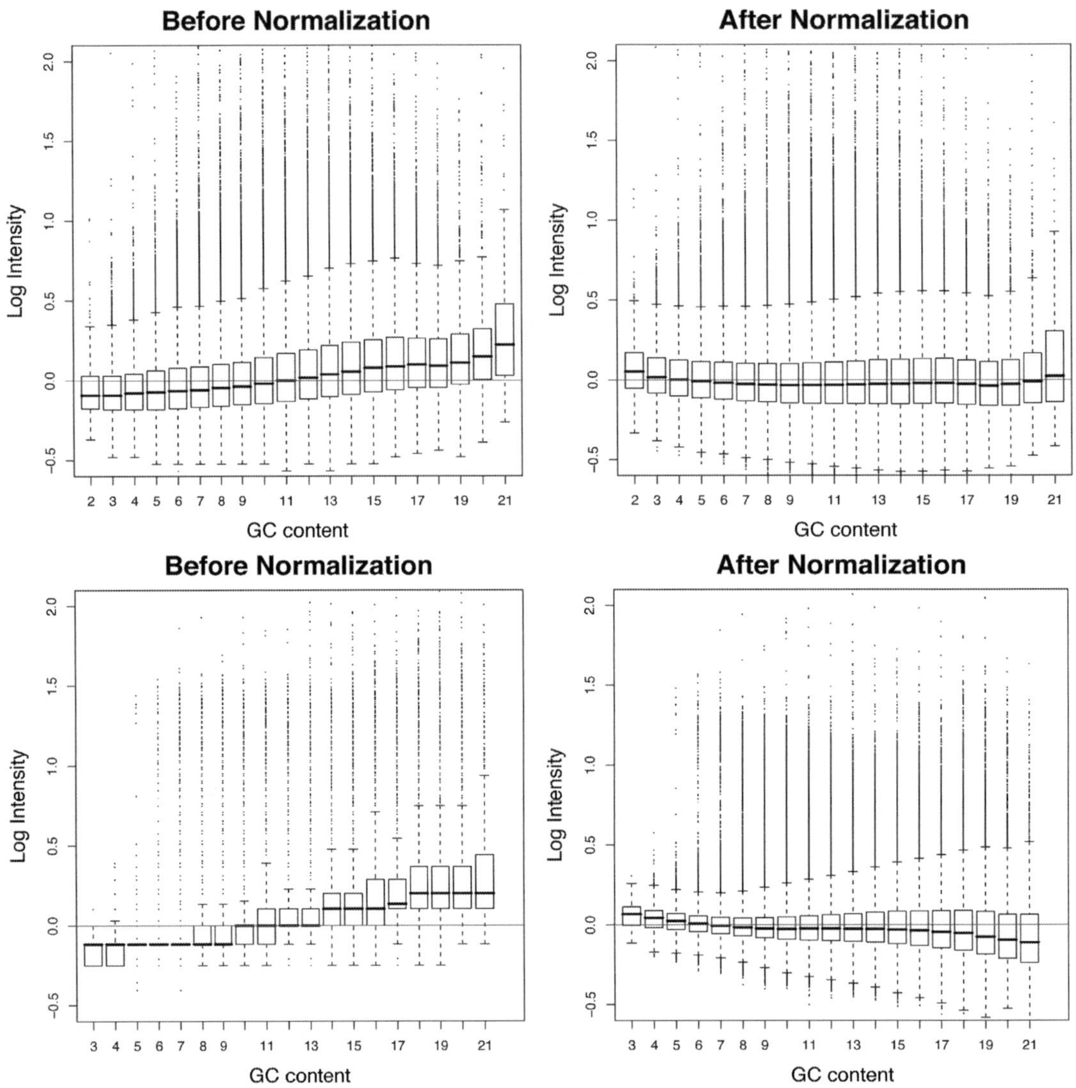

Fig. 9.1. Boxplots of log intensities as a function of GC counts before and after normalization for one control array of the ER data (*top*) and one control array of the spike-in data (*bottom*). The thick line within each *box shows* the median log intensity for all probes with the corresponding number of G's or C's. After normalization the medians are mostly centered around zero.

In comparison, BAC (6), which is built on previous approaches used in gene expression analysis (24–26), uses a Bayesian hierarchical model to identify regions of interest.

In BAC the log transformed measurements are modeled as follows:

$$\begin{aligned} y_{1pr} &= \mu_p + \epsilon_{1pr} \\ y_{2pr} &= \mu_p + \gamma_p + \epsilon_{2pr}, \\ \epsilon_{cpr} &\sim N(0, \lambda_{cp}^{-1}), \end{aligned} \qquad [2]$$

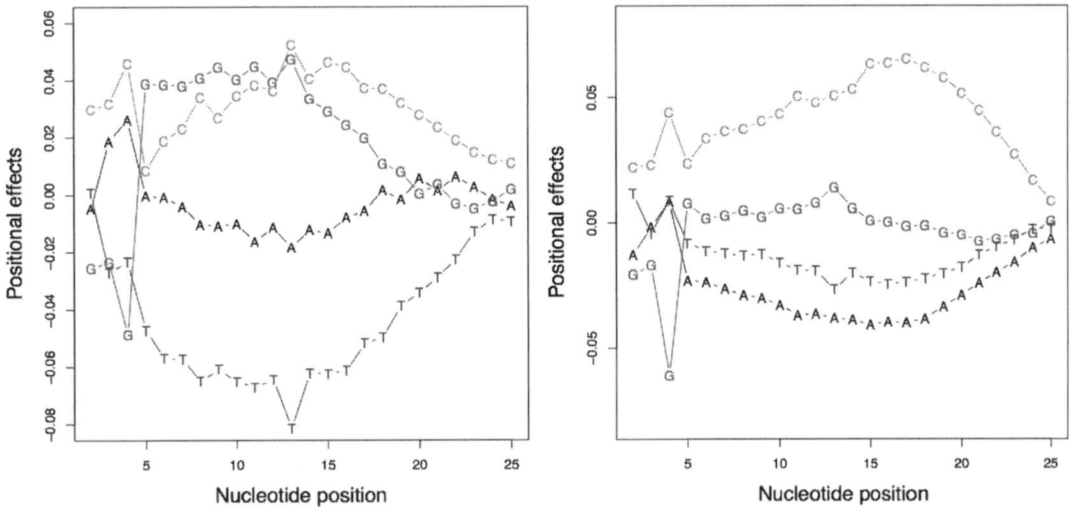

Fig. 9.2. Effect of nucleotide base (A, G, C, T) as a function of its position on the probe sequence for the ER data (*left*) and spike-in data (*right*). G's and C's, especially towards the middle of the probes, have the strongest effect.

where y_{cpr} is the log transformed intensity of probe p from replicate r in condition c with $c=\{1,2\}$ denoting the treatment label equal to one for control and two for IP enriched. In equation [2], μ_p is probe background intensity, and γ_p is a probe enrichment effect, which we expect to be large if probe p is part of a bound region. We model the background as a random effect with Gaussian distribution, namely $\mu_p \sim N(0, \psi^{-1})$ where the variance ψ^{-1} is constant across probes. Even though we would typically normalize our data to remove probe sequence effects (e.g., using MAT), it might still be necessary to include probe specific effects for two main reasons: (i) the MAT sequence normalization model is not perfect and some unexplained residual effects are likely to remain, and (ii) some of the probe-to-probe variation might be due to other (non-sequence specific) factors. To model the fact that enrichment effects can be exactly zero, we use a mixture of a point mass at zero and a Gaussian distribution truncated at zero. BAC takes into account the spatial dependence between probes by allowing the weights of the mixture to be correlated for neighboring probes; see Gottardo et al. (6) for details. BAC also includes an exchangeable prior for the variances, allowing each probe to have a different variance while still achieving some shrinkage. This allows us to regularize empirical variance estimates, which can be very noisy due to the small number of replicates. Finally, non-informative priors are used for all parameters and a simulation technique called Markov chain Monte Carlo is used to estimate the unknown parameters. Among other things, these parameter estimates can be used to compute, for each probe, the probability that the probe belongs to a bound region. The closer the probability is to one, the more evidence there is that the probe belongs to a bound region. Bound regions can then be formed by thresholding these posterior probabilities. A common threshold is 0.5; an FDR-based threshold can also be derived as explained in (27).

We now turn to the ER and spike-in data to evaluate and compare the performance of MAT and BAC. We have applied each method to both datasets, fixing the false discovery rate to 10% (**Table 9.1**). Overall, both MAT and BAC perform relatively well on both datasets as they detect most of the positive controls. On the ER data, BAC performs slightly better as it detects more positive controls. On the spike-in data, we actually know the status of all the regions and we can thus compute the true false discovery rate in addition to the number of positive controls detected. BAC and MAT detect the same number of positive controls, but BAC has a nominal FDR closer to the true FDR (*see* **Note 2**). Finally, for comparison, we have also included the results of MAT and BAC applied to

Table 9.1
Performances of MAT and BAC on the ER and spike-in data. For comparison purposes we have also included the results without normalization

	ER		Spike-in		
	TP	Total	TP	Total	FDR (%)
BAC w/ normalization	73	99	65	72	10
MAT w/ normalization	62	72	65	71	8
BAC w/o normalization	25	83	51	66	23
MAT w/o normalization	83	14084	46	52	12

the unnormalized data. The performance of both methods is clearly inferior. For example, MAT applied to the ER data leads to a huge number of detected regions, most of which are likely false positives.

For the spike-in data, because we know the true status of all the regions, it is also possible to plot a receiver operating characteristic (ROC) curve, which shows the number of true positives versus the number of false positives detected when varying the cut-off of each method. For such an ROC curve, the higher the curve is, the better the performance is. **Figure 9.3** shows that both MAT and BAC are virtually

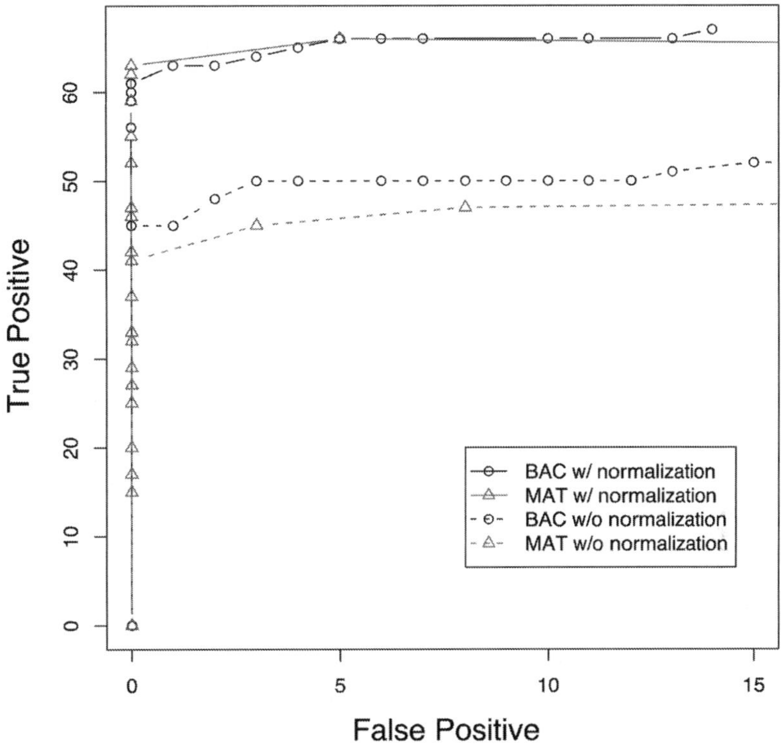

Fig. 9.3. ROC curve for MAT and BAC on the spike-in data.

equivalent when the data are normalized and that both suffer from the lack of normalization. **Figure 9.3** also shows that BAC is slightly better when the data are not normalized (*see* **Note 3**).

4. Notes

1. The normalization implemented in MAT is done for each array separately and uses all the probes on the array to estimate the sequence-specific biases. This is not optimal as probes as part of bound regions do not only measure background but also specific hybridization; this can result in over smoothing for some of the true signals due to enriched regions. To overcome this problem, one could simply replace the least squares estimation by a more robust procedure; see for example (15). In addition, MAT was derived for transcription factor data, but preliminary results on histone modification data suggest that it works relatively well on such data.

2. In the analysis of high-throughput biological discoveries, including ChIP-chip, it is common to use an FDR procedure to account for multiple testing. However, in practice, it can be difficult to get an accurate estimate of the FDR. Based on our experience, the estimation of the FDR is particularly difficult with histone modification data where one expects many enriched regions. In this case, we recommend the use of control regions in order to estimate the FDR. If such control regions are not available, one could simply select a threshold that leads to a reasonable number of enriched regions.
3. In the results shown above, BAC performed slightly better than MAT. This is not surprising because BAC is a more comprehensive modeling approach. This said, BAC is computationally more demanding and users would need to decide whether the improved results are worth the additional computing time. BAC also requires a control sample as well as replicates. This is not the case for MAT, which can be applied to a single array.

Acknowledgments

The author would like to thank Shirley X. Liu, Wei Li, and Evan W. Johnson with whom some of the work presented here originated. The author also thanks Evan W. Johnson for providing the spike-in data.

References

1. Buck, M. J. and Lieb, J. D. (2004) Chip-chip: considerations for the design, analysis, and application of genome-wide chromatin immunoprecipitation experiments. *Genomics* **83**, 349–360.
2. Schena, M., Shalon, D., Davis, R. W. and Brown, P. O. (1995) Quantitative monitoring of gene expression patterns with a complementary DNA microarray. *Science* **270**, 467–470.
3. Lockhart, D. J., Dong, H., Byrne, M. C., Follettie, M. T., Gallo, M. V., Chee, M. S., Mittmann, M., Wang, C., Kobayashi, M., Horton, H. and Brown, E. L. (1996) Expression monitoring by hybridization to high-density oligonucleotide arrays. *Nat. Biotechnol.* **14**, 1675–1680.
4. Cawley, S. E., Bekiranov, S., Ng, H. H., Kapranov, P., Sekinger, E. A., Kampa, D., Piccolboni, A., Sementchenko, V., Cheng, J., Williams, A. J., Wheeler, R., Wong, B., Drenkow, J., Yamanaka, M., Patel, S., Brubaker, S., Tammana, H., Helt, G., Struhl, K. and Gingeras, T. R. (2004) Unbiased mapping of transcription factor binding sites along human chromosomes 21 and 22 points to widespread regulation of noncoding RNAs. *Cell* **116**, 499–509.
5. Keles, S., van der Laan, M. J., Dudoit, S. and Cawley, S. E. (2006) Multiple testing methods for ChIP-chip high density oligonucleotide array data. *J. Comput. Biol.* **13**, 579–613.
6. Gottardo, R., Li, W., Johnson, W. E. and Liu, X. S. (2008) A flexible and powerful Bayesian hierarchical model for ChIP-chip experiments. *Biometrics* **64**, 468–478.

7. Johnson, W. E., Li, W., Meyer, C. A., Gottardo, R., Carroll, J. S., Brown, M. and Liu, X. S. (2006) Model-based analysis of tiling-arrays for ChIP-chip. *Proc. Natl. Acad. Sci. U.S.A.* **103**, 12457–12462.
8. Naef, F. and Magnasco, M. O. (2003) Solving the riddle of the bright mismatches: labeling and effective binding in oligonucleotide arrays. *Phys. Rev. E: Stat. Phys. Plasmas Fluids* **68**, 011906.
9. Wu, Z. and Irizarry, R. A. (2003) Preprocessing of oligonucleotide array data. *Nat. Biotechnol.* **22**, 656–8.
10. Wu, Z. and Irizarry, R. A. (2005) Stochastic models inspired by hybridization theory for short oligonucleotide arrays. *J. Comput. Biol.* **12**, 882–93.
11. Cleveland, W. S. (1979) Robust locally weighted regression and smoothing scatterplots. *J. Am. Stat. Assoc.* **74**, 829–836.
12. Peng, S., Alekseyenko, A. A., Larschan, E., Kuroda, M. I. and Park, P. J. (2007) Normalization and experimental design for Chip-chip data. *BMC Bioinformatics* **8**, 219.
13. Bolstad, B. M., Irizarry, R. A., Astrand, M. and Speed, T. P. (2003) A comparison of normalization methods for high density oligonucleotide array data based on variance and bias. *Bioinformatics* **19**185–93
14. Keles, S. (2007) Mixture modeling for genome-wide localization of transcription factors. *Biometrics* **63**, 10–21.
15. Royce, T. E., Rozowsky, J. S. and Gerstein, M. B. (2007) Assessing the need for sequence-based normalization in tiling microarray experiments. *Bioinformatics* **23**, 988–97.
16. Buck, M. J., Nobel, A. B. and Lieb, J. D. (2005) Chipotle: A user-friendly tool for the analysis of ChIP-chip data. *Genome Biol.* **6**, R97.
17. Benjamini, Y. and Hochberg, Y. (1995) Controlling the false discovery rate: a practical and powerful approach to multiple testing. *J. R. Stat. Soc. Ser. B Stat. Methodol.* **57**, 289–300.
18. Li, W., Meyer, C. A. and Liu, X. S. (2005) A hidden Markov model for analyzing ChIP-chip experiments on genome tiling arrays and its application to p53 binding sequences. *Bioinformatics* **21**, i274–i282.
19. Ji, H. and Wong, W. H. (2005) Tilemap: create chromosomal map of tiling array hybridizations. *Bioinformatics* **21**, 3629–3636.
20. Qi, Y., Rolfe, A., MacIsaac, K., Gerber, G. and Pokholok, D. (2006) High-resolution computational models of genome binding events. *Nat. Biotechnol.* **24**, 963–970.
21. Carroll, J. S., Liu, X. S., Brodsky, A. S., Li, W., Meyer, C. A., Szary, A. J., Eeckhoute, J., Shao, W., Hestermann, E. V., Geistlinger, T. R., Fox, E. A., Silver, P. A. and Brown, M. (2005) Chromosome-wide mapping of estrogen receptor binding reveals long-range regulation requiring the Forkhead protein FoxA1. *Cell* **122**, 33–43.
22. Johnson, D. S., Li, W., Gordon, D. B., Bhattacharjee, A., Curry, B., Ghosh, J., Brizuela, L., Carroll, J. S., Brown, M., Flicek, P., Koch, C., Dunham, I., Bieda, M., Xu, X., Farnham, P., Kapranov, P., Nix, D., Gingeras, T. R., Zhang, X., Holster, H. L., Jiang, N., Green, R., Song, J., McCuine, S., Anton, E., Nguyen, L., Trinklein, N., Ye, Z., Ching, K., Hawkins, D., Ren, B., Scacheri, P. C., Rozowsky, J. S., Karpikov, A., Euskirchen, G. M., Weissman, S., Gerstein, M. B., Snyder, M., Yang, A., Moqtaderi, Z., Hirsch, H., Shulha, H. P., Fu, Y., Weng, Z., Struhl, K., Myers, R. M., Lieb, J. and Liu, X. S. (2008) Systematic evaluation of variability in Chip-chip experiments using predefined DNA targets. *Genome Res.* **18**, 393.
23. Gentleman, R. C., Carey, V. J., Bates, D. M., Bolstad, B. M., Dettling, M., Dudoit, S., Ellis, B., Gautier, L., Ge, Y., Gentry, J., Hornik, K., Hothorn, T., Huber, W., Iacus, S., Irizarry, R. A., Leisch, F., Li, C., Maechler, M., Rossini, A. J., Sawitzki, G., Smith, C., Smyth, G. K., Tierney, L., Yang, J. Y. H. and Zhang, J. (2004) Bioconductor: open software development for computational biology and bioinformatics. *Genome Biol.* **5**, R80.
24. Newton, M. A., Kendziorski, C. M., Richmond, C. S., Blattner, F. R. and Tsui, K. W. (2001) On differential variability of expression ratios: improving statistical inference about gene expression changes from microarray data. *J. Comput. Biol.* **8**, 37–52.
25. Gottardo, R., Pannucci, J. A., Kuske, C. R. and Brettin, T. S. (2003) Statistical analysis of microarray data: a Bayesian approach. *Biostatistics* **4**, 597–620.
26. Gottardo, R., Raftery, A. E., Yeung, K. Y. and Bumgarner, R. E. (2006) Bayesian robust inference for differential gene expression in microarrays with multiple samples. *Biometrics* **62**, 10–18.
27. Newton, M. A., Noueiry, A., Sarkar, D. and Ahlquist, P. (2004) Detecting differential gene expression with a semiparametric hierarchical mixture method. *Biostatistics* **5**, 155–76.

Chapter 10

Use of SNP-Arrays for ChIP Assays: Computational Aspects

Enrique M. Muro, Jennifer A. McCann, Michael A. Rudnicki, and Miguel A. Andrade-Navarro

Abstract

The simultaneous genotyping of thousands of single nucleotide polymorphisms (SNPs) in a genome using SNP-Arrays is a very important tool that is revolutionizing genetics and molecular biology. We expanded the utility of this technique by using it following chromatin immunoprecipitation (ChIP) to assess the multiple genomic locations protected by a protein complex recognized by an antibody. The power of this technique is illustrated through an analysis of the changes in histone H4 acetylation, a marker of open chromatin and transcriptionally active genomic regions, which occur during differentiation of human myoblasts into myotubes. The findings have been validated by the observation of a significant correlation between the detected histone modifications and the expression of the nearby genes, as measured by DNA expression microarrays. This chapter focuses on the computational analysis of the data.

Key words: Chromatin, histone, acetylation, microarray, chromatin immunoprecipitation, single nucleotide polymorphism, database analysis, genome analysis.

1. Introduction

Cellular functions such as proliferation and differentiation are regulated at the transcriptional level and depend on DNA accessibility to determine gene expression. Mechanisms involved in this process include the epigenetic phenomena of DNA methylation and histone modifications (1); disruption of either, being closely linked to aberrant gene expression, may lead to atypical development and/or a potentially malignant transformation (2). While histone modifications include phosphorylation, methylation, ubiquitination, and SUMOylation (3), the most extensively studied modification to date is histone acetylation. However, the exact relation between histone acetylation and gene expression is not yet totally understood (4).

Philippe Collas (ed.), *Chromatin Immunoprecipitation Assays,* Methods in Molecular Biology 567,
DOI 10.1007/978-1-60327-414-2_10, © Humana Press, a part of Springer Science+Business Media, LLC 2009

Histone modifications can be studied in chromatin immunoprecipitation (ChIP) experiments using antibodies that recognize modifications in the side chains of histones (5). ChIP has been combined with DNA expression microarray to map the genome-wide location of modified histones in yeast and flies (6, 7), and quantitative PCR-based ChIP approaches have been used to map histone modification patterns at the β-globin locus in mouse and chicken (8–10). These studies identified associations between domains, genes, regulatory elements, and modification patterns not found in yeast or flies demonstrating that genomes of higher eukaryotes are much more complicated than model systems from lower-order organisms. We have devised a method to analyze histone modifications using a commercially available array, which we applied to elucidate the global relationship between chromatin structure and gene expression in a differentiating myogenic cell line.

We have examined the distribution of permissive chromatin across the human genome through a combination of ChIP and Affymetrix 10 K SNP (single nucleotide polymorphism) microarray (SNP-Array) (11). Traditionally, SNP-Arrays have been used for the simultaneous genotyping of tens of thousands of SNPs through the hybridization of genomic material to an array that contains multiple small nucleotide sequences for each version of the SNP (12, 13); however, this type of array can also be used to detect specific sequences. This study provides an alternative use for the SNP-Arrays: to follow ChIP (ChIP on SNP-Array) with an antibody specific to four acetylated lysine residues of H4, namely Lys5, Lys8, Lys12, and Lys16. We mapped alterations in the pattern of histone H4 (H4) acetylation throughout the entire human genome during muscle cell differentiation from myoblast to myotubes.

More specifically, we have shown that the ChIP on SNP-Array procedure reflects histone modifications associated with gene expression changes (as detected via complementary gene expression analysis with a DNA microarray). Chromatin associated with hyperacetylated H4 is typically relaxed and contains transcriptionally competent genes (14); hence, increases in H4 acetylation detected in some genome locations should be associated with increased transcription from the nearest genes verifying the proposed technique. Accordingly, our experiments clearly indicate that the acetylation status of H4 near the gene promoter region is one of the elements that define the transcriptional competence of a gene.

Upon analysis of the relationship between the ChIP on SNP-Array technique and the gene expression data, we evaluated the hybridization status of the SNP-Array probes according to the hybridization result of the closest DNA expression microarray probe set and the genomic distance between the SNPs and the start of the transcription of the genes related to the DNA microarray probe sets was recorded. The representation of the ratios of SNP-Array probe sets hybridized to non-hybridized at variable distance intervals from the start of the gene starts conclusively showed that there was an association between gene expression and detection of an acetylated histone which was appreciable up to a sort range of 150 Kbases upstream and downstream of gene start of transcription (**Fig. 10.1**). In that range, we found that DNA microarray hybridization (indicating gene expression) was associated to SNP-Array probe set hybridization (indicating histone H4 acetylation) for myoblasts and myotubes with a significant *P* value (<0.0001; Chi-squared test).

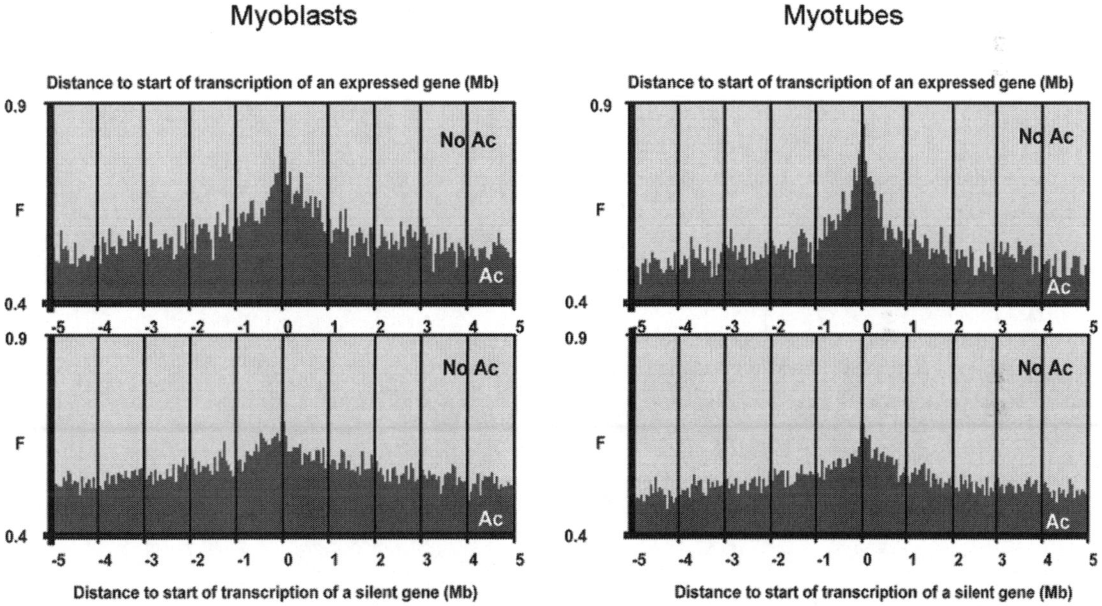

Fig. 10.1. Association between H4 acetylation and gene expression in human myoblasts and myotubes. For all pairs of SNP-Array probes and DNA microarray probesets mapped to the same chromosome, we recorded the state of the probes (hybridized or not) and the distance from the start of transcription of the gene associated to the DNA microarray probeset to the SNP-Array probe. The data was binned in intervals of 50 Kb. Top: fraction of pairs of hybridizing (*dark*) and not-hybridizing (*light*) SNP-Array probes with hybridizing DNA microarray probesets. Bottom: fraction of pairs of hybridizing (*dark*) and not-hybridizing (*light*) SNP-Array probes with non-hybridizing DNA microarray probesets. Left, myoblasts. Right, myotubes.

Having demonstrated that H4 hyperacetylation is associated with the transcriptional start of expressed genes on a global level, we next looked at signal intensities of SNPs in myoblasts and myotubes compared to the intensities of the same probesets in control genotype experiments. In general, we did not find any difference in the overall level of H4 acetylation between the myoblasts and the myotubes (as determined by Western blot, data not shown); however, we did determine that the actual signal intensity of the myotubes was higher than the myoblasts. We classified a SNP probeset as being enriched for acetylated H4 if the ratio of the signal intensities for myoblast/control or the myotubes/control was greater than 1.0. **Figure 10.2** depicts chromosome 10 and is an example of the type of data we acquired using this analysis method. Some regions were enriched in both the developmental stages, for example, those containing the genes *C10orf3* and *C10orf4* (**Fig. 10.2B**). There are also SNPs that are differentially represented. In region 10q24.1, four SNPs located within the same intron of the gene *SORBS1* were associated with acetylated H4 in myotubes only. *SORBS1* is highly expressed in skeletal muscle and is involved in formation of actin stress fibers (15, 16). Within the same region, there was a differential acetylation of H4 associated with SNPs close to the *FER1L3* gene, which is highly expressed in both cardiac and skeletal muscle and is involved in muscle contraction (17). Another gene containing acetylated SNPs, *ADAM12*, is a myogenic factor involved in the fusion of myoblasts into myotubes (18).

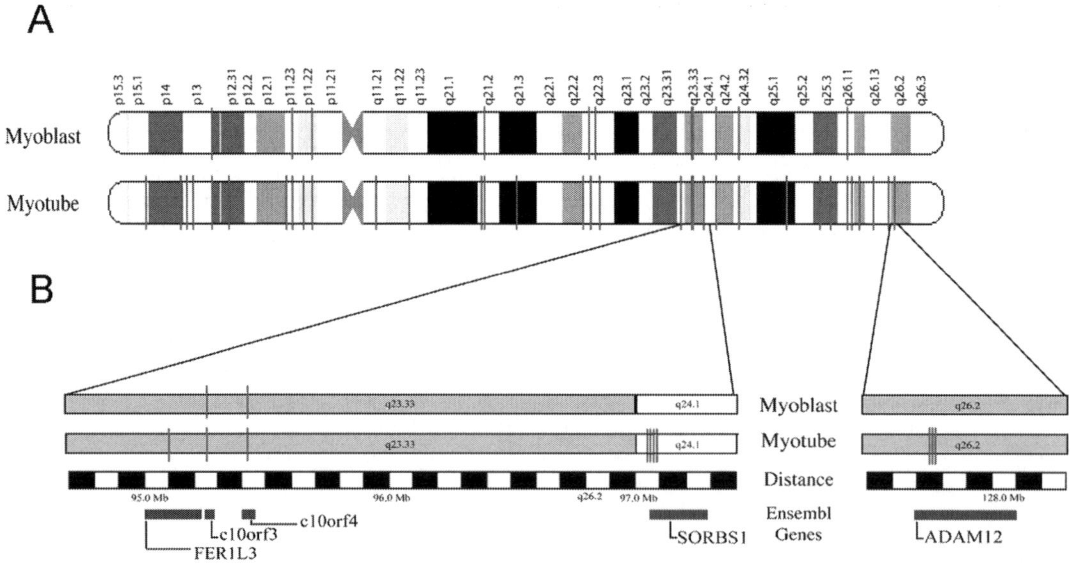

Fig. 10.2. Differential acetylation between myoblasts and myotubes. Longer vertical lines: putative sites for H4 acetylation in myoblasts and myotubes. (**A**) Patterns of acetylated histones in both myoblast and myotubes within chromosome 10. (**B**) Magnification of two regions of the long arm of chromosome 10. *FERL13, SORBS1*, and *ADAM12* are acetylated in myotubes but not in myoblasts. These findings correlate with those in the literature which report that *FERL13* and *SORBS1* (16, 17) are highly expressed in skeletal muscle, and *ADAM12* is essential for the regulation of myoblast fusion (19). Panel A was generated using the Karyoview tool (http://www.ensembl.org/Homo_sapiens/karyoview), whereas panel B was generated using data from the Ensembl Website (http://www.ensemble.org.).

In summary, we have developed and optimized the technique of ChIP on SNP-Array, an extension of the technique of SNP-Array genotyping, expanding the utilization of this technique to the detection and identification of DNA fragments from ChIP. We have illustrated this novel use at the epigenomic level by evaluating the detection of H4 acetylation that correlated to gene expression (as independently detected with DNA expression microarrays) in human myoblasts and myotubes. In addition, another possible use for this technique is the analysis of loss of heterozygosity throughout the entire genome with respect to histone modifications through monitoring the genotypes at each probeset and comparing the genotypes between control arrays and the ChIP on SNP-Arrays.

The ongoing expansion of SNP-Arrays with much larger density implies the ChIP on SNP-Array procedure will result in more precise data enhancing the ability to locate molecular interactions with DNA in addition to providing information on genes for which no probe sets are currently included in standard DNA expression microarrays. Further studies employing the more dense arrays and investigating other histone modifications will contribute to the growing information on the general features of mammalian chromatin structure.

The molecular methods have been described in a previous publication (11). In the reminder of the chapter, we focus and expand on the computational aspects of this work.

2. Materials

2.1. Programs

1. GeneChip DNA Analysis Software 2.0 (Affymetrix), to produce genotype calls from the data files generated by scanning the SNP-Arrays.
2. Genotyping Tools V 1.0 (Affymetrix), for analysis of SNP-Array data.
3. MicroArray Suite 5.0 (Affymetrix), for analysis of gene expression microarray data.
4. Excel^{TM}, for graphical representation (**Fig. 10.1**).

2.2. Databases

1. GeneChip Mapping 10 K library files: Mapping10K_Xba131 (Affymetrix).
2. NCBI human genome version 35 (May, 2004).
3. Ensembl Human v27.35a.1 (http://www.ensembl.org).
4. NCBI SNP database (http://www.ncbi.nlm.nih.gov/SNP).
5. NetAffx (April 12, 2005 release) (Affymetrix).

2.3. Programming Languages and Web Resources

1. Perl (http://www.perl.org) scripts were generated for various computational tasks and for generation of data tables.
2. R programming language and software environment (http://www.r-project.org/).
3. Bioconductor software was used for statistical analysis (http://www.bioconductor.org/). It is based on the R programming language.
4. Karyoview (http://www.ensembl.org/Homo_sapiens/karyoview) was used for graphical representation (**Fig. 10.2A**).
5. The Ensembl web site (http://www.ensemble.org) was used for graphical representation (**Fig. 10.2B**).

3. Methods

Figure 10.3 displays the procedure of data analysis as a flow chart.

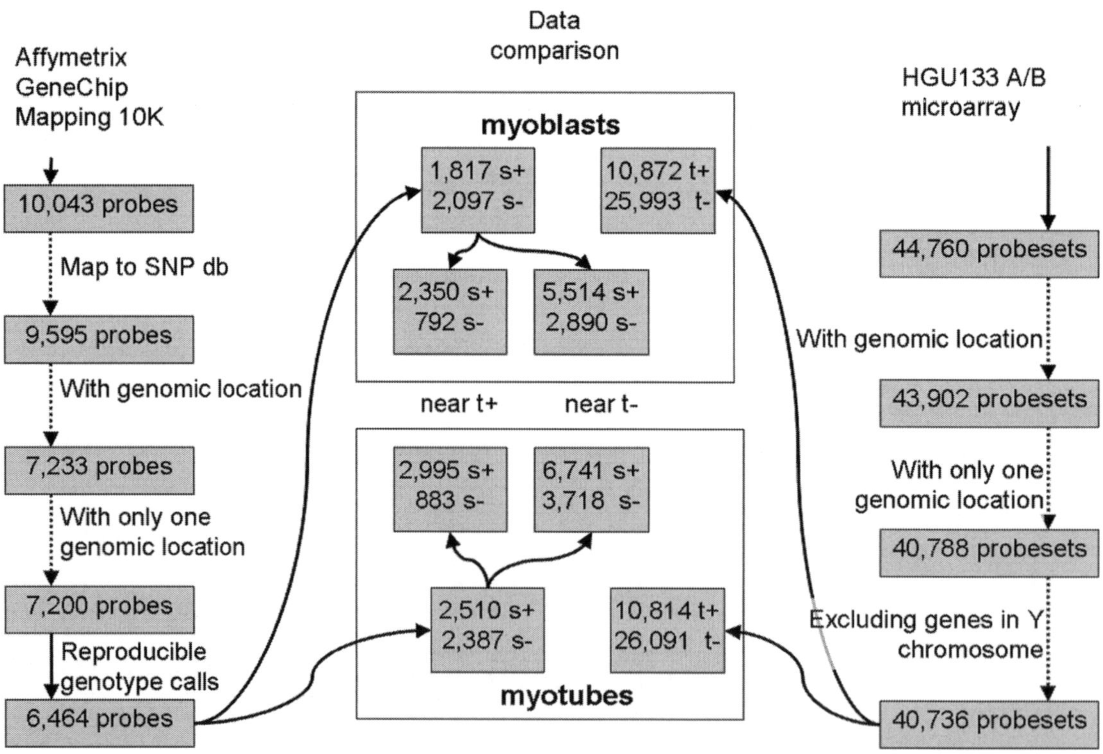

Fig. 10.3. Flow chart of the SNP-Array probe and DNA microarray probe set selection, analysis, and comparison. The boxes labeled myoblasts and myotubes indicate sample specific data. The status of SNP-Array probes is indicated as s+ (hybridized) or s− (non-hybridized). The status of microarray probe sets is indicated as t+ (hybridized) or t− (non-hybridized).

3.1. Scanning

In this study, we used a pre-commercial release version (early access) of the Affymetrix GeneChip Mapping 10 K SNP-Array for the human genome, ax13339. The 10 K SNP-Arrays were scanned with the Affymetrix GeneChip Scanner 3000 using GeneChip Operating System 1.0 (Affymetrix), data files were generated automatically and genotype calls were made automatically by GeneChip DNA Analysis Software 2.0 (Affymetrix). The genetic map used in the analysis was obtained from GeneChip Mapping 10 K library files: Mapping10K_Xba131 and the NCBI 35 version of the human genome (May, 2004) was used in all the analyses. Every SNP on the genechip probe set is characterized by its tscID (The SNP Consortium; see http://snp.cshl.org/) in the Affymetrix annotation (http://www.affymetrix.com). Ensembl Human v27.35a.1 (http://www.ensembl.org) maps the location of each SNP in the genome. The rsID (reference id of the NCBI SNP database, see http://www.ncbi.nlm.nih.gov/SNP) was used to link Affymetrix annotations to Ensembl.

3.2. SNP-Array Probe Set Selection

The Affymetrix GeneChip Mapping 10 K SNP-Array for the human genome, ax13339, contains 10,043 probe sets for the examination of SNPs distributed over the 22 autosomes and the X chromosome. First, we selected the probe sets mapped to an unambiguous position in the human genome (using their tscIDs). Once identified, the probe sets were mapped to rsIDs (NCBI SNP database) and Ensembl (Human v27.35a.1) to obtain their genomic location. It was not possible to map 448 tscIDs to any rsID, 2,362 rsIDs had no attached location, and 33 had multiple locations. The remaining 7,200 probe sets were selected for analysis as they had a unique location. Finally, to further assure SNP data of high quality, control genotyping experiments were performed in triplicate using total human myoblast DNA. Only the 6,464 SNPs for which the same genotyping was obtained at least for two of the three replicates were used for our analysis.

3.3. DNA Microarray Probe Set Selection

The HG U133 A/B microarrays contain 44,760 probe sets for the analysis of mRNA transcripts from genes in the 22 autosomes and in chromosomes X and Y. We selected the probe sets that could be unambiguously mapped to a genomic position according to the gene indicated in the annotations of the probe sets given by NetAffx (April 12, 2005 release). In those annotations, we could not find a genomic location for 858 probe sets, 3,114 had multiple locations, and 52 corresponded to genes in chromosome Y (which is not covered by the SNP-Array employed in this study). Therefore, we selected the remaining 40,736 probe sets, which have a unique genomic location, for analysis (*see* **Note 1**).

3.4. Comparison of SNP-Array Data and DNA Microarray Data

Hybridization of the ChIP products to the SNP microarray produced reproducible results for 3,914 probes (1,817 hybridized and 2,097 not hybridized, s+ and s-, respectively, in **Fig. 10.3**) in a sample of human myoblasts, and for 4,897 probes (2,510 s+ and 2,387 s−) in a sample of human myotubes. To study the relationship between histone H4 hyperacetylation and gene expression, we obtained gene expression data from equivalent human myoblast and myotube samples. Analysis of mRNA cellular transcripts using the Affymetrix HGU133A/B chip set produced reproducible results for 38,865 probesets (10,872 hybridized and 25,993 not hybridized, t+ and t−, respectively, in **Fig. 10.3**) in human myoblasts, and for 36,905 probesets (10,814 t+ and 26,091 t−) in human myotubes (*see* **Note 2**).

Figure 10.3 displays in the middle part the number of SNP-Array probes hybridizing (s+) or not (s−) with a genomic position in a range of 150 Kbases of the corresponding genomic position of gene starts for microarray probe sets detecting their transcripts (*see* **Note 3**).

4. Notes

1. Some of the statistics performed in this project, for example, those displayed in **Fig. 10.1**, required computations comparing elements indicating gene expression (probesets in the DNA expression microarray) and SNP calls (probes in the 10 K SNP chip) that had close genomic positions. Therefore, knowledge of the precise genomic situation of the feature detected (either a SNP or a gene) is also crucial for interpretation of the data in terms of associations between SNP calls and gene expression that is dependent on the relative position of the SNP location respect to the gene. For these reasons, we had to select features in the arrays that would be precisely located, that is, with one and only one genomic location according to the databases analyzed. Some target transcripts and SNPs may have multiple locations because their corresponding sequences in the genomes are duplicated. Generally, this does not imply that those probes are useless; rather, the special requirements of this project imply that those probes with multiple locations must not be used for certain computations. As shown in **Fig. 10.3**, this affects a relatively small number of features. For similar reasons, as the SNP-Array used did not consider probesets in the Y chromosome, we filtered out probe sets in the DNA microarray for that chromosome.

2. The ChIP on SNP-Arrays were performed in biological triplicates and calls of present (P) or absent (A) for a probe set in two of the three replicates were required to deem a probe set hybridized or not hybridized, respectively. The analysis of gene expression was performed in biological triplicates and calls of present (P) or absent (A) for a probe set in two of the three replicates were required to deem a probe set hybridized or not hybridized, respectively. The probes or probe sets with less reproducible results were not used for the comparison of SNP-Array data with gene expression data.
3. In the count of SNP probes localized near gene starts, counting a SNP multiple times is possible since the range used (150 Kbases) is longer than the average distance between human genes (approximately 100 Kbases). This does not introduce a bias in the results because we did not register correlation of gene expression between contiguous genes. Therefore, the effect observed must be due to association between SNP detection and gene expression.

References

1. Jaenisch, R. and Bird, A. (2003) Epigenetic regulation of gene expression: how the genome integrates intrinsic and environmental signals. *Nat. Genet.* **33 Suppl.**, 245–254.
2. Jones, P. A. and Takai, D. (2001) The role of DNA methylation in mammalian epigenetics. *Science* **293**, 1068–1070.
3. Jenuwein, T. and Allis, C. D. (2001) Translating the histone code. *Science* **293**, 1074–1080.
4. Shia, W. J., Pattenden, S. G. and Workman, J. L. (2006) Histone H4 lysine 16 acetylation breaks the genome's silence. *Genome Biol.* **7**, 217.
5. Braunstein, M., Rose, A. B., Holmes, S. G., Allis, C. D. and Broach, J. R. (1993) Transcriptional silencing in yeast is associated with reduced nucleosome acetylation. *Genes Dev.* **7**, 592–604.
6. Bernstein, B. E., Humphrey, E. L., Erlich, R. L., Schneider, R., Bouman, P., Liu, J. S., Kouzarides, T. and Schreiber, S. L. (2002) Methylation of histone H3 Lys 4 in coding regions of active genes. *Proc. Natl. Acad. Sci. USA* **99**, 8695–8700.
7. Schubeler, D., MacAlpine, D. M., Scalzo, D., Wirbelauer, C., Kooperberg, C., van Leeuwen, F., Gottschling, D. E., O'Neill, L. P., Turner, B. M., Delrow, J., Bell, S. P. and Groudine, M. (2004) The histone modification pattern of active genes revealed through genome-wide chromatin analysis of a higher eukaryote. *Genes Dev.* **18**, 1263–1271.
8. Bulger, M., Schubeler, D., Bender, M. A., Hamilton, J., Farrell, C. M., Hardison, R. C. and Groudine, M. (2003) A complex chromatin landscape revealed by patterns of nuclease sensitivity and histone modification within the mouse beta-globin locus. *Mol. Cell Biol.* **23**, 5234–5244.
9. Litt, M. D., Simpson, M., Gaszner, M., Allis, C. D. and Felsenfeld, G. (2001) Correlation between histone lysine methylation and developmental changes at the chicken beta-globin locus. *Science* **293**, 2453–2455.
10. Schneider, R., Bannister, A. J., Myers, F. A., Thorne, A. W., Crane-Robinson, C. and Kouzarides, T. (2004) Histone H3 lysine 4 methylation patterns in higher eukaryotic genes. *Nat. Cell Biol.* **6**, 73–77.
11. McCann, J. A., Muro, E. M., Palmer, C., Palidwor, G., Porter, C. J., Andrade-Navarro, M. A. and Rudnicki, M. A. (2007) ChIP on SNP-chip for genome-wide analysis of human histone H4 hyperacetylation. *BMC Genomics* **8**, 322.
12. Kwok, P. Y. (2001) Methods for genotyping single nucleotide polymorphisms. *Annu. Rev. Genomics Hum. Genet.* **2**, 235–258.
13. Lu, J., McKinsey, T. A., Zhang, C. L. and Olson, E. N. (2000) Regulation of skeletal

myogenesis by association of the MEF2 transcription factor with class II histone deacetylases. *Mol. Cell* **6**, 233–244.

14. Wu, C. (1997) Chromatin remodeling and the control of gene expression. *J. Biol. Chem.* **272**, 28171–28174.
15. Harney, D. F., Butler, R. K. and Edwards, R. J. (2005) Tyrosine phosphorylation of myosin heavy chain during skeletal muscle differentiation: an integrated bioinformatics approach. *Theor. Biol. Med. Model* **2**, 12.
16. Lin, W. H., Huang, C. J., Liu, M. W., Chang, H. M., Chen, Y. J., Tai, T. Y. and Chuang, L. M. (2001) Cloning, mapping, and characterization of the human sorbin and SH3 domain containing 1 (SORBS1) gene: a protein associated with c-Abl during insulin signaling in the hepatoma cell line Hep3B. *Genomics* **74**, 12–20.
17. Davis, D. B., Delmonte, A. J., Ly, C. T. and McNally, E. M. (2000) Myoferlin, a candidate gene and potential modifier of muscular dystrophy. *Hum. Mol. Genet.* **9**, 217–226.
18. Yagami-Hiromasa, T., Sato, T., Kurisaki, T., Kamijo, K., Nabeshima, Y. and Fujisawa-Sehara, A. (1995) A metalloprotease-disintegrin participating in myoblast fusion. *Nature* **377**, 652–656.
19. Lafuste, P., Sonnet, C., Chazaud, B., Dreyfus, P. A., Gherardi, R. K., Wewer, U. M. and Authier, F. J. (2005) ADAM12 and alpha9-beta1 integrin are instrumental in human myogenic cell differentiation. *Mol. Biol. Cell* **16**, 861–870.

Chapter 11

DamID: A Methylation-Based Chromatin Profiling Approach

Mona Abed, Dorit Kenyagin-Karsenti, Olga Boico, and Amir Orian

Abstract

Gene expression is a dynamic process and is tightly connected to changes in chromatin structure and nuclear organization (Schneider, R. and Grosschedl, R., 2007, *Genes Dev.* **21**, 3027–3043; Kosak, S. T. and Groudine, M., 2004, *Genes Dev.* **18**, 1371–1384). Our ability to understand the intimate interactions between proteins and the rapidly changing chromatin environment requires methods that will be able to provide accurate, sensitive, and unbiased mapping of these interactions in vivo (van Steensel, B., 2005, *Nat. Genet.* **37 Suppl**, S18–24). One such tool is DamID chromatin profiling, a methylation-based tagging method used to identify the direct genomic loci bound by sequence-specific transcription factors, co-factors as well as chromatin- and nuclear-associated proteins genome wide (van Steensel, B. and Henikoff, S., 2000, *Nat. Biotechnol.* **18**, 424–428; van Steensel, Delrow, and Henikoff, 2001, *Nat. Genet.* **27**, 304–308). Combined with other functional genomic methods and bioinformatics analysis (such as expression profiles and 5C analysis), DamID emerges as a powerful tool for analysis of chromatin structure and function in eukaryotes. DamID allows the detection of the direct genomic targets of any given factor independent of antibodies and without the need for DNA cross-linking. It is highly valuable for mapping proteins that associate with the genome indirectly or loosely (e.g., co-factors). DamID is based on the ability to fuse a bacterial Dam-methylase to a protein of interest and subsequently mark the factor's genomic binding site by adenine methylation. This marking is simple, highly specific, sensitive, inert, and can be done in both cell culture and living organisms. Below is a short description of the method, followed by a step-by-step protocol for performing DamID in *Drosophila* cells and embryos. Due to space limitations, the reader is referred to recent reviews that compare the method with other profiling techniques such as ChIP-chip as well as protocols for performing DamID in mammalian cells (NSouthall, T. D. and Brand, A. H., 2007, *Nat. Struct. Mol. Biol.* **14**, 869–871; Orian, A., 2006, *Curr. Opin. Genet. Dev.* **16**, 157–164; Vogel, M. J., Peric-Hupkes, D. and van Steensel, B. 2007, *Nat. Protoc.* **2**, 1467–1478).

Key words: DamID, gene regulation, chromatin, transcription, nuclear organization, genomics, *Drosophila*.

1. Introduction

To monitor dynamic changes in chromatin and nuclear organization (1, 2), we describe below a step-by-step protocol for performing DamID chromatin profiling.

Philippe Collas (ed.), *Chromatin Immunoprecipitation Assays,* Methods in Molecular Biology 567,
DOI 10.1007/978-1-60327-414-2_11, © Humana Press, a part of Springer Science+Business Media, LLC 2009

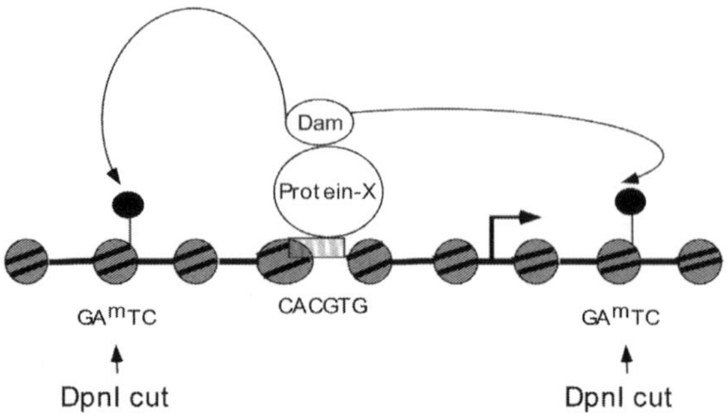

Fig. 11.1. The DamID method. Binding of the Dam-Fusion proteins to its cognate binding sites – for example CACGTG (*dashed box*) – results in flanking DAM methylation (*black circle*). Subsequently, the methylated flanked fragment is isolated from the genomic DNA using DpnI digest. Chromatin is represented as *gray circles*.

To perform a DamID profiling experiment, a bacterial DNA adenine methylase (DAM) is fused to the protein of interest (**Fig. 11.1**). Trace amounts of the chimeric protein are expressed in cells or as a transgene in animals. DNA binding of the chimeric protein results in local methylation in the vicinity of binding sites on adenine nucleotides within the Dam recognition sequence (GA^{m}TC). Subsequently, GA^{m}TC methylated DNA fragments are isolated using DpnI digest, which cleaves specifically GA^{m}TC. Considering that GATC sequences are frequently present in the genome (on average every 0.2–2.5 kb), the fragments isolated contain regions near by or within genes in addition to the binding site itself (**Fig. 11.1**). To account for accessibility and non-specific Dam binding, a DamID experiment is performed as a comparison between the relative binding of Protein X-Dam chimeric protein to that of a free Dam protein. Isolated 0.2–2.5 kb DpnI genomic fragments from Dam-Only (reference) and Dam-X-Fusion (experimental) are directly labeled with Cy3 and Cy5 dyes and hybridized directly to a cDNA/EST or genomic tiling microarray (3–6). The Dam methylation in eukaryotes is transcriptionally as well as developmentally inert, and therefore is ideal for network analysis in vivo. Indeed DamID was used to map the binding site of sequence-specific transcription factor networks, and to monitor co-factors recruitment (7–12). It is powerful for studying heterochromatin-associated proteins as well proteins required for nuclear organization and dynamics (13–18). DamID can also be used to evaluate recruitment to a single gene of interest using a Southern blot approach (4, 19, 20). DamID is not limited to *Drosophila* and has been used to map proteins in *Arabidopsis thaliana* and mammalian genomes (21–23). In this chapter we describe a simple procedure to perform DamID using *Drosophila*

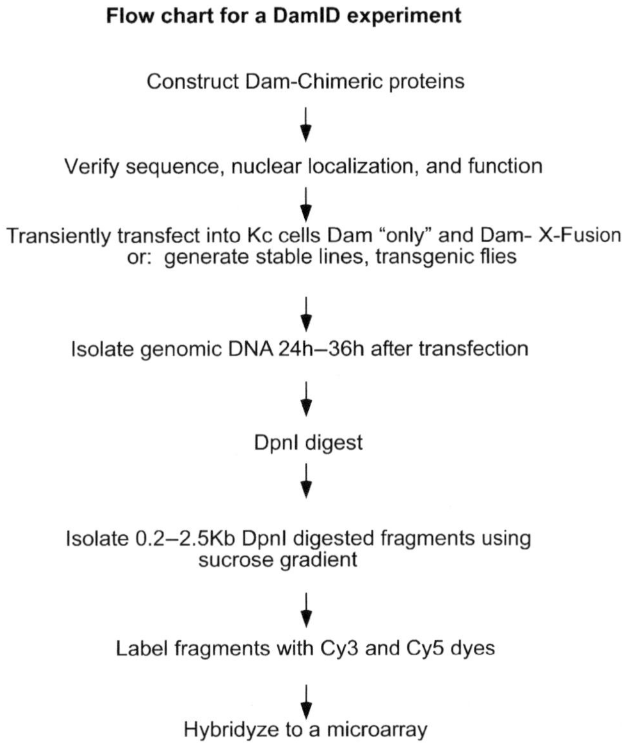

Fig. 11.2. Design and flow-chart for a DamID experiment.

Kc_{167} $_{cells}$ and Dam-transgenic *Drosophila melanogaster* embryos using a sucrose gradient (**Fig. 11.2**). We also included protocols for constructing Dam-fusions proteins, transfection of *Drosophila* cells, and isolation of genomic DNA from large quantities of *Drosophila* embryos. While we have tried to be as conclusive as possible, an excellent DamID source can be found at: http://research.nki.nl/Vansteensellab/, which contains technical information, published DamID data sets, and answers to frequently asked questions.

2. Materials

All materials should be of high molecular and analytic grade.

2.1. Construction of Dam-Fusion Expression Vectors

1. pNDamMyc and pCMycDam expression vectors. Vectors can be obtained from the Van Steensel laboratory (for academic and non-profit use). A complete list of vectors; their sequences, maps and cloning strategies are available for download from the Van Steensel lab (see above link).

2. Full-length cDNA encoding the protein of interest.

2.2. Electroporation of Kc Cells

1. HyQ-SFX-Insect MP (#SH30350.03, HyClone) supplemented with 20 mM L-glutamine.
2. 100 × 20 mm^2 tissue culture plates (Falcon).
3. 0.4 cm gap electroporation cuvettes (Bio-Rad).
4. Dam expression vectors (pNDamMyc (*see* **Note 1**), a vector encoding the Dam-fusion protein of interest) and a heat shock (hs)-Casper GFP vector (transfection control). All constructs should be prepared with a high-quality Plasmid Maxi Kit (such as #12163, Qiagen) or by $CsCl_2$ purification.
5. Bio-Rad Gene Pulser II/Capacitance Extender II Electrophoresis System (Bio-Rad), or a similar cell electroporator.
6. Tissue culture grade sterile tips and pasture pipettes, as well as 15 and 50 mL plastic tubes.

2.3. Purification of Genomic DNA from Transfected Kc Cells for DamID Labeling

1. $T_{10}E_{10}$ buffer: 10 mM Tris-HCl, pH 7.5, 10 mM EDTA.
2. $T_{10}E_{0.1}$ buffer: 10 mM Tris-HCl, pH 7.5, 0.1 mM EDTA.
3. TENS buffer: 10 mM Tris-HCl, pH 7.5, 10 mM EDTA, 100 mM NaCl, 0.5% SDS. Store solutions 1–3 at room temperature (RT).
4. TENS/K solution: 200 μg/mL proteinase K (#03-115-887, Roche Diagnostics) in TENS. Prepare freshly before use and keep at room temperature.
5. Buffer-saturated phenol:chloroform:isoamylalcohol (25:24:1) saturated with 10 mM Tris-HCl pH 8.0, 1 mM EDTA.
6. 3 M Na-Acetate (NaAc), pH 5.2.
7. DNase-free RNaseA (10 mg/mL).

2.4. Purification of Genomic DNA from Fly Embryos for DamID Labeling

1. Yeast paste. Dissolve baking yeast in water to form paste. Keep at room temperature or 4°C. Prepare freshly every 2 days.
2. Household bleach.
3. 1 M Tris-base, pH 9.0.
4. 0.5 M EDTA.
5. 5 M NaCl.
6. 50% sucrose, filtered.
7. 20% SDS.
8. Proteinase K, 20 mg/mL stock.
9. Phenol:chloroform:isoamylalcohol.
10. 3 M NaAc, pH 5.2.
11. DNase-free RNase A (10 mg/mL; #R5503, Sigma).

12. Homogenizing buffer: 0.1 M Tris-HCl, pH 9.0, 0.1 M EDTA, 0.1 M NaCl, 5% sucrose. Store at 4°C.
13. 3 mL glass homogenizer fitted with pestle A (tight).
14. Embryo collection sieves (#052-006, 230 × 260 mm^2, Whatman Biometra)
15. 15 cm embryo collection plates ("grape plates")
16. Population cage containing 100–200 fly bottles.

2.5. DpnI Digestion of Genomic DNA

1. DpnI (New England Biolabs).
2. Restriction buffer No. 4 (New England Biolabs; supplied with DpnI).
3. DNase-free RNase A (10 mg/mL; #R5503, Sigma).

2.6. Sucrose Gradient Fractionation

1. 5% sucrose sol.: 5% sucrose, 10 mM Tris-HCl, pH 7.5, 10 mM EDTA, 150 mM NaCl.
2. 30% sucrose sol: 30% sucrose, 10 mM Tris-HCl, pH 7.5, 10 mM EDTA, 150 mM NaCl, a dash of Bromophenol-Blue crystals to give the solution a bit of color. Filter each solution through a 0.22 μm filter and keep sterile at 4°C.
3. 3 M NaAc, pH 5.2
4. Ultra-Clear™ Tubes (14 × 89 mm^2, #BC-344059, Beckman).
5. Gradient mixer with a peristaltic pump.
6. Ultra centrifuge with a SW40-Ti swing-out rotor.
7. 1% agarose gel.
8. Wide-spectrum DNA ladder.

2.7. Labeling of DpnI Methylated DNA

1. BioPrime DNA labeling kit (Invitrogen).
2. PCR grade dNTPs (#28-4065-51, Amersham).
3. 10X dNTP Genomic labeling mix: 1.2 mM each dATP, dGTP and dTTP, 0.6 mM dCTP, 10 mM Tris-HCl pH 8.0, 1 mM EDTA.
4. Yeast tRNA (# 15401-011, Invitrogen); 5 mg/mL stock.
5. Cy3-dCTP (PA53021, Amersham); 1 mM stock.
6. Cy5-dCTP (PA55021, Amersham), 1 mM stock.
7. 25 μg competitor DNA, i.e., the plasmid encoding the Dam-fusion protein that was used to transfect the Kc cells.
8. Strataclean Resin (#400714, Stratagene).
9. Glycogen (Roche).
10. Poly [dA]-Poly [dT] 1 mg/mL stock (#P9764-25UN, Sigma).
11. Microcon YM-30 filters (#42410, Millipore).

12. 20X SSC.
13. Hybridization oven set at 55°C.
14. 37°C heating block or water bath.

3. Methods

3.1. Construction of Dam Expression Vectors

1. Clone the gene of interest in frame into the multiple cloning sites (MCS) of both pNDamMyc and pCMycDam expression vectors (*see* **Notes 2, 3**). The ORF of the gene of interest can be cloned upstream of the Myc-tag (EQKLISEEDL, 9E10) in the pCMycDam vector. Similarly, the gene of interest can be cloned downstream of the Myc tag in the pNDamMyc vector using its MCS. In both cases the short Myc-tag serves as a linker between the protein of interest and the Dam, and could be used for detection of the chimeric protein.
2. We recommend that the sequence, proper expression, and nuclear localization of the chimeric protein be verified prior to performing the DamID experiment.

3.2. Electroporation of Kc Cells

1. One 90% confluent 100 × 20 mm dish ($\sim 1 \times 10^8$ cells) is required per transfection. A 1:10 split of sub-confluent Kc cells growing in SFX supplemented with L-glutamine will provide this appropriate cell density after 48 h at 25°C (*see* **Notes 4, 5**). The protocol described below is for a single plate transfection. Note that five starting plates (five independent transfections for each construct) are required for DamID analysis of a single protein.
2. Resuspend cells and pool in a 15 mL sterile tube. Spin at 1,000*g* for 3 min, aspirate supernatant, and resuspend cell pellet in 0.81 mL SFX-glutamine.
3. Mix 10 μg of the expression vector with the cell suspension and transfer to a 0.4 cm gap electroporation cuvette.
4. Electroporator setup: turn the capacitance rotary switch to "high capacitance", set the voltage at 0.25 kV and high capacitance at 1. A good electroporation should yield a time constant in the range of 16–22.
5. In the hood, carefully remove the cell suspension from the cuvette while avoiding the upper layer of foam and cell debris. Split the cell suspension evenly (380–400 μL) to two 100 × 20 mm dishes supplemented with 10 mL SFX-glutamine.
6. Grow cells at 25°C for approximately 36–48 h before continuing to the DNA purification and labeling stages (*see* **Notes 6, 7**).

7. If the transfection is intended for analysis of the nuclear localization of the Dam-fusion protein by immunofluorescence or for Western blot analysis, heat shock induction of the protein is required. Heat shock is carried out by incubating the cells at 37°C for 1 h and subsequently 6 h recovery period at 25°C (*see* **Note 6**).

3.3. Purification of Genomic DNA from Transfected Kc Cells for DamID Labeling

1. Collect cells from 10 plates of transfected Kc cells into two 50 mL tubes. Spin at 1,500*g* for 3 min in a tabletop centrifuge.
2. Remove the supernatant and pool the pellets in 7 mL ice-cold T_{10} E_{10} by gently pipetting up and down.
3. Squirt in 7.5 mL freshly prepared room temperature TENS/K. Gently invert the tube a few times to induce sufficient mixing.
4. Incubate the tube at 55°C for 2 h in a hybridization oven with gentle shaking. Mix gently after 30 min and return to oven.
5. Add 15 mL buffer-saturated phenol:chloroform:isoamylalcohol and mix gently by inverting the tube. Spin for 20 min at 2,200*g* at RT.
6. Gently transfer the supernatant to a clean tube and add 15 mL isopropanol and 1.5 mL of 3 M NaAc, pH 5.2.
7. Mix gently until DNA forms a large spool. Carefully remove the DNA spool using a large pipetting tip and drain it gently on the side of the tube. Continue transferring the DNA and draining it on the side of a set of clean Eppendorf tubes in order to further assist the drying process.
8. Transfer DNA to a clean Eppendorf tube and add 0.3 mL of $T_{10}E_{10}$ and 20 μg DNase-free RNase. Incubate at 37°C for 30 min. Mix the DNA gently by pipetting up and down using a blue tip, which has been cut at the tip. Return the DNA to 37°C and incubate overnight. Important: The DNA must be completely dissolved before the next step.
9. Add 0.3 mL TENS/K and gently mix by pipetting up and down with a blue tip. Incubate tube at 55°C for 2 h.
10. Add 0.6 mL phenol:chloroform:isoamylalcohol, mix gently, and spin 15 min 10,000*g* in a tabletop centrifuge (*see* **Note 8**).
11. Transfer the supernatant to a clean Eppendorf tube. Add 60 μL 3 M NaAc, pH 5.2, and 0.6 mL isopropanol. Carefully mix by gently inverting the tube a few times.
12. Spool the DNA onto a yellow tip and briefly dip into an Eppendorf tube with 70% ethanol in order to remove the salt.

13. Transfer DNA into a clean Eppendorf and dissolve it by incubating the DNA with 500 μL $T_{10}E_{0.1}$ for several hours at 37°C. Pipette up and down a few times with a blue tip to dissolve the DNA. At this stage the tubes can be incubated overnight at 37°C. Important: Only go to the DpnI digestion step (**Section 3.5**) if the DNA is completely dissolved in solution.

3.4. Purification of Genomic DNA from Fly Embryos for DamID Labeling

1. Set up a population cage with approximately 100 bottles of flies (*see* **Notes 9–12**).
2. Synchronize flies by changing the embryo collection plates twice over 1 h.
3. Collect embryos of the appropriate age (for example 4–6 h for early developmental stages) by washing the embryos off the collection plate with water and a paintbrush into an embryo collection sieve.
4. Wash the embryos thoroughly with water and dry off the sieve using a paper towel ("Kimwipes™").
5. Place the collection chamber in a household bleach and dechorinate embryos for 2 min. Embryos should be thoroughly immersed in the bleach.
6. Wash embryos well with distilled water until there is no trace of bleach (*see* **Note 13**).
7. Place ~500 μL of embryos in homogenizer tube on ice.
8. Add 1 mL ice-cold homogenizing buffer to the embryos and grind well with a tight glass pestle, while keeping the embryos on ice.
9. Transfer to an Eppendorf tube and immediately add 25 μL 20% SDS and 5 μL of 20 mg/mL proteinase K. Mix gently.
10. Incubate for 2 h at 55°C while mixing gently every 30 min.
11. Add an additional 25 μL of 20% SDS and 8 μL proteinase K, mix gently, and incubate for 3 h at 55°C.
12. Add 25 μL 20% SDS.
13. Spin down debris for 10 min at maximum speed at RT.
14. Discard the upper phase of lipid layer and keep supernatant.
15. Add 1 volume of phenol:chloroform:isoamylalcohol and mix gently by inverting.
16. Spin for 4 min at 10,000*g* at RT and transfer supernatant to a new Eppendorf tube.
17. Precipitate the genomic DNA with 1 volume of isopropanol and 0.1 volume of NaAc, pH 5.2.

18. Mix gently and spin down for 1 min at 10,000*g* at RT. Do not spin longer to avoid DNA shearing.
19. Air-dry the DNA pellet (2–3 min).
20. Resuspend genomic DNA in 0.3 mL $T_{10}E_{10}$buffer with 20 μg DNase-free RNase (Roche, 10 mg/mL stock).
21. Follow same protocol for DamID in KC cells (continue from **Section 3.3**, Step 9).

3.5. DpnI Digestion of Genomic DNA

1. Set up the following digest:
   i. 400 μL DNA
   ii. 120 μL 10X buffer 4
   iii. 0.5 μL DNase-free RNase
   iv. 640 μL DDW
   v. 40 μL DpnI.
2. Mix gently by pipetting up and down with a blue tip and incubate for 16 h at 37°C.
3. After the designated incubation period, add an additional 10 μL DpnI, and further incubate for 2 h. The DNA should be less viscous at this stage.
4. Determine DNA concentration using a Hoechst fluorometer/ NanoDrop™.

3.6. Sucrose Gradient Fractionation

1. Rinse an Ultra-Clear™ open-top Beckman tube (14 × 89 mm) with water to remove any dust and dry completely.
2. Using the gradient maker, make a gradient composed of 5.6 mL each of 5 and 30% buffered sucrose solutions. It is easiest to fill the tubes slowly (low pump pressure) from the bottom using a glass capillary.
3. Layer 1 mL of the Digested DNA on top of the gradient with great caution not to disturb the gradient layers using a blue tip (set aside the remaining 200 μL of digested DNA as input for the analysis of the gradients fractions).
4. Load the gradient onto a SW40-Ti swing-out rotor and balance the rest of the tubes with water (*see* **Note 14**).
5. Run the gradient according to the following settings:
   i. Speed: 25,000 rpm
   ii. Temperature: 20°C
   iii. Time: 16 h
   iv. Deceleration: setting 9 (slow deceleration).
6. Carefully collect 0.4 mL fractions using a blue tip from the surface of the gradient reserving each fraction separately in an individual Eppendorf tube. Run 20 μL of each fraction on a

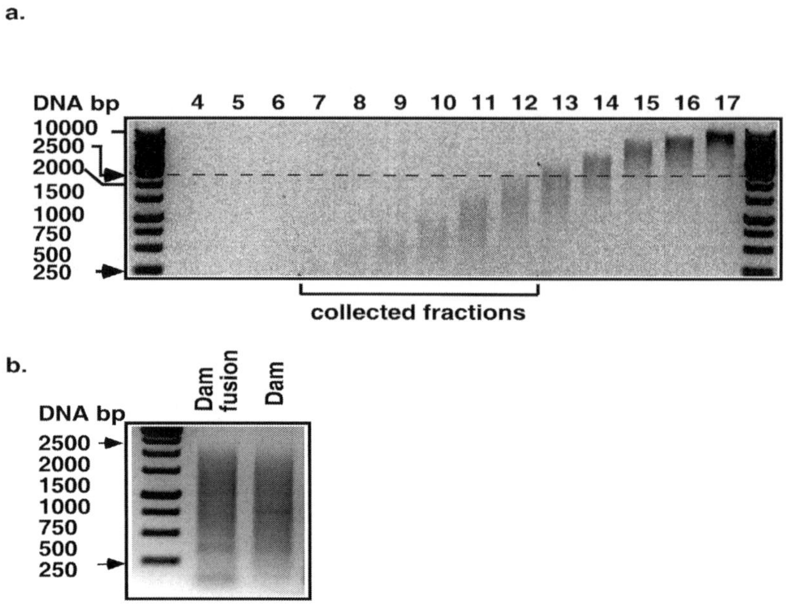

Fig. 11.3. Purification of methylated DNA fragments. **(A)** Post Dpn I digested genomic DNA was resolved over a 5–30% sucrose gradient fractionation. 20 μL sample from each fraction was run on a 1% agarose gel. Fractions # 7–11 containing DNA fragments at the size of 0.2–2.5 kb were combined, and subsequently precipitated. **(B)** Analysis of 3 μL on a 1% agarose gel of 0.2–2.5 kb dam-methylated fragment from pooled fractions prior to labeling.

1% agarose gel beside a large spectrum DNA ladder. Collect the desired fractions (<2.5 kb) typically found in fractions 6–12 (**Fig. 11.3A**).

7. Pool and mix the fractions that contain the DNA fragments smaller than 2.5 kb.
8. Distribute 0.7 mL amounts of the mix over Eppendorf tubes containing 0.7 mL isopropanol and 70 μL 3 M NaAc, pH 5.2, and mix well. Let the DNA precipitate for 1.5 h at –20°C, but no longer than 2 h.
9. Spin the tubes for 20 min at 12,000*g* at 4°C using a table centrifuge. Remove supernatant and wash the pellet in 1 mL 70% ethanol. Spin at 12,000*g* for 5 min.
10. Carefully and completely remove the supernatant. Air-dry the pellet, and re-dissolve and pool the DNA in 50 μL $T_{10}E_{0.1}$(total volume after pooling).
11. Measure the DNA concentration using a Hoechst fluorometer/NanoDrop™. Typically the total yield of methylated fragments (<2.5 kb) is ~15–25 μg. Store DNA at –20°C.

3.7. Labeling of DpnI Methylated DNA

3.7.1. Preparing the Competitor DNA

1. Set up the following reaction:
   i. 50 μg competitor DNA (Dam-Fusion Plasmid)
   ii. 10 μL DpnI (5 units per μg DNA)
   iii. 20 μL 10X NEB buffer 4 (provided with DpnI)
   iv. DDW is added to a final volume of 200 μL.
2. Incubate the reaction at 37°C for 2 h or overnight.
3. Remove the enzyme by adding 5 μL well mixed StrataClean beads to the above reaction and incubate at RT for 2 min while mixing from time to time.
4. Spin at top speed for 2 min and transfer supernatant to a clean Eppendorf tube.
5. Add to the supernatant 20 μL 3 M NaAc, pH 5.2, 550 μL 100% cold ethanol, and 3 μL glycogen, and incubate at –70°C for 20 min.
6. Spin at full speed in a cold table centrifuge for 20 min.
7. Remove supernatant and wash pellet with 70% cold ethanol. Re-spin the tube at top speed for 5 min. Air-dry and resuspend in at 5 μg/μL (~10 μL) $T_{10}E_{0.1}$buffer.

3.7.2. Labeling DNA for Microarray Hybridization

1. Set up the following reaction in a PCR tube (reactions below are intended for hybridization to a 12 k spotted array. Adjustments should be made according to array geometry and size):
   i. 2 μg of DNA pooled fragments from either "experimental" or "reference" (Dam-Fusion or Dam fragments; *see* **Note 15**).
   ii. Bring the DNA to a total volume of 42 μL with DDW (included in BioPrime kit).
   iii. 40 μL 2.5X random primer/reaction buffer mix (Bio-Prime kit).
2. Incubate the reaction at 95°C for 5 min and remove onto ice immediately afterwards.
3. Set up the following reaction on ice:
   i. 82 μL of the above DNA reaction
   ii. 10 μL of genomic 10X dNTP mix (*see* **Note 16**)
   iii. 6 μL Cy5-dCTP or Cy3-dCTP
   iv. 4 μL Klenow Fragment (provided with the BioPrime kit).
4. Incubate at 37°C for 2 h.
5. After incubation period, return the tube onto ice, and stop the reaction by adding 5.5 μL 0.5 M EDTA, pH 8.0.
6. Add 400 μL $T_{10}E_{0.1}$to the stopped labeling reaction and transfer to a Microcon YM-30 filter.
7. Spin at 8,000*g* for 10 min.
8. Invert the filter and place in a clean collection tube.

9. Spin for 1 min at 8,000*g* to recover the probe (the probe should be approximately 20–40 μL in volume).
10. Combine the "experimental" and "reference" purified labeled DNA fragments in a clean Eppendorf tube and add the following:
    i. 50 μg of the pre-digested competitor DNA (*see* **Note 17**)
    ii. 200 μg yeast tRNA
    iii. 40 μg poly [dA]-Poly [dT] (*see* **Note 18**)
    iv. 400 μL $T_{10}E_{0.1}$.
11. Concentrate the probe with a Microcon YM-30 filter as mentioned in Steps 6–9.
12. Adjust the volume of the probe mixture to 30 μL with $T_{10}E_{0.1}$ and add 6 μL 20X SSC (to a final volume of 36 μL and concentration of 3.4X SSC).
13. Protect from light, and preferably hybridize to array of choice, or otherwise keep frozen at –20°C.

4. Notes

1. Only the "empty" pNDamMyc vector should be used to express the Dam-Myc protein since it contains an initiating Methionine, and a stop codon 15 amino acids after the Myc tag.
2. It is impossible to predict which of the two expression vectors will produce a Dam-fusion protein that successfully migrates into the nucleus and binds the specific DNA sites within the chromatin. Therefore, we recommend that the gene of interest is cloned into both vectors and that nuclear localization of the Dam-fusion protein in Kc cells is sequentially analyzed by means of staining with anti-Myc tag.
3. When cloned into the pNDamMyc vector, the Dam-Myc protein is fused to the N-terminus of the protein of interest. Therefore, when cloning the gene of interest into the pNDamMyc vector, ensure that the start codon has been removed from the gene and that it contains a stop codon at the end of the sequence. Similarly, when using the pCMyc-Dam vector, the Dam-Myc protein is fused to the C-terminus of the protein of interest. Therefore, ensure that the gene sequence contains an intact start codon and that the inset's stop codon has been removed.
4. All equipment must be sterilized and stages should be carried out in a cell culture flow hood under sterile conditions.

5. It may be necessary to optimize various parameters such as cell growth conditions, plasmid concentrations, and electroporation field strength and pulse length when using different Kc sublines.
6. To prevent non-specific Dam saturation the DamID experiments are performed under the control of the pCasper-hs (heat shock) promoter, but in the absence of heat shock. This leaky promoter expresses trace amounts of the chimeric protein that are sufficient for target tagging yet at the same time are not detected by immunofluorescence or Western blot.
7. It is necessary to supplement the SFX growth media with 20 mM L-glutathione, as this significantly increases the transfection efficiency. The transfection efficiency should be monitored by expression of a heat shock induced GFP protein. We favorably use a pCasper hs-GFP vector for this purpose.
8. This step should be repeated if supernatant does not appear to be clear.
9. The adult flies used for embryo collections are generally most productive for a period of 3–7 days after emerging when kept in good conditions.
10. To express only trace amounts of the DamID chimeric proteins use UAS-Dam flies without mating (crossing) them to the Gal4 driver.
11. We had good results with generating UAS-Dam fusions but were not able to generate viable hs-CaSaper based transgenic flies.
12. For performing experiments testing for factors recruitment during early embryogenesis and germ cells make sure to generate and clone the chimeric proteins using the UASp vector.
13. Embryos can be stored at this point for a long period of time in a saran wrap at –80°C.
14. The sucrose gradient should be managed with great care throughout all the steps in order to prevent mixture of the layers.
15. "Experimental" DNA refers to the labeled DNA fractions that were obtained from the transfection of the plasmid encoding the Dam-fusion protein, and "reference" DNA refers to the DNA purified from pNDamMyc or pCMycDam transfected cells. Label the "experimental" and "reference" DNA with a different florescent probe.
16. Do not use the dNTP mix provided with the kit. Instead, prepare the 10X dNTP mix with PCR grade dNTPs, which can be purchased separately.
17. The unlabeled competitor DNA competes with the labeled transfected vector to avoid background artifacts.

18. Poly [dA]-Poly [dT] blocks hybridization to polyA tails of cDNA array elements.

Acknowledgments

We thank Dr. Susan Parkhurst for protocols and advice. We are grateful to Dr. Bas Van-Steensel, the inventor of the DamID method for sharing protocols, and support of the DamID community. We thank Dr. Tom Schultheiss for reading this manuscript. DK is supported at the Technion by a fellowship from the Lady Davis Foundation. AO is supported by the German-Israeli foundation (GIF 936-273), ISF-F.I.R.S.T. 1215-07 grant and a Human Frontier Science Program CDA (0048/2005).

References

1. Schneider, R. and Grosschedl, R. (2007) Dynamics and interplay of nuclear architecture, genome organization, and gene expression. *Genes Dev.* **21**, 3027–3043.
2. Kosak, S. T. and Groudine, M. (2004) Form follows function: The genomic organization of cellular differentiation. *Genes Dev.* **18**, 1371–1384.
3. van Steensel, B. (2005) Mapping of genetic and epigenetic regulatory networks using microarrays. *Nat. Genet.* **37 Suppl**, S18–24.
4. van Steensel, B. and Henikoff, S. (2000) Identification of in vivo DNA targets of chromatin proteins using tethered dam methyltransferase. *Nat. Biotechnol.* **18**, 424–428.
5. van Steensel, B., Delrow, J. and Henikoff, S. (2001) Chromatin profiling using targeted DNA adenine methyltransferase. *Nat. Genet.* **27**, 304–308.
6. NSouthall, T. D. and Brand, A. H. (2007) Chromatin profiling in model organisms. *Nat. Struct. Mol. Biol.* **14**, 869–871.
7. Orian, A. (2006) Chromatin profiling, DamID and the emerging landscape of gene expression. *Curr. Opin. Genet. Dev.* **16**, 157–164.
8. Vogel, M. J., Peric-Hupkes, D. and van Steensel, B. (2007) Detection of in vivo protein–DNA interactions using DamID in mammalian cells. *Nat. Protoc.* **2**, 1467–1478.
9. Greil, F., Moorman, C. and van Steensel, B. (2006) DamID: mapping of in vivo protein–genome interactions using tethered DNA adenine methyltransferase. *Methods Enzymol.* **410**, 342–359.
10. Orian, A., van Steensel, B., Delrow, J., Bussemaker, H. J., Li, L., Sawado, T., Williams, E., Loo, L. W., Cowley, S. M., Yost, C., Pierce, S., Edgar, B. A., Parkhurst, S. M. and Eisenman, R. N. (2003) Genomic binding by the *Drosophila* Myc, Max, Mad/Mnt transcription factor network. *Genes Dev.* **17**, 1101–1114.
11. Choksi, S. P., Southall, T. D., Bossing, T., Edoff, K., de Wit, E., Fischer, B. E., van Steensel, B., Micklem, G. and Brand, A. H. (2006) Prospero acts as a binary switch between self-renewal and differentiation in *Drosophila* neural stem cells. *Dev. Cell* **11**, 775–789.
12. Bianchi-Frias, D., Orian, A., Delrow, J. J., Vazquez, J., Rosales-Nieves, A. E. and Parkhurst, S. M. (2004) Hairy transcriptional repression targets and cofactor recruitment in *Drosophila*. *PLoS Biol.* **2**, e178.
13. Greil, F., van der Kraan, I., Delrow, J., Smothers, J. F., de Wit, E., Bussemaker, H. J., van Driel, R., Henikoff, S. and van Steensel, B. (2003) Distinct HP1 and Su(var)3-9 complexes bind to sets of developmentally coexpressed genes depending on chromosomal location. *Genes Dev.* **17**, 2825–2838.

14. de Wit, E., Greil, F. and van Steensel, B. (2007) High-resolution mapping reveals links of HP1 with active and inactive chromatin components. *PLoS Genet.* **3**, e38.
15. Tolhuis, B., de Wit, E., Muijrers, I., Teunissen, H., Talhout, W., van Steensel, B. and van Lohuizen, M. (2006) Genome-wide profiling of PRC1 and PRC2 Polycomb chromatin binding in *Drosophila* melanogaster. *Nat. Genet.* **38**, 694–699.
16. Pindyurin, A. V., Moorman, C., de Wit, E., Belyakin, S. N., Belyaeva, E. S., Christophides, G. K., Kafatos, F. C., van Steensel, B. and Zhimulev, I. F. (2007) SUUR joins separate subsets of PcG, HP1 and B-type lamin targets in *Drosophila*. *J. Cell Sci.* **120**, 2344–2351.
17. Pickersgill, H., Kalverda, B., de Wit, E., Talhout, W., Fornerod, M. and van Steensel, B. (2006) Characterization of the *Drosophila* melanogaster genome at the nuclear lamina. *Nat. Genet.* **38**, 1005–1014.
18. de Wit, E., Braunschwieg, U., Greil, F., Bussemaker, H. and van Steensel, B. (2008) Global chromatin domain organization of the *Drosophila* genome. *PLoS Genetics* **4**, e1000045.
19. Grewal, S. S., Li, L., Orian, A. Eisenman, R. N. and Edgar, B. A. (2004) Myc-dependent regulation of ribosomal RNA synthesis during *Drosophila* development. *Nat. Cell Biol.* 7, 295–302.
20. Song, S., Cooperman, J., Letting, D. L., Blobel, G. A. and Choi, J. K. (2004) Identification of Cyclin D3 as a direct target of E2A using DamID. *Mol. Cell Biol.* **24**, 8790–8802.
21. Reddy, K. L., Zullo, J. M., Bertolino, E. and Singh, H. (2008) Transcriptional repression mediated by repositioning of genes to the nuclear lamina. *Nature* **452**, 243–247.
22. Zhang, X., Germann, S., Blus, B. J., Khorasanizadeh, S., Gaudin, V. and Jacobsen, S. E. (2007) The Arabidopsis LHP1 protein colocalizes with histone H3 Lys27 trimethylation. *Nat. Struct. Mol. Biol.* **14**, 869–871.
23. Germann, S., Juul-Jensen, T., Letarnec, B. and Gaudin, V. (2006) DamID, a new tool for studying plant chromatin profiling in vivo, and its use to identify putative LHP1 target loci. *Plant J.* **48**, 153–163.

Chapter 12

Chromosome Conformation Capture (from 3C to 5C) and Its ChIP-Based Modification

Alexey Gavrilov, Elvira Eivazova, Iryna Pirozhkova, Marc Lipinski, Sergey Razin, and Yegor Vassetzky

Abstract

Chromosome conformation capture (3C) methodology was developed to study spatial organization of long genomic regions in living cells. Briefly, chromatin is fixed with formaldehyde in vivo to cross-link interacting sites, digested with a restriction enzyme and ligated at a low DNA concentration so that ligation between cross-linked fragments is favored over ligation between random fragments. Ligation products are then analyzed and quantified by PCR. So far, semi-quantitative PCR methods were widely used to estimate the ligation frequencies. However, it is often important to estimate the ligation frequencies more precisely which is only possible by using the real-time PCR. At the same time, it is equally necessary to monitor the specificity of PCR amplification. That is why the real-time PCR with TaqMan probes is becoming more and more popular in 3C studies. In this chapter, we describe the general protocol for 3C analysis with the subsequent estimation of ligation frequencies by using the real-time PCR technology with TaqMan probes. We discuss in details all steps of the experimental procedure paying special attention to weak points and possible ways to solve the problems. A special attention is also paid to the problems in interpretation of the results and necessary control experiments. Besides, in theory, we consider other approaches to analysis of the ligation products used in frames of the so-called 4C and 5C methods. The recently developed chromatin immunoprecipitation (ChIP)-loop assay representing a combination of 3C and ChIP is also discussed.

Key words: 3C, ChIP-loop assay, 4C, 5C, TaqMan probes, real-time PCR, chromatin, genome spatial organization.

1. Introduction

It becomes increasingly evident that spatial organization of the eukaryotic genome plays an important role in regulation of gene activity. Hence, it is very important to have a reliable experimental approach permitting to find out whether two remote genomic

Philippe Collas (ed.), *Chromatin Immunoprecipitation Assays,* Methods in Molecular Biology 567,
DOI 10.1007/978-1-60327-414-2_12, © Humana Press, a part of Springer Science+Business Media, LLC 2009

sites interact with each other in the nuclear space. In model systems, such as plasmid constructs or phage DNA, it is sometimes possible to answer this question using electron microscopy. With the development of fluorescent in situ hybridization (FISH), it became possible to determine location of specific genomic elements in the nuclear space and thus to find out if these elements co-localize or not. However, this approach is far from being precise. Localization of two signals in the same voxel does not necessarily mean that the corresponding genomic elements interact with each other. Additional information can be obtained by using fluorescence resonance energy transfer (FRET). This approach was successfully used to study protein–protein interactions. However, the possibility of using FRET in the studies of the spatial organization of the genome has not yet been demonstrated even in model systems. Thus the so-called 3C technology is presently the only experimental approach proven to permit identification of distant genomic regions interacting with each other in the nuclear space.

The basic principle of the 3C protocol is shown in **Fig. 12.1**. Cells are treated with formaldehyde to cross-link proteins to other proteins nearby and DNA. After lysis of nuclei by SDS and solubilization of proteins that were not cross-linked, the resulting DNA–protein network is subjected to cleavage by a restriction enzyme(s), which is followed by ligation at a low

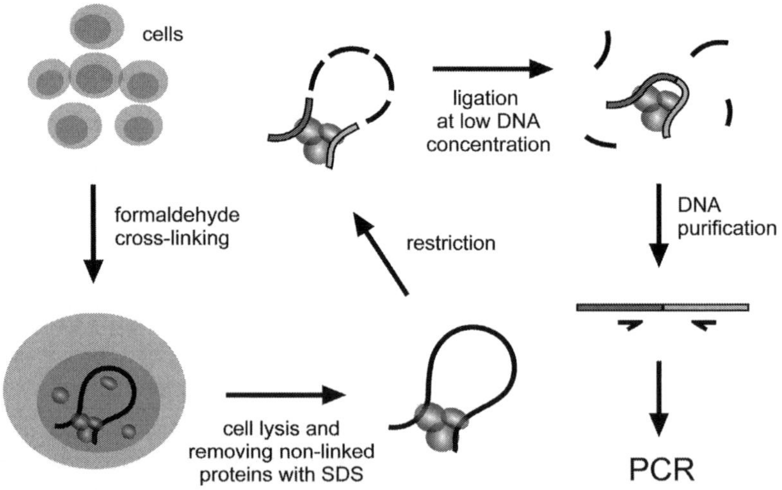

Fig. 12.1. A scheme representing the main principles of 3C technology. Cells are treated with formaldehyde and lysed. Non-linked proteins are removed by SDS, and cross-linked chromatin is digested with a restriction enzyme(s), followed by ligation at a low DNA concentration. Ligation products are analysed by PCR (with one primer pair one of four possible ligation products is examined). Two interacting restriction fragments are shown as framed light lines.

DNA concentration. Under such conditions, ligations between cross-linked DNA fragments are strongly favored over ligations between random fragments. After ligation, the cross-links are reversed, and ligation products are detected and quantified by polymerase chain reaction (**Fig. 12.1**). The cross-linking frequency of two specific restriction fragments, as measured by the amount of corresponding ligation product, is proportional to the frequency with which these two genomic sites interact. Thus, 3C analysis provides information about the spatial organization of chromosomal regions in vivo (1, 2).

Developed on the yeast system (1), 3C technology was then adopted to analyze spatial organization of genomic loci in higher eukaryotes. Successfully analyzed were both relatively small genomic domains such as mouse interferon gamma gene domain (25 Kb) (3), the mouse immunoglobulin kappa (Ig) gene domain (30 Kb) (4), chicken alpha globin gene domain (40 Kb) (5), and longer areas such as the T-helper type 2 cytokine locus (140 Kb), (6) mammalian alpha and beta globin gene domains (up to 200 Kb) (2, 7–10), and others. 3C was also used to detect *trans* interactions between functionally related elements located on different chromosomes (11, 12). In these and other studies, it was clearly demonstrated that spatial structure of genomic domains dynamically changed upon activation/repression of gene expression and other process taking place in cell nucleus.

In this chapter, we discuss a protocol for 3C analysis using real-time PCR with TaqMan probes. In the Section 1.1 and Section 1.2 below we also give an overview of methods based on the 3C technology, namely the 4C, 5C and chromatin immunoprecipitation (ChIP)-loop assays.

1.1. 4C and 5C

Nowadays, 3C technology is getting more and more widespread, and is gradually a routine method to study interactions of genomic elements. In parallel, derivative methods adopting the same idea are being developed. In this way, improving technology of DNA micro-arrays and quantitative DNA sequencing resulted in appearance of the so-called 4C and 5C methods adopted for full genome screening of interaction partners for some selected genome site. Differences between 3C, 4C, and 5C concern only ligation product analysis.

4C was independently developed in two variants differing in names but not in abbreviations. The first one is designated as circular chromosome conformation capture. Ligation products first are amplified by PCR. The strategy is aimed at amplification of circular DNA molecules originated from cross-ligation of both ends of cross-linked restriction fragments (**Fig. 12.2A**). Two PCR primers are designed to anneal at the opposite ends of a restriction fragment of interest, facing outwards. In such a way, all DNA fragment ligated with the fragment of interest at both ends are

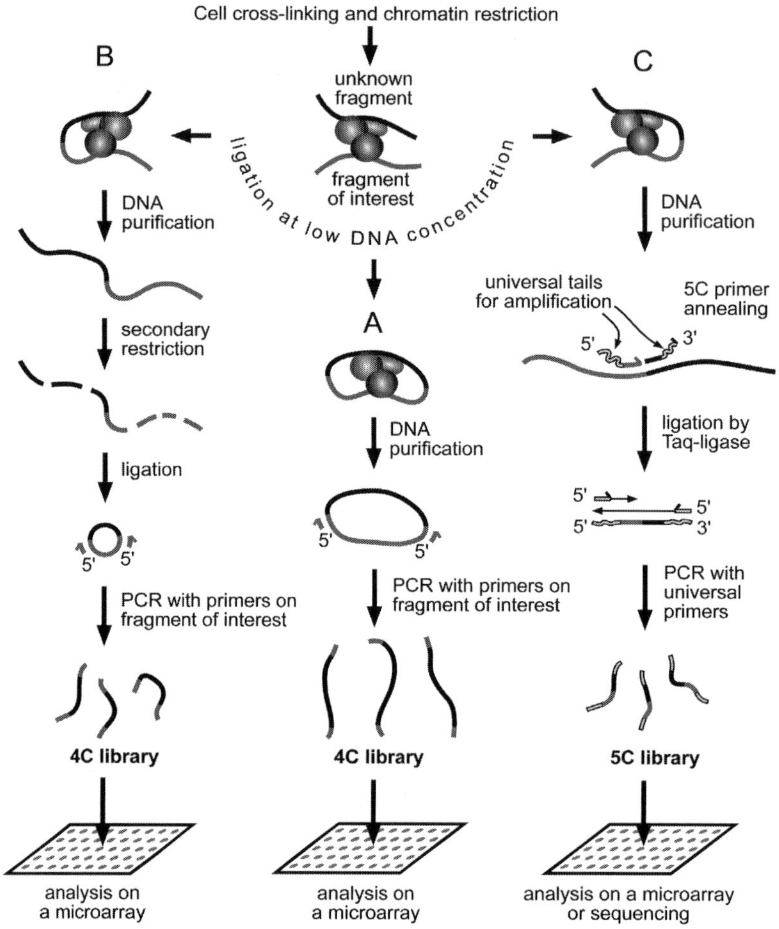

Fig. 12.2. An outline of the 4C and 5C technologies. The procedures are performed as usual until ligation products are obtained. A fragment of interest is shown as a light line. Its interaction partner, unknown and supposed to be established, is shown as a black line. (**A**) Circular chromosome conformation capture (4C). Circular ligation products are used as templates for PCR amplifications with primers annealing at the opposite sides of the fragment of interest and facing outwards. Resulting 4C library is analyzed using a DNA micro-array technique. (**B**) Chromosome conformation capture on chip 4C. Ligation products are digested with a frequently cutting restriction enzyme and re-ligated to form small DNA circles. The ones containing junction of the fragment of interest and its interaction partner are amplified with primers specific to the fragment of interest to form a 4C library that is analysed on a DNA micro-array. (**C**) Chromosome conformation capture carbon copy (5C). Ligation products are mixed with special 5C primers that anneal across ligated junction and are ligated by *Taq* ligase. 5C primers contain universal tails for amplification that serve for amplification of resulting ligation products with universal primers. The 5C DNA library is analysed on a DNA micro-array or by quantitative DNA sequencing.

amplified, and the resulting 4C DNA library representing cross-link environment of a fragment of interest is analysed by the DNA micro-array technology (13).

The second variant of 4C is designated as chromosome conformation capture on chip. In this very similar technique, ligation products are digested with a frequently cutting secondary restriction enzyme and then ligated to form small circular DNA molecules that are amplified with primers facing outwards, specific to the restriction fragment of interest (**Fig. 12.2B**). The resulting 4C DNA library is analyzed by DNA micro-array (chip) technology (14).

5C designates chromosome conformation capture carbon copy. Ligation products in this case are mixed with special primers that are designed to anneal at the very ends of restriction fragments, ones facing outwards and the others inwards, so that an end of each primer covers exactly a half of a restriction site. In such a way, outward and inward primers anneal tail-to-head across ligated junction of definite ligation products and then ligated with *Taq* ligase (**Fig. 12.2C**). Additionally, these primers contain universal tails for amplification. Such amplification having been done, resulting 5C DNA library is analyzed using either micro-arrays or quantitative DNA sequencing. The 3C library determines which 5C ligation products are generated and how frequently. As a result, the 5C library is a quantitative "carbon copy" of a part of the 3C library, as determined by the collection of 5C primers (15).

1.2. ChIP-Loop Assay

Chromatin immunoprecipitation and chromosome conformation capture methods operate with the same principle of fixing DNA–protein contacts in vivo, but are meant to address different issues. The first gives information of which proteins bind to one or another genomic site. The second is aimed to show which genomic sites interact in the nuclear space. While ChIP data may be frequently interpreted without any concerns on DNA spatial organization, 3C data is regarded to some degree as incomplete as long as it is not supplemented with the knowledge of which proteins are involved in interactions of sites of a locus under study. That is why many studies involving 3C analysis have been assisted by ChIP experiments (4, 8, 10).

Recently, a method was developed to, at the same time, allow determining which genomic sites interact and suggesting candidate proteins mediating the interaction. This method was called a ChIP-loop assay (16). It is a combination of 3C and ChIP and is performed as follows. Cells are fixed with formaldehyde, lysed, and the cross-linked chromatin is purified of free proteins by urea gradient ultracentrifugation (16). Purified cross-linked chromatin is digested with a restriction enzyme and subjected to precipitation with specific antibodies including standard for ChIP steps of preclearing with protein-A/G beads, incubation with specific (or preimmune) antibodies, and final washing of the beads. Then the beads with precipitated chromatin are resuspended in ligation buffer, and the chromatin is ligated by T4 DNA ligase. Ligation products are then purified and analyzed as in usual 3C experiments (**Fig. 12.3**) (16–18).

Thus, ChIP-loop assay allows segregating from a panel of tested proteins those that may take part in DNA loop organization. However, it should be understood that protein being cross-linked

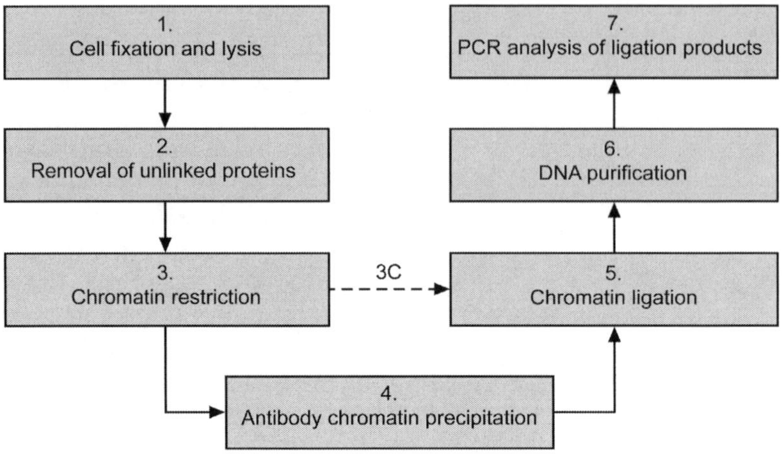

Fig. 12.3. The main steps of ChIP-loop assay. In 3C analysis the antibody precipitation step is omitted, which is shown by an arrow with a dotted stem.

to interacting DNA fragments is not sufficient for assuming protein participation in DNA loop formation; the protein may bind DNA nearby interacting sites but do not mediate the interaction. To that end, additional experiments may be helpful, for example blocking protein expression and examining if the characteristic spatial configuration of the DNA region under study is lost (19). Nevertheless in some aspects, ChIP-loop assay provides a better insight than 3C and ChIP do when used apart. It concerns the situation when a positive ChIP signal originates from a cell subpopulation where the locus examined has a linear configuration, whereas a positive 3C signal corresponds to another cell subpopulation in which the protein does not bind to the corresponding DNA sites.

2. Materials

2.1. Cell Fixation and Lysis

1. Equipment for cell culturing.
2. Materials necessary for single-cell suspension preparation.
3. PBS/FBS: PBS supplemented with 10% fetal bovine serum (FBS) (if fixation is not carried out in cell growth medium).
4. Fix solution: solution of formaldehyde in the cell suspension buffer.
5. 2.5 M glycine.
6. PBS.
7. Cell lysis buffer: 10 mM Tris-HCl (pH 8.0), 10 mM NaCl, 0.2% NP-40, fresh protease inhibitor cocktail.

2.2. Treatment of DNA Cross-Linked to Proteins with Restriction Endonuclease(s)

1. Highly concentrated restriction enzyme(s) of choice.
2. 10X restriction buffer.
3. 20% SDS (w/v).
4. 20% Triton-X-100 (v/v).

2.3. Ligation of DNA Cross-Linked to Proteins

1. 20% Triton-X-100 (v/v).
2. T4 DNA ligase.
3. 10X ligation buffer.

2.4. Cross-Link Reversion and DNA Purification

1. Proteinase K.
2. RNase A.
3. Phenol (pH 8.0).
4. Chloroform.
5. 3 M sodium acetate (pH 5.2).
6. Ethanol.
7. 10 mM Tris-HCl (pH 7.5).

2.5. TaqMan Real-Time PCR Analysis of Ligation Products

1. Primers.
2. TaqMan probes.
3. dNTPs.
4. 10X *Taq* polymerase buffer.
5. Hot-start *Taq* DNA polymerase.

3. Methods

3.1. Cell Fixation and Lysis

See **Note 1** about fixation principle.

1. Prepare single-cell suspension containing 1×10^7 cells in 2–8 mL of growth medium or PBS/FBS (*see* **Note 2**).
2. Add freshly made fix solution to obtain a final volume of 10 mL and formaldehyde concentration of 2% and incubate for 10 min at room temperature with slow agitation (*see* **Note 3**).
3. Stop fixation by adding 2.5 M glycine to a concentration of 0.125 M and cool the sample on ice.
4. Harvest the cells by centrifugation for 5–10 min at 200–300*g* and 4°C, wash with 10 mL of cold PBS and harvest again.
5. Resuspend the cell pellet in 5 mL of cold lysis buffer and incubate for 10 min on ice to release nuclei (*see* **Note 4**).
6. Harvest the nuclei by centrifugation for 5 min at 600*g* at 4°C (*see* **Note 5**).

3.2. Treatment of DNA Cross-Linked to Proteins with Restriction Endonuclease(s)

See **Note 6** about choosing appropriate restriction enzymes and buffers.

1. Suspend the nuclear pellet in 0.5 mL of 1.2X restriction buffer.
2. Add 20% SDS to a final concentration of 0.3% and incubate at 37°C for 1 h with vigorous shaking (for example, 1,400 rpm on a temperature-controlled shaker, Eppendorf) (*see* **Note 7**).
4. Add 20% Triton-X-100 to a concentration of 1.8% and incubate at 37°C for 1 h with shaking (1,400 rpm) to sequester the SDS.
5. Add 400–1,500 units of a highly concentrated restriction enzyme and carry out restriction overnight at 37°C with shaking (1,400 rpm) (*see* **Note 8**).
6. Inactivate the enzyme by addition of 20% SDS to a concentration of 1.3% and incubation at 65°C for 20 min (*see* **Note 9**).

3.3. Ligation of DNA Cross-Linked to Proteins

1. Mix the "restriction" solution with 7 mL of 1X ligation buffer in a 50 mL tube (*see* **Note 10**).
2. Add 20% Triton-X-100 to a final concentration of 1% and incubate at 37°C for 1 h with shaking (for example, 400 rpm in a bacterial incubator, the tube set upright) to sequester the SDS.
3. Add 100 units of T4 DNA ligase and incubate first for 4–5 h at 16°C and then for 30 min at room temperature with slow agitation.

3.4. Cross-Link Reversion and DNA Purification

1. Reverse cross-links by overnight incubation of the whole sample (~8 mL) at 65°C in the presence of 300 μg of proteinase K.
2. Add 300 μg of RNase A and digest the RNA at 37°C for 30–45 min.
3. Extract the solution successively with 7 mL of phenol, phenol–chloroform, and chloroform in a 15 mL tube. Centrifugation at each step is performed for 10 min at 2,000–3,000*g* and room temperature.
4. Mix the solution with the same volume of pure water in a 50 mL tube (*see* **Note 11**).
5. Add 3 M sodium acetate (pH 5.2) to a concentration of 0.2 M, 2 volumes of 96% ethanol and incubate overnight at –70°C for DNA precipitation.
6. Precipitate the DNA by centrifugation for 1 h at 3,200*g* and 4°C (*see* **Note 12**).

7. Wash the DNA pellet with 10 mL of cold 70% ethanol and centrifuge for 20 min at 3,200*g* and 4°C.
8. Dry the DNA pellet and dissolve in 150 μL of 10 mM Tris-HCl (pH 7.5) carefully washing the tube bottom (*see* **Note 13**).

3.5. TaqMan Real-Time PCR Analysis of Ligation Products

1. Design primers and TaqMan probes for analysis of ligation products (*see* **Note 14**). Annealing temperatures of primers and TaqMan probes should be 56–60°C and 68–70°C, respectively, and the size of PCR products should be within the range of 50–250 bp.
2. Prepare a reaction mixture that should contain in a final volume of 20 μL a DNA matrix, 1X PCR buffer, 0.5 μM of each primer, 0.25 μM of TaqMan probe, 0.2 mM of each dNTP, and 0.75 unit of hot-start Taq DNA polymerase. The PCR is carried out as follows: initial denaturation for 5 min at 94°C; 50–60 cycles of 15 s at 94°C, 60 s at 60°C, plate read. As a matrix, use a 3C template in parallel with a random ligation template (*see* **Notes 15–17**).
3. Determine a relative amount of corresponding ligation product in a 3C template.

3.6. Random Ligation Matrix Preparation

1. Set up a 100 μL restriction reaction including 5–10 μg of BAC (YAC), 1X restriction buffer and 25–50 units of a restriction enzyme(s) used for 3C analysis, and digest the DNA for 3 h at 37°C.
2. Extract the solution successively with one volume of phenol, phenol:chloroform, and chloroform (use centrifugation for 3 min at 12,000*g* and room temperature at each step).
3. Add 3 M sodium acetate (pH 5.2) to a concentration of 0.2 M, 2 volumes of 96% ethanol and incubate for at least 1 h at –70°C for DNA precipitation.
4. Precipitate the DNA by centrifugation for 15 min at 12,000*g*.
5. Wash the DNA pellet with 0.5 mL of 70% ethanol and centrifuge for 5 min at 12,000*g*.
6. Dissolve the DNA pellet in 50 μL of 1X ligation buffer, add 20 units of T4 DNA ligase, and incubate for 4 h at 16°C.
7. Dilute the solution with 1 volume of pure water and repeat phenol–chloroform extraction and ethanol precipitation as described above.
8. Dilute the DNA pellet in 100 μL of 10 mM Tris-HCl (pH 7.5).
9. In the same way digest and religate pure genomic DNA of organism under study and use it for adjusting DNA concentration in BAC (YAC) random ligation templates (*see* **Note 15**).

3.7. 3C Control Experiments

3.7.1. Control of Restriction Efficiency

Usually restriction does not go to completion because of severity of reaction conditions. Efficiency of digestion as a rule does not exceed 80–90%. Besides, it may differ from site to site because of blocking of sites by the proteins cross-linked nearby or due to some other reasons, and if so it will influence the amount of specific ligation products. Mixed digestion of 3C templates with two different restriction enzymes producing compatible DNA ends may cause additional problems. Therefore, the efficiency of digestion should be checked for different sites throughout the locus under study. For this purpose, Southern blot analysis or PCR-stop analysis of restriction products can be used.

3.7.2. Control of Measurement of Cross-Linking Frequencies by the Amount of Ligation Products

Measuring of cross-linking efficiency by the amount of the corresponding ligation product seems to depend on a condition of the two DNA ends whose ligation is regarded – their lengths, integrity of ligation sites, and presence of cross-linked proteins. These properties determine mobility of cross-linked DNA fragments and their ability to reach each other, as well as accessibility of the cohesive ends to DNA ligase. So, it is desirable to repeat 3C experiments with primers designed to anneal at the other side of restriction fragments or/and with another restriction enzyme and to see whether the results are similar.

3.7.3. Control of Quality and Quantity of a 3C Template

If 3C analysis is carried out on different types of cells, it should be taken into consideration that quality and quantity of 3C DNA may vary depending on the cell type: differences in internal cellular conditions may cause variations in efficiency of fixation, restriction and ligation as well as simply in degree of integrity of DNA subjected to degradation by cellular enzymes. So, as an internal control of quality of experimental procedures, 3C analysis is performed on a locus that can reasonably be assumed to have similar spatial organization in all cell types used. Housekeeping genes transcribed at the same level in different cell types are usually selected as such control loci (2, 4, 6). It is recommended that the control locus is located far from loci that are known to have different transcriptional status in the cells under study. It is better to perform 3C analysis for several pairs of restriction fragments of a control locus: if differences (or absence of differences) in cross-linking frequencies observed in different cell types are reproduced independently of a fragment combination, then the results can be thought reliable (5). If so, cross-linking frequencies measured for different fragment combinations within a cell type are averaged and the resulting figure is considered as the relative cross-linking value of 1.

Another way of getting such an internal control is based on the assumption that adjacent restriction fragments must exhibit similar cross-linking frequencies independently of a cell type and gene activity (4). Thus, a cross-linking frequency of adjacent restriction fragments can be used to normalize the data of 3C analysis obtained for different cell types. In this case, too, it is recommended to measure cross-linking frequencies of several pairs of fragments and average the results. Again it should be checked that differences (or absence of differences) observed in different cell types are reproduced independently of a fragment pair.

3.7.4. Mock Controls

Mock controls are usually done by omitting one of three important steps of the analysis – fixation (when formaldehyde is not added), restriction (a restriction enzyme is not added), or ligation (ligase is not added). In the first case amplification products, if any, must be detected in noticeably smaller amounts comparing with the normal experiment; in the second and third cases there must not be amplification products at all.

3.8. Interpretation of 3C Data

Data of 3C analysis are usually represented as a graph of dependence of relative cross-linking frequency of an anchor restriction fragment (i.e., the fragment bearing an anchor primer) and other fragments on position of these fragments in a locus (**Fig. 12.4**). Cross-linking frequencies are then conceived as frequencies with which genomic sites interact in nuclear space. However, it should be taken into consideration that different nucleoprotein complexes may be

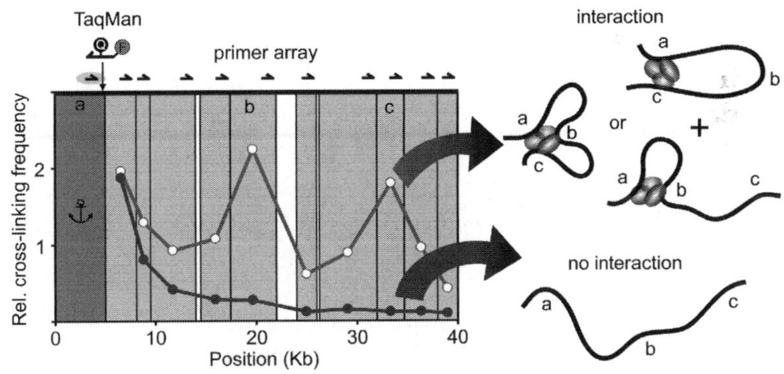

Fig. 12.4. Representation and interpretation of 3C data. A hypothetical case is illustrated. The graph shows the dependence of relative cross-linking frequency of one selected restriction fragment (anchor fragment, dark shadowing) and other fragments (test fragments, light shadowing) on position of the test fragments in the locus. Black vertical lines show positions of restriction sites. A primer array used for PCR is shown above the graph. The anchor primer is marked by an oval. Next to it the site of TaqMan probe annealing is shown. The results presented on the graph are interpreted as shown from the right. The curve with unfilled points reflects the situation when site "a" interacts with sites "b" and "c". However, this curve does not answer the question whether it is simultaneous interaction ("a+b+c") or only superposition of "a+b" and "a+c" interactions. The curve with filled points reflects the situation when the locus has a linear configuration.

composed of different proteins and DNA sequences and have different structure. So efficiency of formaldehyde fixation of one complex may not be just the same as efficiency of fixation of another complex even if these two complexes are characterized by similar lifetime.

To interpret adequately the results of 3C analysis it is necessary to take into account some more important moments. First, restriction fragments located nearby in DNA sequence or adjacent to fragments containing really interacting sites are situated close to each other in nuclear space and may get casually cross-linked. This results in appearance of ligation products that do not correspond to specific spatial interactions. Ligation products of even very remote in DNA sequence fragments are usually detected in the amounts exceeding random ligation background. The closer in DNA sequence fragments are, the higher cross-linking frequency these fragments demonstrate independently of their involvement in the formation of chromatin hubs, and the more difficult it is to discriminate between a specific interaction and an accidental one. So, the results of 3C analysis become less and less reliable with the decrease in distance between analyzed restriction fragments.

Second, as we mentioned before, restriction usually does not go to completion. Hence some fragments may get to be in one nucleoprotein complex as a result of their not being cut from cross-linked fragments. Obviously, such "false" cross-linking concerns mainly the fragments that are situated in the neighborhood of interacting fragments but do not participate in the interaction themselves. Cross-linking frequencies measured for these fragments are thus overestimated, while the other cross-linking frequencies are underestimated because of incomplete restriction. As a result, peaks of 3C curves smear.

Additional problems may be caused by inequality of digestion efficiency for different sites of a locus under study. The simplest way to correct for this is to compare each cross-linking frequency with digestion efficiencies for the corresponding restriction sites.

If digestion efficiency is low, the situation is additionally complicated when the anchor fragment or test fragment or both of them are followed by one or several small restriction fragments. If not digested from the analyzed DNA fragments, these short fragments may give rise to additional (longer) amplicons. PCR signals from amplicons of different size are summarized and cannot be discriminated when real-time PCR is employed. As a result, corresponding cross-linking frequencies may be overestimated.

Moreover, 3C analysis allows estimating only relative probabilities of interaction of different restriction fragments within the area under study. The possibility of existence of an alternating three-dimensional configuration of a studied locus in the same cells should be taken into account. In addition, it should be considered that three-dimensional organization of this locus can differ in two copies of homologous chromosomes in the same cells and also in different cells present in the population.

4. Notes

1. Cross-linking of protein and DNA, such as they are in living cells, is performed by treatment of cells with formaldehyde that reacts with amino and imino groups of DNA and proteins forming DNA–protein and protein–protein links.
2. Fixation is performed directly in growth medium or after some manipulations needed; e.g., for separation of cells of interest from other cells or cell dissociation if tissue is operated. Some protocols propose to perform fixation on isolated nuclei (1), but we think that the less influences cells undergo before fixation, the less probability that the specific spatial organization of chromatin is disturbed.
3. Formaldehyde concentration and time of incubation may be varied to get more or less fixed chromatin.
4. The process of lysis can be monitored by staining cells with trypan blue.
5. The nuclear pellet can be frozen in liquid nitrogen and stored at –70°C.
6. When selecting a restriction enzyme, the following should be considered:
   (a) A size of a restriction fragment to be analyzed should not be very small (less than 0.1 Kb) or very big (more than 15 Kb).
   (b) DNA ends produced must not be blunt.
   (c) It is acceptable to digest DNA with more than one restriction enzyme if resulting DNA ends are compatible (i.e., can be cross-ligated).
   (d) Because fixed chromatin but not pure DNA is subjected to treatment with restriction enzymes, SDS is used to disperse chromatin via removal of unlinked proteins and provide access of the enzyme to DNA. SDS is then sequestered by Triton-X-100, but some restriction enzymes do not work well or do not work at all in the resulting solution. *Hind* III, *Eco*RI, and *Bgl* II work correctly; *BamH* I, *Spe*I, *Pst*I, and *Nde*I work with lesser efficiency (20); enzymes preferring low-salt buffers, for example *Sac*I, as a rule do not work. Moreover, when

low-salt restriction buffer is used, the nuclei tend to form an insoluble pellet. The efficiency of restriction of cross-linked templates with the selected enzyme(s) should be tested experimentally (*see* **Section 3.7.1**).

7. It is possible to decrease concentration of SDS if it helps to achieve better efficiency of restriction. However, the concentration of SDS should not be lower than 0.1%. Otherwise the solubilization of non-cross-linked proteins will be incomplete.
8. Even very intense shaking (1,400 rpm) does not affect enzyme work, but on the contrary provides proper mixing the suspension which, as a rule, is quite heterogeneous until the cross-link reversion step.
9. Most cross-links are thought to be preserved after incubation at 65° C for such a short period of time as 20 min (Dr. Erik Splinter, personal communication).
10. Dilution in the ligation buffer gives DNA concentration of about several ng per μL that is enough to provide preferably intramolecular ligation (ligation between cross-linked DNA fragments) (20).
11. Dilution by water before ethanol precipitation is aimed to reduce precipitation of DTT, a component of the ligation buffer, during centrifugation (Dr. Erik Splinter, personal communication). Replacement of DTT by β-mercaptoethanol in the ligation buffer might solve this problem, but it still remains to check whether ligase works normally in such buffer.
12. It was shown that centrifugation at 3,200*g* is sufficient to precipitate 3C DNA. Perform centrifugation at that force using a cellular centrifuge if you do not have at your disposal a suitable high-speed centrifuge to carry out DNA precipitation from 50 mL of solution.
13. The DNA pellet may be spread on the conical bottom of a 50 mL tube and invisible.
14. Ligation products are usually present in 3C templates in such low amounts that it is more reliable to quantify them by real-time PCR rather than by semi-quantitative PCR methods using ethidium bromide staining or radioactively labeled primers (10, 19, 21). Primers for PCR are designed to anneal at the same ends (left or right) of restriction fragments and face outwards. In this way head-to-head ligation products are analyzed. One primer is selected as an anchor primer which is sequentially used to carry out PCR with all other primers. Unidirectionality of the primers eliminates the possibility of generating PCR products because of partial digestion and subsequent ligation through circularization. It is desirable

that any primer combination is operable. Therefore, all primers should have similar annealing temperature and should not form strong homo- and heterodimers. To design such a primer set is not at all a simple task. A possibility remains that one primer pair or another will not work well producing, for example, considerable background of non-specific PCR products. That is why it is preferable to carry out real-time PCR using sequence-specific DNA probes, such as TaqMan, rather than non-specific DNA dyes (SYBRgreen, etc.).

A TaqMan probe is an oligonucleotide that is designed to anneal between PCR primers and bears a fluorescent dye at the 5' end and a quencher dye inside or at the 3' end. When Taq polymerase meets with the probe, the enzyme cleaves it, which results in separation of the fluorescent and quencher dyes and thus in increase of fluorescence (*see* **Fig. 12.5D**). In the 3C assay a TaqMan probe is designed to anneal downstream to the anchor primer within the same restriction fragment (**Fig. 12.4**). In such

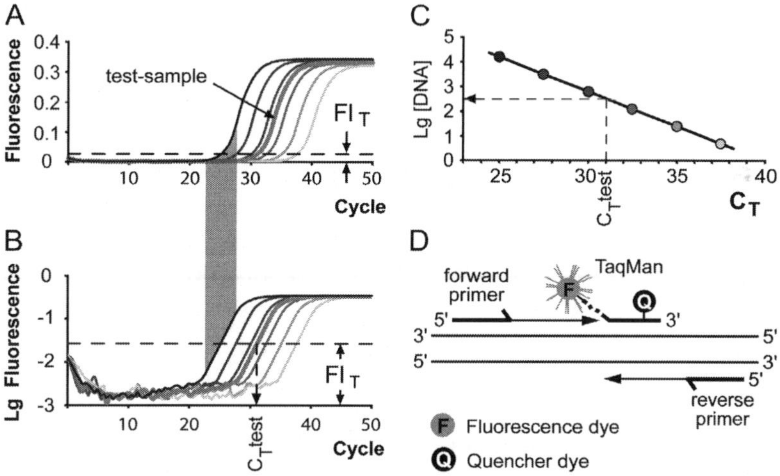

Fig. 12.5. Determination of DNA quantity by real-time PCR with TaqMan probes. A generalized experiment is discussed. (**A**) and (**B**). The dependences of the intensity of reporter fluorescence on a number of amplification cycles; A – normal scale, B – logarithmic scale. The six thin curves of different colour intensity illustrate the results of six real-time PCR amplifications carried out on five-fold dilutions of a standard DNA of known quantity. The thick curve corresponds to the test-sample with unknown DNA quantity that is supposed to be measured. The shadowing shows an exponential phase of the fluorescence growth for the PCR carried out on non-diluted standard template (**A**). This region is seen as a linear part of the curve when the logarithmic scale is used (**B**). A dotted line indicates the fluorescence level corresponding to the middle of the exponential phase (threshold fluorescence, Fl_T). An amplification cycle by which the threshold fluorescence is achieved is called threshold cycle C_T. The threshold cycle for the sample with unknown DNA quantity is designated as test C_T ($C_T$test). (**C**) A calibration curve showing dependence of a threshold cycle on a common logarithm of the start DNA quantity. The points representing successive dilutions of the DNA standard are indicated by decreasing of intensity of filling of the corresponding circles on the calibration curve. Applying $C_T$test to the graph gives the relative amount of DNA in the sample. (**D**) The principle of real-time PCR with TaqMan probes. A TaqMan probe bears a fluorescence and a quencher dye and anneals between PCR primers. Cleavage of the probe by *Taq*-polymerase results in the separation of the fluorophore from the quencher, and consequently in a fluorescence increase.

a way, the same TaqMan probe can be used for all PCR with the given anchor primer. The direction of the TaqMan probe is recommended to be opposite to the anchor primer (19).

15. To judge an amount of specific DNA by the amount of the amplification product, once again it should be taken into account that different primer pairs may work with different efficiency. To correct for this, all primer pairs should be tested on the matrix in which for each primer pair there is an equal amount of the target DNA sequence to amplify. An equimolar mix of all possible ligation products can be used as such a matrix. Originally, in the yeast system, this kind of matrix was prepared by restriction and religation of pure genomic DNA (1). But genomes of higher eukaryotes are frequently thousands times larger, and restriction and religation of the whole genome will result in appearance of uncountable amount of ligation product variants, each variant being present in minute amounts very difficult to amplify. The problem is solved by using DNA of bacterial or yeast artificial chromosomes bearing the locus of interest. The BAC (or YAC) is digested with a selected restriction enzyme(s) and, after inactivation or removal of the restriction enzyme(s), religated at a high DNA concentration (hundreds of ng per μL) to allow intermolecular ligation (20, 22). Another way to prepare a random ligation matrix is to amplify the DNA fragments bearing restriction sites of interest, purify the amplification products, mix them in equimolar amounts, digest, and religate (20). This way seems to be much more time-consuming.

The PCR is performed, individually for each primer pair, on series of successive dilutions of the random ligation matrix (we usually used six five-fold dilutions of BAC starting with 1,250 pg per reaction and finishing with 0.4 pg), and the dependence of intensity of reporter fluorescence on a cycle of amplification is plotted (**Fig. 12.5A**). For each dilution a threshold cycle C_T is determined, the cycle at which the fluorescence achieves some selected value within an exponential phase of fluorescence growth, for example corresponding to the middle of this phase. It is easier to operate with the graph on a logarithmic scale of fluorescence. In that case, the exponential phase is seen as a linear part of the curve (**Fig. 12.5B**). After C_Ts have been determined, the calibration curve is plotted representing the dependence of logarithm of DNA quantity on C_T. Theoretically, such a curve must be a descending line (**Fig. 12.5C**). In parallel, PCR is performed on several dilutions of a 3C template (for example, 5, 15, and 45 ng of DNA per reaction), and the rate of PCR product accumulation, in terms of C_T, is applied to the calibration curve in order to determine a relative amount of the corresponding ligation product in the 3C template (**Fig. 12.5C**). The values obtained should not exceed the frames of the calibration curve. The digested and religated pure genomic DNA is added to each reaction with the random ligation matrix so that final DNA concentration is adjusted to the amount of DNA used for PCR with 3C templates (for example, 20 ng (20)), because PCR amplification is influenced by the amount of genomic DNA present in 3C templates. To be sure that the calibration curve is reliable, one may check if the values obtained for any two dilutions of 3C templates indeed differ as much as the amounts of DNA in these two dilutions.

16. To maintain conditions of amplification, all components of real-time PCR, including water, should be aliquoted and stored at –20°C.
17. Each PCR is carried out in a triple or quadruple repeat and corresponding results are averaged.

References

1. Dekker, J., Rippe, K., Dekker, M. and Kleckner, N. (2002) Capturing chromosome conformation. *Science* **295**, 1306–1311.
2. Tolhuis, B., Palstra, R. J., Splinter, E., Grosveld, F. and de Laat, W. (2002) Looping and interaction between hypersensitive sites in the active beta-globin locus. *Mol. Cell* **10**, 1453–1465.
3. Eivazova, E. R. and Aune, T. M. (2004) Dynamic alterations in the conformation of the *Ifng* gene region during T helper cell differentiation. *Proc. Natl. Acad. Sci. U.S.A.* **101**, 251–256.
4. Liu, Z. and Garrard, W. T. (2005) Long-range interactions between three transcriptional enhancers, active V gene promoters, and a 3' boundary sequence spanning 46 kilobases. *Mol. Cell. Biol.* **25**, 3220–3231.
5. Gavrilov, A. A. and Razin, S. V. (2008) Spatial configuration of the chicken α-globin gene domain: immature and active chromatin hubs *Nucleic Acids Res* **36**, 4629–40.
6. Spilianakis, C. G. and Flavell, R. A. (2004) Long-range intrachromosomal interactions in the T helper type 2 cytokine locus. *Nat. Immunol.* **5**, 1017–1027.
7. Palstra, R. J., Tolhuis, B., Splinter, E., Nijmeijer, R., Grosveld, F. and de Laat, W. (2003) The beta-globin nuclear compartment in development and erythroid differentiation. *Nat. Genet.* **35**, 190–194.
8. Vakoc, C., Letting, D. L., Gheldof, N., Sawado, T., Bender, M. A., Groudine, M., Weiss, M. J., Dekker, J. and Blobel, G. A. (2005) Proximity among distant regulatory elements at the beta-globin locus requires GATA-1 and FOG-1. *Mol. Cell* **17**, 453–462.
9. Zhou, G. L., Xin, L., Song, W., Di, L. J., Liu, G., Wu, X. S., Liu, D. P. and Liang, C. C. (2006) Active chromatin hub of the mouse alpha-globin locus forms in a transcription factory of clustered housekeeping genes. *Mol. Cell Biol.* **26**, 5096–5105.
10. Vernimmen, D., De Gobbi, M., Sloane-Stanley, J. A., Wood, W. G. and Higgs, D. R. (2007) Long-range chromosomal interactions regulate the timing of the transition between poised and active gene expression. *EMBO J.* **26**, 2041–2051.
11. Spilianakis, C. G., Lalioti, M. D., Town, T., Lee, G. R. and Flavell, R. A. (2005) Interchromosomal associations between alternatively expressed loci. *Nature* **435**, 637–645.
12. Ling, J. Q., Li, T., Hu, J. F., Vu, T. H., Chen, H. L., Qiu, X. W., Cherry, A. M. and Hoffman, A. R. (2006) CTCF mediates interchromosomal colocalization between Igf2/H19 and Wsb1/Nf1. *Science* **312**, 269–272.
13. Zhao, Z., Tavoosidana, G., Sjölinder, M., Göndör, A., Mariano, P., Wang, S., Kanduri, C., Lezcano, M., Sandhu, K. S., Singh, U., Pant, V., Tiwari, V., Kurukuti, S. and Ohlsson, R. (2006) Circular chromosome conformation capture (4C) uncovers extensive networks of epigenetically regulated intra- and interchromosomal interactions *Nat. Genet.* **38**, 1341–1347.

14. Simonis, M., Klous, P., Splinter, E., Moshkin, Y., Willemsen, R., de Wit, E., van Steensel, B. and de Laat, W. (2006) Nuclear organization of active and inactive chromatin domains uncovered by chromosome conformation capture-on-chip (4C). *Nat. Genet.* **38**, 1348–1354.

15. Dostie, J., Richmond, T. A., Arnaout, R. A., Selzer, R. R., Lee, W. L., Honan, T. A., Rubio, E. D., Krumm, A., Lamb, J., Nusbaum, C., Green, R. D. and Dekker, J. (2006) Chromosome Conformation Capture Carbon Copy (5C): a massively parallel solution for mapping interactions between genomic elements. *Genome Res.* **16**, 1299–1309.

16. Horike, S., Cai, S., Miyano, M., Cheng, J. F. and Kohwi-Shigematsu, T. (2005) Loss of silent-chromatin looping and impaired imprinting of DLX5 in Rett syndrome. *Nat. Genet.* **37**, 31–40.

17. Cai, S., Lee, C. C. and Kohwi-Shigematsu, T. (2006) SATB1 packages densely looped, transcriptionally active chromatin for coordinated expression of cytokine genes. *Nat. Genet.* **38**, 1229–1230.

18. Kurukuti, S., Tiwari, V. K., Tavoosidana, G., Pugacheva, E., Murrell, A., Zhao, Z., Lobanenkov, V., Reik, W. and Ohlsson, R. (2006) CTCF binding at the H19 imprinting control region mediates maternally inherited higher-order chromatin conformation to restrict enhancer access to Igf2. *Proc. Natl. Acad. Sci. U.S.A.* **103**, 10684–10689.

19. Splinter, E., Heath, H., Kooren, J., Palstra, R. J., Klous, P., Grosveld, F., Galjart, N. and de Laat, W. (2006) CTCF mediates long-range chromatin looping and local histone modification in the beta-globin locus. *Genes Dev.* **20**, 2349–2354.

20. Splinter, E., Grosveld, F. and de Laat, W. (2004) 3C technology: Analyzing the spatial organization of genomic loci in vivo. *Methods Enzymol.* **375**, 493–507.

21. Wurtele, H. and Chartrand, P. (2006) Genome-wide scanning of HoxB1-associated loci in mouse ES cells using an open-ended Chromosome Conformation Capture methodology. *Chromosome Res.* **14**, 477–495.

22. Dekker, J. (2006) The 3 C's of Chromosome Conformation Capture: controls, controls, controls. *Nat. Methods* **3**, 17–21.

Chapter 13

Determining Spatial Chromatin Organization of Large Genomic Regions Using 5C Technology

Nynke L. van Berkum and Job Dekker

Abstract

Spatial organization of chromatin plays an important role at multiple levels of genome regulation. On a global scale, its function is evident in processes like metaphase and chromosome segregation. On a detailed level, long-range interactions between regulatory elements and promoters are essential for proper gene regulation. Microscopic techniques like FISH can detect chromatin contacts, although the resolution is generally low making detection of enhancer–promoter interaction difficult. The 3C methodology allows for high-resolution analysis of chromatin interactions. 3C is now widely used and has revealed that long-range looping interactions between genomic elements are widespread. However, studying chromatin interactions in large genomic regions by 3C is very labor intensive. This limitation is overcome by the 5C technology. 5C is an adaptation of 3C, in which the concurrent use of thousands of primers permits the simultaneous detection of millions of chromatin contacts. The design of the 5C primers is critical because this will determine which and how many chromatin interactions will be examined in the assay. Starting material for 5C is a 3C template. To make a 3C template, chromatin interactions in living cells are cross-linked using formaldehyde. Next, chromatin is digested and subsequently ligated under conditions favoring ligation events between cross-linked fragments. This yields a genome-wide 3C library of ligation products representing all chromatin interactions in vivo. 5C then employs multiplex ligation-mediated amplification to detect, in a single assay, up to millions of unique ligation products present in the 3C library. The resulting 5C library can be analyzed by microarray analysis or deep sequencing. The observed abundance of a 5C product is a measure of the interaction frequency between the two corresponding chromatin fragments. The power of the 5C technique described in this chapter is the high-throughput, high-resolution, and quantitative way in which the spatial organization of chromatin can be examined.

Key words: Chromosome conformation capture, chromatin looping, chromatin structure, long-range gene regulation, high-throughput.

1. Introduction

Tremendous efforts are underway to annotate all genes and other functional elements within the human genome as well as the genomes of several model organisms such as *C. elegans* and

Philippe Collas (ed.), *Chromatin Immunoprecipitation Assays*, Methods in Molecular Biology 567,
DOI 10.1007/978-1-60327-414-2_13, © Humana Press, a part of Springer Science+Business Media, LLC 2009

D. melanogaster (e.g., 1). These large-scale efforts will ultimately result in linear maps of genes and elements within the genome. Although these maps will provide crucial information about gene content and regulatory potential encoded within genomes, these maps will not directly reveal which regulatory elements regulate each gene. Identification of functional relationships between regulatory elements and target genes is complicated because elements such as enhancers, repressors, and insulators can be located at large genomic distances from their cognate target genes, and in some cases can even be located on other chromosomes (2, 3).

During the last couple of years it has been demonstrated that distant regulatory elements regulate genes through direct physical associations with target genes resulting in the formation of chromatin loops (e.g., 4, 5). Thus, the spatial organization of chromosomes plays a critical role in bringing together functionally related genomic elements. This further implies that mapping spatial organization of genomes will be a powerful approach to determine which (distant) regulatory elements regulate any given gene.

The spatial organization of chromosomes likely plays roles in many other nuclear processes as well. Most notably, the formation of condensed metaphase chromosomes involves organization of chromosomes in topologically reproducible compact shapes, which is essential for faithful chromosome segregation. During interphase, chromosomes are less condensed and appear diffuse and unorganized. However, recent studies have revealed that the interphase nucleus also displays significant spatial organization (6, 7). For instance, the three-dimensional arrangement of chromosomes and positioning of genes with respect to each other and to sub-nuclear structures such as the nuclear envelope is correlated with gene activity. However, currently it is unknown whether the three-dimensional organization of the nucleus directly affects gene expression, or whether it is a downstream consequence of gene expression. Comprehensive studies of the spatial organization of chromosomes during interphase and metaphase may provide new insights into these long-standing questions.

The spatial organization of chromosomes can be studied using microscopic methods to visualize the locations of genes inside individual cells. Although modern imaging technologies allow detailed analysis of positioning of specific loci inside living cells, the resolution of light microscopy is in many cases not sufficiently high to detect looping interactions between promoters and enhancers. To overcome these issues we developed the chromosome conformation capture (3C) technology that allows high-resolution analysis of chromatin looping events and chromosome conformation in general (8).

3C is used to measure physical interaction frequencies between small chromatin fragments in vivo (8–10). The method uses formaldehyde cross-linking to capture chromatin interactions in living cells (**Fig. 13.1**). Chromatin is then digested and

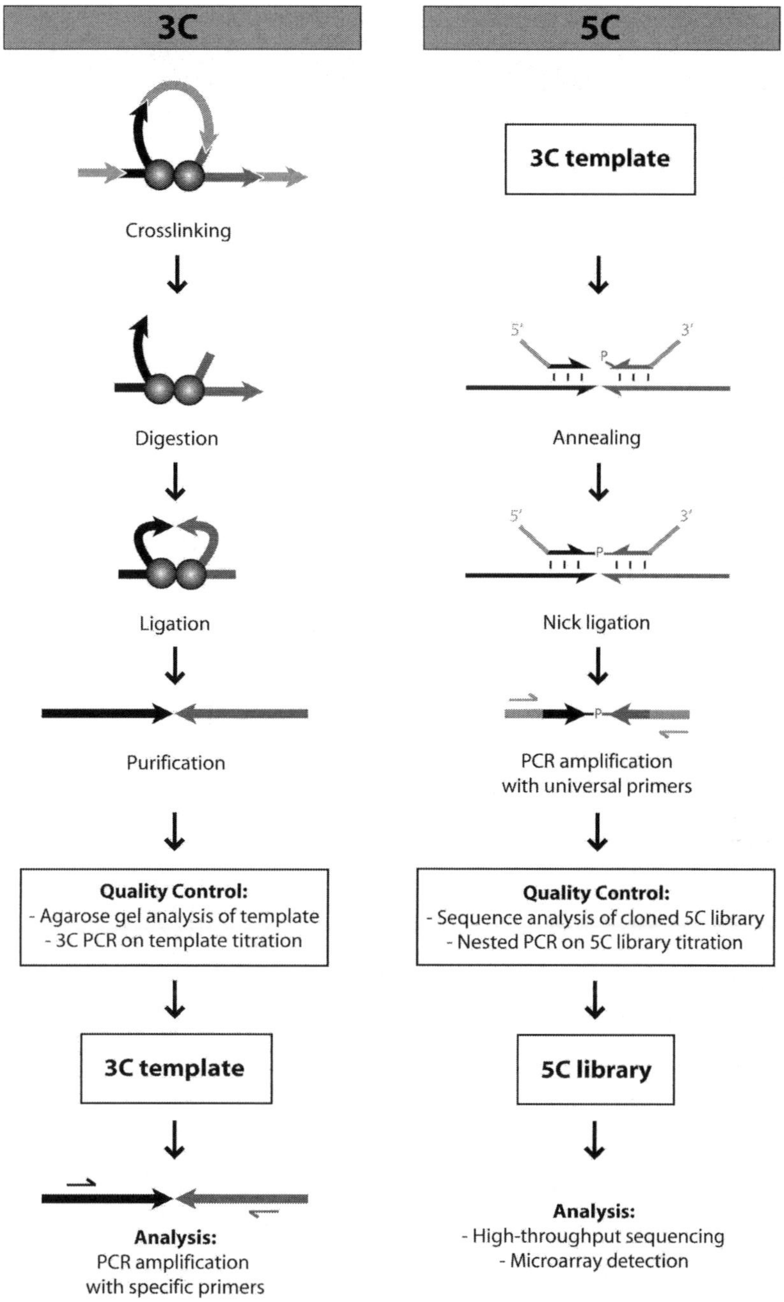

Fig. 13.1. Overview of the 3C (*left*) and 5C (*right*) methods. In 3C, chromatin contacts are cross-linked with formaldehyde. The DNA is digested with a restriction enzyme and is subsequently ligated under conditions favoring intramolecular ligation. After purification, the 3C template is run on gel and titrated in a 3C PCR experiment for quality control. Conventional 3C analysis is done in a one-to-one fashion by performing semi-quantitative PCR using specific primers for individual restriction fragments. The 5C method involves multiplex ligation mediated amplification of 5C primers detecting 3C ligation products. The quality of a 5C library is checked by nested PCR and by sequencing of individual cloned products. DNA interaction frequencies are examined in a many-to-many fashion by either microarray analysis or deep sequencing of the 5C library.

subsequently ligated under very dilute conditions to form intramolecular ligation products of cross-linked restriction fragments specifically. After ligation, cross-links are reversed and the DNA is purified. The resulting 3C template is a library of genomic ligation products reflecting all chromatin interactions that occur throughout the entire genome. Individual ligation products are detected through semi-quantitative PCR using pairs of primers that recognize specific combinations of ligated restriction fragments. The abundance of a particular ligation product in the 3C library and hence the yield of PCR product is in proportion to the nuclear interaction frequency between the two corresponding restriction fragments.

The 3C method has been proven to be a powerful tool to detect long-range interactions that are involved in gene regulation, both in *cis* and in *trans*(3, 11–14). The limitation of this technique is that interactions are analyzed in a "one-to-one" manner. Analysis of large numbers of interactions is time-consuming and labor intensive. Hence, 3C is most appropriate to study interactions between candidate loci located in a relatively small genomic region of up to a few hundred kilobases.

Recently, we have adapted 3C to allow for high-throughput and comprehensive analyses of interaction networks between large numbers of genomic elements (15, 16). The resulting "3C-carbon copy" or "5C" technique combines 3C with highly multiplexed ligation mediated amplification (LMA) and thereby permits millions of interactions to be tested simultaneously in a "many-to-many" fashion. In the 5C method, the relative abundance of the ligation products is detected by forward and reverse 5C primers that are designed to anneal directly up or downstream, respectively, of the newly formed restriction site in a 3C ligation product (**Fig.13.1**). After annealing to the 3C template, the primers are ligated by *Taq* DNA ligase, which specifically ligates nicked DNA. The ligated primer pairs form copies of the unique ligation junctions that characterize 3C ligation products present in the original 3C library, hence the name "3C carbon copy" or 5C. LMA allows for very high levels of multiplexing because thousands of forward and reverse primers can be combined to detect millions of unique chromatin interactions in a single assay. Using common tails on the 5C primers, all 5C ligation products can be simultaneously amplified with universal primers. The resulting product is a 5C library that can be subsequently analyzed by either deep-sequencing or microarray analysis.

Under ideal conditions, the abundance of a 5C product in the 5C library directly reflects the frequency with which the two corresponding chromatin segments interact in the nucleus. However, the efficiency of formation of 5C products can be biased due to differences in 5C primer annealing efficiency and PCR amplification of 5C ligation products. These biases are minimized by careful design of 5C primers so that they are all of equal length and all have identical melting temperatures. Any remaining technical biases can be corrected for by using a so-called control 5C library. A control 5C library is generated by performing 5C with a special control 3C library as template. The control 3C library is composed of randomly ligated fragments of the region of interest. As a result, every possible ligation product will be equally represented and any differences in abundance of 5C products in the 5C control library generated with the control 3C library will be due to annealing and amplification differences between 5C primers. Any biases in 5C library composition due to primer differences are removed by dividing the signal for each ligation product in the 5C library by the signal of the corresponding product in the control 5C library. This ratio is a quantitative measure for the interaction frequency of the two corresponding DNA fragments in the nucleus. These quantitative results make the 5C technique extremely powerful.

The 5C method can be used for different types of large-scale studies. The type of study will determine the design of a 5C experiment, because the combination of forward and reverse 5C primers defines the interactions that can be measured in the assay. For example, 5C can be used to determine a profile of chromatin interactions between one or a few fragments of interest and all other fragments within a large genomic domain. This approach can be used to discover the elements involved in regulation of one or a few specific genes. In this case, reverse primers are designed for the fragments containing the transcription start sites of the genes and forward primers are designed for all other fragments within the genomic domain of interest. Other studies can be focused on the identification of the global chromatin conformation of a specific region by determining dense networks of interaction frequencies between every pair of restriction fragments in that region. For this type of analysis, forward and reverse 5C primers are designed in an alternating manner for consecutive restriction fragments within the region of interest. Both types of data generated by 5C will give invaluable information about the spatial organization of chromatin and will provide new insights into the elements and mechanisms involved in long-range gene regulation.

2. Materials

2.1. Generation of a 3C Template

1. Deionized autoclaved water for use in all solutions.
2. 7×10^7–1×10^8 mammalian cells grown under appropriate conditions.
3. Cell culture medium.

4. 37% (v/v) formaldehyde (*see* **Note 1**). This product is flammable, can cause skin burns, is an eye irritant, and is toxic by inhalation. Therefore, formaldehyde should be handled with protective gear in a chemical fume hood.
5. 2.5 M glycine. Store at room temperature.
6. Lysis buffer: 10 mM Tris-HCl, pH 8.0, 10 mM sodium chloride, 0.2% (v/v) Igepal CA-630. Store at 4°C.
7. Protease inhibitor cocktail (Sigma P8340).
8. Dounce homogenizer (pestle A).
9. 10X restriction buffer (NEB).
10. Restriction enzyme (NEB) (*see* **Note 2**).
11. 1 and 10% (w/v) SDS. Store at room temperature.
12. 10% (v/v) Triton X-100. Store at room temperature.
13. 10X T4 ligation buffer: 500 mM Tris-HCl, pH 7.5, 100 mM magnesium chloride, 100 mM DTT. Store at –20°C (*see* **Note 3**).
14. 10 mg/mL bovine serum albumin (BSA).
15. 100 mM ATP. Store at –20°C.
16. T4 DNA ligase (300 cohesive end units/μL) (Invitrogen, cat. no. 15224-025).
17. 10 mg/mL proteinase K. Dissolve in 1X TE buffer, pH 8.0, aliquot, and store at –20°C.
18. Saturated phenol, pH 8.0. This product is a toxic and corrosive material. Wear protective gear and handle in a chemical fume hood. Store at 4°C.
19. Phenol pH 8.0:chloroform (1:1). This product is a toxic and corrosive material. Wear protective gear and handle in a chemical fume hood. Store at 4°C.
20. 1X Tris-EDTA (TE), pH 8.0: 10 mM Tris-HCl, pH 8.0: 1 mM EDTA, pH 8.0. Store at room temperature.
21. 3 M sodium acetate, pH 5.2. Store at room temperature.
22. 10 mg/mL of DNase-free, RNase A (Sigma, cat. no. R6513). Dissolve in water, aliquot, and store at –20°C.
23. 10X Tris-borate-EDTA (TBE) buffer: 0.89 M Tris base, 0.89 M boric acid, 0.02 M EDTA, pH 8.0.
24. 10 mg/mL ethidium bromide. This product should be regarded as mutagenic to man and could be carcinogenic. Therefore, ethidium bromide should be handled with protective gear.
25. 4X DNA loading buffer: 10% (w/v) ficoll, 0.17% (w/v) xylene cyanol (*see* **Note 4**).

2.2. Generation of a Control 3C Template

1. LB medium, pH 7.0: 1% (w/v) bacto-tryptone, 0.5% (w/v) bacto-yeast extract, 1% sodium chloride.
2. Large-construct DNA purification kit (Qiagen, cat. no. 12462).

2.3. Quality Control of 3C and Control Templates

1. 10X PCR buffer: 600 mM Tris-H_2SO_4, pH 8.9, 180 mM ammonium sulfate. Store at –20°C.
2. 50 mM magnesium sulfate. Store at –20°C.
3. dNTP mix (25 mM each) (Invitrogen, cat. no. 10297-018).
4. 80 μM 3C template titration primers (*see* **Note 5**). Dissolve in 1X TE buffer, pH 8.0, and store at –20°C.
5. 5 U/μL *Taq* DNA polymerase (NEB, cat. no. M0267L).

2.4. Preparation of a 5C Primer Pool

1. 10 U/μL T4 polynucleotide kinase (PNK) (NEB, cat. no. M0201S).
2. 10X T4 polynucleotide kinase reaction buffer (PNK buffer) (NEB, cat. no. M0201S): 700 mM Tris-HCl, pH 7.6, 100 mM magnesium chloride, 50 mM DTT.
3. 10 mM ATP. Store at –20°C.

2.5. Preparation of a 5C and Control Library

1. 1 μg/μL Salmon sperm DNA (SSD). Dilute in 1X TE, pH 8.0, aliquot, and store at –20°C (*see* **Note 6**).
2. 10X 5C annealing buffer (NEBuffer 4, NEB): 200 mM Tris-acetate, pH 7.9, 500 mM potassium acetate, 100 mM magnesium acetate, 10 mM DTT.
3. 10X *Taq* DNA ligase buffer (NEB, cat. no. B0208S): 200 mM Tris-HCl pH 7.6, 250 mM potassium acetate, 100 mM magnesium acetate, 100 mM DTT, 10 mM NAD^+, 1% (v/v) Triton X-100.
4. 40 U/μL *Taq* DNA ligase (NEB, cat. no. M0208S).
5. 10X PCR buffer II (Applied Biosystems, cat. no. N8080243).
6. 25 mM magnesium chloride.
7. Ampli *Taq* Gold DNA polymerase (Applied Biosystems, cat. no. N8080243).
8. 80 μM T3 (5′ TATTAACCCTCACTAAAGGGA 3′) and T7 (5′ TAATACGACTCACTATAGCC 3′) PCR primers (*see* **Note 7**).
9. MinElute PCR purification kit (Qiagen, cat. no. 28004).

2.6. Quality Control of a 5C Library: Nested PCR

1. 80 μM nested PCR primers (*see* **Note 8**).

2.7. Quality Control of a 5C Library: Cloning and Sequence Analysis

1. Zero Blunt TOPO PCR Cloning Kit (Invitrogen, cat. no. K2800-20).
2. 40 mg/mL Xgal in dimethylformamide (US Biological, cat. no. X1000-05).
3. QIAprep Spin Miniprep kit (Qiagen, cat. no. 27106).
4. –21 M13 sequencing primer (5′ TGTAAAACGACGGC-CAGT 3′).
5. Sequencing reagents.

3. Methods

3.1. Generation of a 3C Template

1. Grow between 7 × 10^7 and 1 × 10^8 cells under the preferred conditions in the appropriate medium.
2. Aspirate the medium and add 22.5 mL of fresh medium to the cells (*see* **Note 9**).
3. To cross-link chromatin, add 625 µL of 37% formaldehyde to obtain a 1% final concentration. Mix gently and incubate at room temperature for 10 min. Gently rock the plates every 2 min.
4. Stop the reaction by adding 1.25 mL of 2.5 M glycine. Mix gently and incubate at room temperature for 5 min, followed by incubation on ice for at least 15 min to stop cross-linking completely (*see* **Note 10**).
5. Scrape the cells from the plates with a cell scraper and transfer the cells to a 250 mL conical tube.
6. Centrifuge the cross-linked cells at 450*g* for 10 min. Discard the supernatant (*see* **Note 11**).
7. Mix 2 mL of ice-cold lysis buffer with 200 µL of protease inhibitor cocktail and add it to the cell pellet. Resuspend well and let the suspension sit on ice for at least 15 min to let the cells swell (*see* **Note 12**).
8. Dounce homogenize the cells by stroking 10 times, followed by incubation on ice for 1 min and subsequently stroking for another 10 times.
9. Transfer the suspension to two 1.7 mL centrifuge tubes, spin at 2,000*g* at room temperature for 5 min.
10. Discard the supernatant and wash each pellet with 1 mL cold 1X restriction buffer. Centrifuge at 2,000*g* for 5 min at room temperature.
11. Repeat the wash step in Step 10.

12. Discard the supernatant, resuspend each pellet in 500 μL of 1X restriction buffer, and pool the suspensions (*see* **Note 13**). Divide the supsension into 50 μL aliquots in 1.7 mL centrifuge tubes (~22 tubes) (*see* **Note 14**).
13. Add 312 μL of 1X restriction buffer per tube.
14. To remove the proteins that are not directly cross-linked to the DNA, add 38 μL of 1% SDS per tube. Mix well while avoiding air bubbles (*see* **Note 15**), incubate at 65°C for 10 min (*see* **Note 16**) and place tubes back on ice.
15. Quench the SDS by adding 44 μL of 10% Triton X-100 to each tube. Mix well but avoid air bubbles.
16. Add 400 U of restriction enzyme per tube, mix well, and digest the DNA overnight at the manufacturer recommended temperature (*see* **Note 17**).
17. Inactivate the restriction enzyme by adding 86 μL of 10% SDS and incubate at 65°C for 30 min (*see* **Notes 16** and **18**).
18. Meanwhile, prepare a ligation master mix containing $N\times$ (745 μL 10% Triton X-100, 745 μL 10X ligation buffer, 80 μL 10 mg/mL BSA, 80 μL 100 mM ATP, and 5960 μL water) for N tubes. Aliquot 7.61 mL ligation master mix in 15 mL conical tubes and put on ice.
19. Transfer the digestion product from Step 17 into each 15 mL conical tube. To ligate the DNA fragments, add 10 μL of T4 DNA ligase per tube. Mix by inverting the tubes several times, spin down shortly, and incubate at 16°C for 2 h.
20. Add 50 μL of 10 mg/mL proteinase K per tube. Mix by inverting the tubes multiple times, spin down shortly, and reverse the cross-linking by incubating at 65°C overnight.
21. Add 50 μL of 10 mg/mL proteinase K per tube and incubate at 65°C for an additional 2 h (*see* **Note 19**).
22. Pool two ligation mixtures into one clean 50 mL conical tube (~11 tubes). Add 20 mL of phenol, pH 8.0, to each tube, vortex for 2 min and spin the tubes at 2,500*g* for 10 min at room temperature.
23. Transfer the aqueous upper phase to a clean 50 mL conical tube (*see* **Note 20**). Add 20 mL of phenol pH 8.0:chloroform per tube, vortex each tube for 1 min, and spin the tubes at 2,500*g* for 10 min at room temperature.
24. Pool the aqueous phases into four 50 mL conical tubes (*see* **Note 21**). Adjust the volume to 50 mL per tube with 1X TE, pH 8.0 (*see* **Note 22**) and transfer the DNA solutions to four 250 mL screw-cap centrifuge tubes that are suitable for high-speed spinning.

25. Add 5 mL of 3 M sodium acetate, pH 5.2, per tube, mix and add 125 mL 100% ice-cold ethanol per tube. Mix gently and precipitate the DNA by incubating at –80°C for at least 1 h (*see* **Note 23**).
26. Pellet the DNA by spinning at 10,000*g* for 20 min at 4°C.
27. Discard the supernatant. Dissolve each pellet in 500 μL of 1X TE buffer, pH 8.0, and pool the DNA solutions.
28. Wash each tube with an additional 500 μL of 1X TE buffer, pH 8.0, and pool it with the DNA solution from the previous step. Mix and aliquot 500 μL into eight fresh 1.7 mL centrifuge tubes.
29. Add 500 μL of phenol pH 8.0:chloroform to each tube, vortex for 1 min, and spin at 18,000*g* for 5 min at room temperature.
30. Transfer 450 μL of each upper aqueous phase to a fresh 1.7 mL centrifuge tube and repeat Step 29.
31. Transfer 400 μL of each upper aqueous phase to fresh 1.7 centrifuge tubes (*see* **Note 21**). Add 40 μL of 3 M sodium acetate, pH 5.2, and vortex briefly. Add 1 mL of 100% ethanol, mix gently, and precipitate the DNA by placing the tubes at –80°C for at least 30 min.
32. Pellet the DNA by spinning at 18,000*g* for 20 min at 4°C.
33. Aspirate the supernatant and wash the pellets with 1 mL of 70% ethanol. Make sure the pellet is resuspended well to allow the salt in the pellet to dissolve into the ethanol. Spin at 18,000*g* for 15 min at 4°C.
34. Repeat Step 33 at least five times or until the volume of the pellet does not decrease anymore (*see* **Note 24**).
35. Remove the last traces of 70% ethanol and air-dry the pellets briefly.
36. Dissolve all pellets in a total volume of 1 mL of 1X TE buffer, pH 8.0.
37. To degrade RNA, add 1 μL of 10 mg/mL of (DNase-free) RNase A and incubate at 37°C for 15 min.
38. Prepare a 0.8% agarose/0.5X TBE gel containing 0.5 μg/mL ethidium bromide.
39. Load 0.1, 0.2, and 0.4 μL of the 3C template and 150 ng of a molecular weight standard (**Fig. 13.2A**).
40. After running the gel, estimate the 3C template concentration by comparing the intensities to the molecular weight standard (*see* **Note 25**).
41. Aliquot the 3C template and store it up to at least 2 years at –20°C.

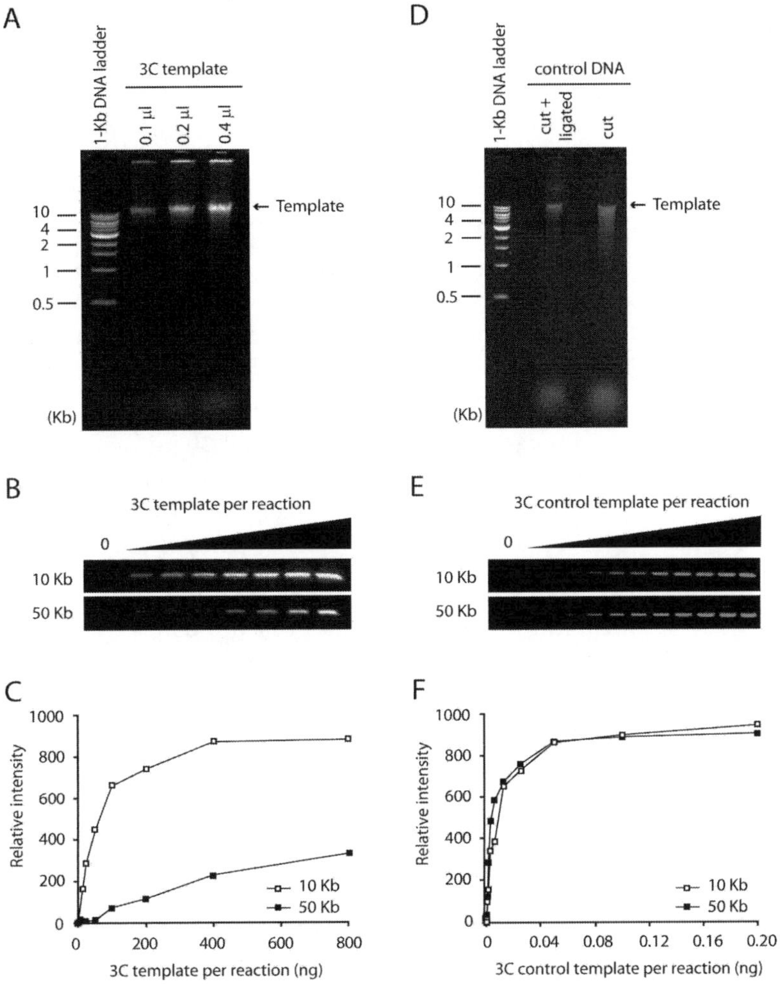

Fig. 13.2. Quality control of 3C and control templates. (**A**) Increasing amounts of 3C template were resolved on a 0.8% agarose gel. Typically a 3C template runs as a rather tight band larger than 10 kb. A DNA smear indicates poor ligation efficiency. (**B**) Agarose gel analysis and (**C**) quantification of a 3C template titration in a 3C PCR. Increasing amounts of 3C template were analyzed with two 3C primer pairs. One primer pair interrogates an interaction between fragments that are close (10 kb) to each other in the linear genome. The other primer pair examines the interaction between two more distant (50 kb) fragments. A titration curve should demonstrate a linear increase at the beginning of the curve and should reach a plateau with increasing amounts of 3C template. The curve of the PCR product representing the interaction between two adjacent restriction fragments should be above the curve of the PCR product testing the interaction between two, more distant fragments. (**D**) Digested and religated BAC DNA were resolved on a 0.8% agarose gel. Digested control BAC DNA usually runs as a smear on the gel. Religated control DNA should run as a band above 10 kb and the smear should be mostly gone. (**E**) Agarose gel analysis and (**F**) quantification of a control 3C template titration in a 3C PCR. Increasing amounts of control 3C template were analyzed with the same 3C primer pairs as for the 3C template. The titration curves of both PCR products should look more similar compared to the curves of a 3C template, indicating that all interactions in the control 3C template are represented equally.

3.2. Generation of a Control 3C Template

1. Select one or more BAC clones that cover the genomic region of interest with the least possible overlap between the clones and leaving the lowest number of gaps (*see* **Note 26**).
2. Purify the BAC DNA from 500 mL overnight cultures using the large-construct DNA purification kit.
3. If more than one BAC clone is used, mix the different clones in equimolar amounts (*see* **Note 27**).
4. Estimate the DNA concentration by running 1 μL of digested BAC DNA and a molecular weight standard of known concentration on a 0.8% agarose/0.5 TBE gel containing 0.5 μg/mL ethidium bromide. The concentration of the BAC DNA should be between 50 and 100 ng/ μL.
5. Prepare the following reaction: 20 μg BAC DNA, 160 μL 10X restriction buffer, 800 U restriction enzyme, and add water till 1.6 mL. Mix well and digest the BAC DNA overnight at the manufacturer recommended temperature (*see* **Note 28**).
6. Split the samples in 4X 400 μL in fresh 1.7 mL centrifuge tubes.
7. Add 500 μL of phenol pH 8.0:chloroform, vortex for 30 s, and spin at 18,000*g* for 5 min at room temperature.
8. Transfer the aqueous phases to fresh 1.7 mL centrifuge tubes. Add 40 μL of 3 M sodium acetate, pH 5.2, and vortex briefly. Add 1 mL of 100% ethanol, mix gently, and precipitate the DNA by placing the tubes at –20°C for at least 15 min.
9. Pellet the DNA by centrifuging at 18,000*g* for 20 min at 4°C.
10. Wash the pellets with 1 mL of 70% ethanol and spin at 18,000*g* for 15 min at 4°C.
11. Resuspend each pellet in 161 μL water and dissolve the BAC DNA by incubating at 37°C for 15 min.
12. Take 4 μL of each tube and keep that separate for future gel analysis.
13. Prepare the following ligation reactions: 157 μL digested BAC DNA, 20 μL T4 ligation buffer, 2 μL 10 mg/mL BSA, 2 μL 100 mM ATP, and 19 μL T4 DNA ligase. Incubate at 16°C overnight.
14. Inactivate the T4 DNA ligase by incubating the reactions at 65°C for 15 min.
15. Add 200 μL phenol pH 8.0:chloroform to each tube, vortex for 30 s, and spin at 18,000*g* for 5 min at room temperature.
16. Transfer the aqueous phases to fresh 1.7 mL centrifuge tubes and repeat the phenol pH 8.0:chloroform extraction once.
17. Transfer the aqueous phases to fresh 1.7 mL centrifuge tubes. Add 200 μL chloroform, vortex for 30 s, and spin at 18,000*g* for 5 min at room temperature.

18. Pool two aqueous phases in one fresh 1.7 mL centrifuge tube rendering two tubes, each containing ~ 350 μL.
19. Add 35 μL of 3 M sodium acetate, pH 5.2, and vortex briefly. Add 875 μL of 100% ethanol, mix gently, and precipitate the DNA by placing the tubes at –20°C for at least 15 min.
20. Pellet the DNA by centrifuging at 18,000*g* for 20 min at 4°C.
21. Wash the pellet with 1 mL of 70% ethanol and spin at 18,000*g* for 15 min at 4°C.
22. Aspirate the supernatant and air-dry the pellet briefly.
23. Resuspend both pellets in a total volume of 200 μL of 1X TE buffer, pH 8.0. Dissolve the DNA by incubating at 37°C for 15 min.
24. Prepare a 0.8% agarose/0.5X TBE gel containing 0.5 μg/mL ethidium bromide.
25. Load 1 μL of the control 3C template, 4 μL of digested BAC DNA from Step 12 and a molecular weight standard of known concentration (**Fig. 13.2D**).
26. After running the gel, estimate the control 3C template concentration by comparing the intensities to the molecular weight standard (*see* **Note 29**).
27. Aliquot the control 3C template and store it up to at least 2 years at –20°C.

3.3. Quality Control of 3C and Control Templates

1. Prepare 8–12 two-fold serial dilutions of the 3C template starting with 250 ng/μL and ending with a "no template" control. Do the same for the control 3C template starting with a dilution of around 5 ng/μL. The minimum volume of each dilution is 8 μL. Every dilution is used for two separate PCR reactions each containing a specific pair of PCR primers designed to detect a 3C interaction between fragments that are either nearby or far apart on the linear genome (*see* **Note 5**).
2. PCR reactions are set up as follows: 4 μL 3C template dilution, 2.5 μL 10X PCR buffer, 2 μL 50 mM magnesium sulfate, 0.2 μL 25 mM dNTP mix, 0.125 μL 80 μM 3C primer 1, 0.125 μL 80 μM 3C primer 2, 0.2 μL 5 U/μL *Taq* DNA polymerase, and 15.85 μL water.
3. Amplify the DNA products using the following PCR parameters: 1 cycle 5 min at 95°C; 35 cycles 30 s at 95°C followed by 30 s at 65°C followed by 30 s at 72°C; 1 cycle 30 s at 95°C followed by 30 s at 65°C followed by 8 min at 72°C.
4. Add 8 μL of 4X DNA loading buffer to each PCR reaction and mix by pipetting. Analyze 14 μL of each sample on a 1.5% agarose/0.5X TBE gel containing 0.5 μg/mL ethidium bromide.

5. Quantify the PCR products using a gel documentation system and plot the quantity of PCR product versus the amount of 3C template used as input material (**Fig. 13.2C–F**) (*see* **Note 30**).

3.4. 5C Primer Design and Preparation of a 5C Primer Pool

1. Two types of 5C primers should be designed: forward 5C primers that anneal directly upstream of the restriction site of the 3C ligation product and reverse 5C primers that anneal exactly downstream of it (**Fig. 13.3**). Usually, the 5′ half of the restriction site is included at the 3′ end of the forward primer and the 3′ half of the restriction site is incorporated at the 5′ end of the reverse primer (*see* **Notes 31** and **32**).

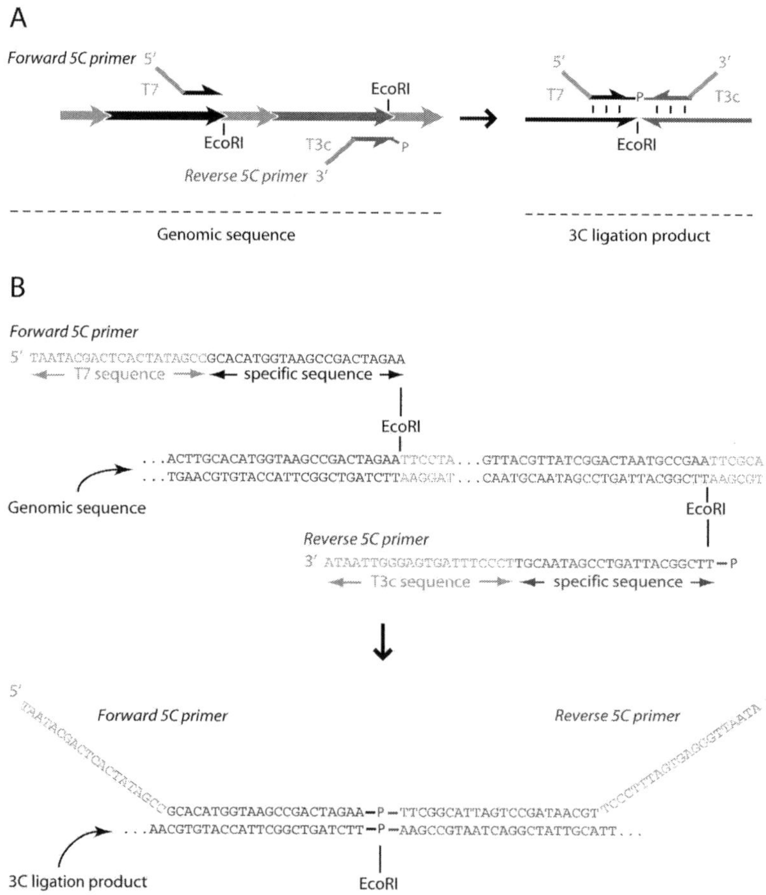

Fig. 13.3. 5C primer design. (**A**) Diagram and (**B**) example of 5C primer design. Both forward and reverse 5C primers are designed on the 5′ end of a restriction site in the genomic sequence and include three bases of the restriction site. Note, the forward and reverse 5C primers are designed on opposite strands in the genomic sequence (**A** *left*, **B** *top*), but anneal to the same strand in the 3C template (**A** *right*, **B** *bottom*). Universal tails are added to the specific sequence of the 5C primers. The T7 sequence is added to the 5′ end of forward 5C primers. The complement of the T3 sequence (T3c) is added to the 3′ end of reverse 5C primers. Only reverse 5C primers are phosphorylated at the 5′ end.

2. Only one primer (either a forward or a reverse) is designed per restriction fragment. 5C can only detect a 3C ligation product recognized by a combination of a forward primer and a reverse primer. Therefore, one should carefully decide for which fragments forward primers should be designed and for which fragments reverse primers.
3. Typically, the sequence specific part of the primers is around 40 nucleotides long. The annealing temperature should be adjusted to ~72°C. Primers with a too low annealing temperature should be discarded. If the annealing temperature of a primer is too high, specific nucleotides should be removed from the 5′ end of forward primers and from the 3′ end of reverse primers till the annealing temperature is ~72°C. The removed nucleotides should be replaced by random nucleotides, so that the total length of each primer remains exactly the same (*see***Note 33**).
4. It is recommended to exclude primers that anneal to repetitive sequences in the genome because they are likely to generate excessively large amounts of ligation products .
5. To ensure the ligation of 5C products, all reverse primers should be modified with a phosphate group at the 5′ end of the primer. This can be done either during synthesis of the individual primers or by phosphorylating a pool of reverse primers with PNK (see below for protocol).
6. All 5C ligation products are simultaneously amplified with two universal PCR primers. Therefore, common tails should be added to both the 5′ end of 5C forward primers and the 3′ end of 5C reverse primers (**Fig.13.3**). Typically, the T7 sequence is used for the forward primer tails and the complementary T3 promoter sequence is applied to the reverse primer tails.
7. Order the primers as 50 μM stock solutions in 1X TE, pH 8.0.
8. Pool all forward primers in equimolar amounts (*see* **Note 34**).
9. Make a separate pool of all reverse primers.
10. If the reverse primers were synthesized without a 5′ phosphate group, perform the following phosphorylation reaction: add 10 μL PNK buffer, 10 μL 10 mM ATP, and 10 μL PNK to 70 μL reverse primer pool; incubate at 37°C for 30 min; inactivate PNK by incubating the sample at 65°C for 10 min.
11. Add the forward primers to the reverse primers in a way that all individual primers are present at equimolar amounts.
12. Aliquot the 5C primer pool and store it at -20°C.

3.5. Preparation of a 5C and Control Library

1. Prepare 12 reactions each containing an amount of 3C template that corresponds to ~100,000 genome copies (*see* **Notes 35, 36**). These 12 reactions include 10 5C reactions for making a 5C library, 1 "no primer" control and 1 "no ligase" control.
2. Adjust the total DNA quantity in each reaction to 1.5 μg with 1 μg/μL SSD.
3. Add 1 μL of the appropriate 5C primer pool dilution to each 5C annealing reaction except the "no primer" control (*see* **Note 37**).
4. Add 2 μL of 5C annealing buffer and adjust the final volume to 20 μL with water.
5. Denature the 3C template and primers by incubating the samples at 95°C for 5 min.
6. Anneal the primers to the 3C template by incubating the samples at 48°C for 16 h.
7. Add 20 μL 1X *Taq* ligase buffer to the "no ligase" control.
8. Add 20 μL 1X *Taq* ligase buffer containing 10 units of *Taq* DNA ligase to all other samples.
9. Mix by pipetting and continue the incubation at 48°C for 1 h.
10. Inactivate the reactions by incubating the samples at 65°C for 10 min.
11. Split each reaction into 4X 6 μL and set up the following PCR reactions: 6 μL 5C ligation product, 2.5 μL 10X PCR buffer II, 1.8 μL 25 mM magnesium chloride, 0.2 μL 25 mM dNTP mix, 0.5 μL 80 μM T7, 0.5 μL 80 μM T3, 0.225 μL Ampli *Taq* Gold DNA polymerase and 13.275 μL water. Include a water control for the PCR (*see* **Note 38**).
12. Amplify the DNA products using the following PCR parameters: 1 cycle 9 min at 95°C; 24 cycles 30 s at 95°C followed by 30 s at 60°C followed by 30 s at 72°C; 1 cycle 30 s at 95°C followed by 30 s at 60°C followed by 8 min at 72°C.
13. Pool the PCRs from the samples for the 5C library.
14. Pool the PCRs from the "no ligase" control and make a separate pool of the PCRs from the "no primer" control.
15. Run aliquots from the 5C library, the "no ligase" and "no primer" controls and the water control on a 2% agarose/0.5X TBE gel containing 0.5 μg/mL ethidium bromide.
16. If the control lanes are empty and a clean single band is observed for the 5C library, continue with purifying the 5C library using the MinElute PCR Purification Kit.
17. Analyze serial dilutions of the purified 5C library and a molecular weight standard of known concentration on a 2% agarose/0.5X TBE gel containing 0.5 μg/mL ethidium bromide.

18. After running the gel, estimate the 5C library concentration by comparing the intensity to the molecular weight standard.
19. Aliquot the 5C library and store it at –20°C.

3.6. Quality Control of a 5C Library: Nested PCR

1. Prepare 8–12 two-fold serial dilutions of the 5C library starting with 2–10 ng/μL and ending with a water (no template) control. Do the same for the control 5C library. The minimum volume of each dilution is 12 μL. Every dilution is used for two separate PCR reactions each containing a specific pair of PCR primers. Each pair of PCR primers is designed to detect a 5C ligation product representing an interaction between fragments that are either nearby or far apart on the linear genome (*see* **Note 8**).
2. Set up the following PCR reactions for each dilution: 6 μL 5C library dilution, 2.5 μL 10X PCR buffer, 2 μL 50 mM magnesium sulfate, 0.2 μL 25 mM dNTP mix, 0.125 μL 80 μM 5C nested primer 1, 0.125 μL 80 μM 5C nested primer 2, 0.2 μL 5 U/μL *Taq* DNA polymerase, and 13.85 μL water.
3. Amplify the DNA products using the following PCR parameters: 1 cycle 5 min at 95°C; 20 cycles 30 s at 95°C followed by 30 s at 56°C followed by 30 s at 72°C; 1 cycle 30 s at 95°C followed by 30 s at 56°C followed by 8 min at 72°C.
4. Add 8 μL of 4X DNA loading buffer to each PCR reaction and mix by pipetting. Analyze 14 μL of each sample on a 2% agarose/0.5X TBE gel containing 0.5 μg/mL ethidium bromide.
5. Quantify the PCR products using a gel documentation system and plot the relative quantity of PCR product versus the amount of 5C library used as input material (**Fig. 13.4**) (*see* **Note 39**).

3.7. Quality Control of a 5C Library: Cloning and Sequence Analysis

1. To ensure that none of the 5C primers is misbehaving in the assay, take an aliquot of the purified 5C library and clone the PCR products using the Zero Blunt TOPO PCR cloning kit.
2. Inoculate 100 colonies and isolate the plasmids containing the 5C ligation products using a Qiaprep Spin Miniprep kit.
3. Sequence the inserts using the –21 M13 sequencing primer and analyze the sequences (*see* **Note 40**).
4. The 5C library is ready for deep sequencing or microarray analysis, if the results from the nested PCR show that the 5C library is a reliable copy of the 3C template and when the sequence analysis of the 5C library does not identify problematic primers. We recommend acquiring specific instructions regarding sample preparations for the particular microarray platform or sequencing platform that will be used.

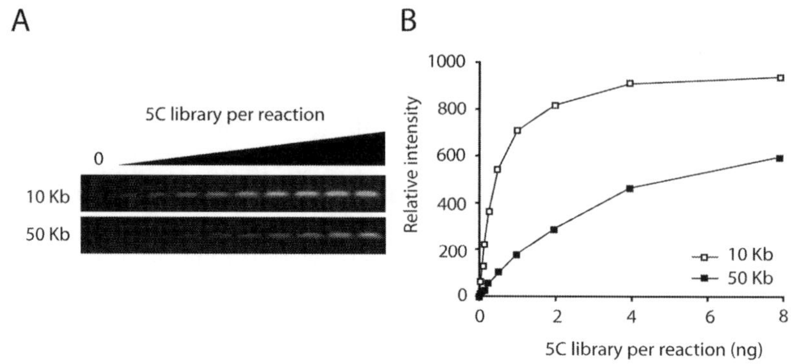

Fig. 13.4. Quality control of a 5C library by nested PCR. (**A**) Agarose gel analysis and (**B**) quantification of a 5C library titration in a nested PCR. Increasing amounts of 5C library were analyzed with two nested primer pairs. One primer pair interrogates an interaction between fragments that are close (10 kb) to each other in the linear genome. The other primer pair examines the interaction between two more distant (50 kb) fragments. A titration curve should demonstrate a linear increase at the beginning of the curve and should reach a plateau with increasing amounts of 5C library. The curve of the PCR product testing the interaction between two adjacent restriction fragments should be above the curve of the PCR product testing the interaction between two, more distant fragments. Note, when using a control 5C template, these two titration curves should be more similar.

3.8. Normalization and Analysis of the Results

Both the number of sequence hits and the intensity on the microarray are a measure for the amount of 5C ligation product in the 5C library. The abundance of specific 5C ligation products in the 5C library is proportionate to the frequency with which the two corresponding restriction fragments interact inside the nucleus. Interaction frequencies are calculated by dividing the amount of a specific 5C ligation product in the 5C library by the amount of the same 5C product in the control 5C library. This ratio is a direct measure for the interaction incidence, and is normalized for any differences in primer efficiency and 5C ligation product amplification efficiency. However, the ratio remains an arbitrary unit, meaning that only the frequencies obtained within a single 5C experiment can be directly compared. To be able to compare interaction frequencies from different 5C experiments, one has to use a common internal control to normalize the different datasets. This internal control can be a set of interactions of which the frequencies are expected to be the same in the different 5C experiments. Including a set of interactions between fragments located within a conserved gene desert region on human chromosome 16 have been used successfully to normalize 5C experiments (15) (*see* **Note 41**).

Due to the many-to-many setup, the 5C data is typically presented in a matrix (**Fig. 13.5**). The color of each cell represents the measured amount of 5C ligation product formed by the corresponding forward and reverse 5C primers on the x- and y-axes. The resulting 5C heatmap can be regarded as a collection of 3C

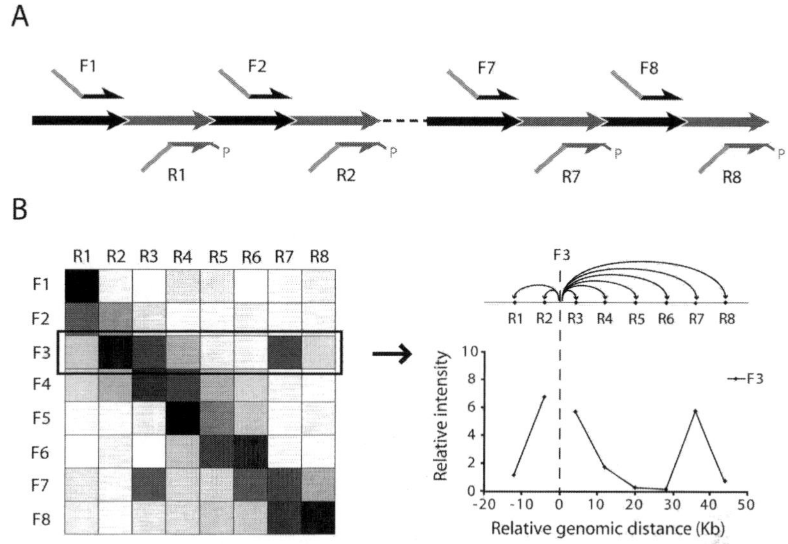

Fig. 13.5. Expected results from a 5C experiment using an alternating scheme. (**A**) Diagram of a 5C experiment using alternating forward and reverse 5C primers for consecutive restriction fragments. (**B**) Expected results are shown in a heatmap, where every box represents an interaction between two fragments corresponding to the primers indicated to the left and the top. The gray level of each box is a measure of interaction frequency as determined by 5C. White corresponds to no interaction, whereas black corresponds to a high-interaction frequency. Interactions between neighboring fragments are usually strong and are represented by a diagonal across the heatmap. High interaction frequencies that are off the diagonal represent long range interactions. Data from individual column or row can be translated into a 3C graph for that particular primer.

graphs. Consequently, the analysis of data from a 5C experiment is similar to analysis of 3C data (17). 3C graphs of individual 5C primers can be obtained by plotting the values of its corresponding column or row. Only background interactions are detected in the absence of a looping interaction. The interaction frequencies of background interactions are inversely correlated to the genomic distance between the restriction fragments. In the presence of a specific long-range interaction, a local peak on top of the background interactions will be detected.

4. Notes

1. Formaldehyde older than 6 months–1 year will result in less efficient cross-linking and should not be used.
2. There are several criteria to take into consideration when selecting the restriction enzyme. Most importantly, the enzyme should cut efficiently under the conditions of the

3C protocol. The use of *Bam*HI is not recommended because it has proven to be less efficient under the specific 3C conditions. One can choose enzymes that recognize a 6-mer palindromic sequence and cut every ~4 kb (e.g., *Eco*RI, *Hind* III, *Bgl* II) or choose enzymes that cut more frequently (e.g., *Mse* I). The choice of enzyme will also depend on the desired resolution and the distribution of the restriction sites within the region of interest.

3. Avoid multiple freeze–thaw cycles of buffers containing DTT. It is best to store these solutions in aliquots.
4. The use of bromophenol blue is not recommended, as this dye will run at the same position as the PCR products and will, therefore, interfere with DNA quantification.
5. For quality control, 3C templates should be titrated in a 3C PCR experiment. Typically, two head-to-head 3C ligation products are measured by semi-quantitative PCR and agarose gel quantification. One tested interaction should be between two adjacent or nearby restriction fragments (10 kb) and one should be between two, more distant fragments (50–80 kb). The 3C primers have to be designed unidirectionally along the linear genome and 50–150 bp upstream of the 3′ end of the predicted restriction fragment (8). Generally, the 3C primers are 28–32 bp long, have a GC content of approximately 50% and should preferably be unique as determined by BLAST.
6. Since the 5C technology is very sensitive to contaminations, we recommend dividing all reagents mentioned in **Section 2.6** into single use aliquots.
7. The PCR primers should be adjusted if other tail sequences were used in the design of the 5C primers. The primers should also be modified dependent on the detection method used in the assay. For microarray analysis, we recommend using a Cy3 labeled reverse primer to generate labeled antisense 5C products. For sequence analysis, the 5C libraries should be amplified with 5′-phosphorylated primers to allow ligation of linkers used for subsequent sequencing. We advise to consult your microarray or sequencing facility to decide on the proper modification for either analysis method.
8. For quality control, 5C libraries should be titrated in a nested PCR experiment. Typically, two 5C ligation products are measured by semi-quantitative PCR and agarose gel quantification. One 5C ligation product should correspond to an interaction between two adjacent or nearby restriction fragments (~10 kb) and one to an interaction between two more distant fragments (50–80 kb). The forward nested primer should be designed starting at the first specific

nucleotide at the 3′ end of the common tail. The same holds true for the reverse nested primer on the reverse strand of the 5C ligation product. Generally, the 5C nested primers are 19–22 bp long and preferentially have a GC content of approximately 50%.

9. Cells grown in suspension should be pelleted gently by centrifuging at 300*g* for 10 min. Next, the cells should be resuspended in 45 mL fresh culture medium and the amounts of formaldehyde and glycine in the following two steps should be doubled.
10. It is essential to cross-link the cells for exactly 10 min. Shorter incubation times will result in lower detection signals of chromatin interactions, whereas longer incubation times will cause too many cross-links resulting in reduced digestion efficiency.
11. The experiment can be paused at this point by incubating the cell pellet on dry ice for 20 min and storing the pellet at –80°C. Pellets can be stored at –80°C for at least 1 year.
12. If fewer cells were used for cross-linking, the volumes in this 3C template protocol should be adjusted accordingly.
13. The cell lysate should not be viscous. Viscous lysates are caused by insufficient cross-linking due to using formaldehyde that is too old (*see* **Note 1**).
14. Try to avoid sedimentation of the suspension to prevent uneven distribution of the chromatin.
15. Air bubbles will make it more difficult at a later stage to quench the SDS with Triton.
16. Longer incubation at 65°C will cause reversal of the cross-linking of the chromatin interactions and should be avoided.
17. The mixture should appear slightly granular.
18. The solution should be clear at this stage.
19. The second proteinase K digestion step increases the 3C template yield after phenol:chloroform extraction.
20. Both the phenol and the aqueous phase can appear cloudy. The DNA accumulates close to the interface during the first extraction. Take off as much material as possible without transferring any interface material.
21. The supernatant and interface should both be clear at this stage. If not, perform another phenol pH 8.0:chloroform extraction.
22. Dilution of the solution helps to reduce the amount of salt precipitating in the next step.
23. The tubes can also be left at –80°C overnight and the protocol can be continued the next day.

24. PCR reactions with 3C templates are inhibited by residual salt in the template. Thorough desalting is required. Desalting columns are not recommended because they often "size fractionate" the DNA and can change the nature of the samples.
25. Typically, the 3C template concentration is around 200–250 ng/μL. The 3C template should run as a fairly tight band of more than 10 kb. A DNA smear indicates poor ligation efficiency and material trapped in the wells indicates incomplete DNA digestion. Very little RNA should be present.
26. It is recommended to use a control 3C template of randomly ligated fragments to create a control 5C library. This is used to correct for differences in annealing efficiencies of the 5C primers and slight amplification biases of the different 5C ligation products in a normal 5C library. For small genomes (e.g., yeast), a control 3C template can be generated by digesting and randomly ligating whole genomic DNA (10). However, the complexity of the ligation mixture resulting from a larger genome (e.g., human) becomes too high to reliably detect individual ligation products. Therefore, a control template should be made by using equimolar amounts of a set of minimally overlapping BAC clones that covers the genomic region of interest.
27. To determine accurate relative concentrations, perform a real-time quantitative PCR with primers recognizing a common BAC vector region. Alternatively, BAC DNA can be digested and quantified on an agarose gel.
28. The volume of the added restriction enzyme should not exceed 10% of the total volume, because the glycerol in the enzyme storage buffer will inhibit enzymatic activity.
29. Typically, the control 3C template concentration is around 100 ng/μL. Digested BAC DNA appears as a smear of DNA fragments with the larger bands migrating around 10 kb. Ligated BAC DNA should appear as a tight smear migrating just above 10 kb.
30. A titration curve should demonstrate a linear increase at the beginning of the curve and should reach a plateau with increasing amounts of 3C template. The curve of the PCR product testing the interaction between two adjacent restriction fragments should be above the curve of the PCR product testing the interaction between two, more distant fragments. This validates that the relative amounts of 3C ligation products in the 3C template are consistent with the actual interaction frequencies of the chromatin fragments in vivo. In contrast, both curves will be much more similar for a control 3C template. If excess salt is still present in the 3C template

sample, the curve will show an irregular linear phase and/or appear biphasic and it is recommended to re-precipitate the 3C template and to extensively wash the DNA pellet with 70% ethanol. The 3C template should be titrated again after re-precipitation.

31. It is recommended to first read the complete protocol before starting with the 5C primer design.
32. Importantly, forward and reverse primers will anneal to the same strand of the 3C ligation product formed by a head-to-head ligation of two restriction fragments. Thus, the forward and reverse primers are designed on the 3′ ends of restriction fragments so that they will anneal to different strands on the regular genomic sequence. This approach will prevent detection of partial digestion products and of ligation products resulting from self-circularization of restriction fragments. The latter occurs very frequently in the generation of the 3C library.
33. 5C is much less sensitive to fluctuations in primer efficiency than regular PCR because all amplicons are equal in size and are amplified using a single universal PCR primer pair. In addition, they are designed with equal annealing temperatures. However, it is recommended to make a 3C control template to correct for any differences in annealing efficiencies of the 5C primers and slight amplification biases of the different 5C ligation products in a normal 5C library (*see* **Section3.2**).
34. The total primer concentration will remain 50 μM. To calculate the individual primer concentration, divide the total primer concentration by the number of primers in the pool.
35. For the human genome the amount corresponding to ~100,000 genome copies is ~400 ng. The amount of control 3C template used in the 5C reaction should be determined by a titration experiment (*see* **Note 36**).
36. We recommend performing a (control) 3C template titration experiment before preparing the (control) 5C library. Prepare serial dilutions of the 3C template. Include a "no ligase" control with the highest 3C template concentration as well as a "no template" control and perform the 5C reactions according to the protocol. A titration curve should demonstrate a linear increase at the beginning of the curve and should reach a plateau with increasing amounts of 3C template.
37. Typically, 1 fmol of each individual 5C primer is added to the reaction. We recommend performing a primer titration experiment before preparing the 5C library. Prepare serial dilutions of the 5C primer pool. Include a "no ligase" control with the highest primer concentration as well as a "no primer" control at the end and perform the 5C reactions according to

the protocol using 100,000 genome copies of 3C template per reaction. A titration curve should demonstrate a linear increase at the beginning of the curve and should reach a plateau with increasing amounts of 5C primer pool. The optimal primer concentration is the concentration that corresponds to the point where the curve just reached the plateau. A peak in the curve could be an indication that the amplification primers are quenched in the PCR reaction. This can be resolved by decreasing the number of PCR cycles or by increasing the amount of T7 and T3 primers.

38. To avoid any artifacts of primers annealing to each other and subsequently being ligated by residual activity of *Taq* ligase, it is essential to set up the PCR reactions immediately after Step 10 and not to store the ligation reactions at 4°C.

39. A titration curve should demonstrate a linear increase at the beginning of the curve and should reach a plateau with increasing amounts of 5C library. The curve of the PCR product testing the interaction between two adjacent restriction fragments should be above the curve of the PCR product testing the interaction between two, more distant fragments. This confirms that the 5C ligation products in the 5C library form an accurate copy of the 3C template and hence the interaction frequencies in vivo.In contrast, both curves should be more similar for a control 5C library.

40. Pay attention to individual 5C primer sequences that are overrepresented in the cloned inserts. This could indicate that this primer is somehow misbehaving in the assay. 5C ligation products corresponding to interactions of nearby fragments are expected to be slightly overrepresented in a 5C library. However, in a control 5C library every 5C ligation product should be equally present. If problematic primers are identified by sequence analysis, a new 5C primer pool should be made without the troublesome primers.

41. Including an internal control has consequences for the experimental setup. First of all, BAC clones covering the region of the internal control are added in equimolar amounts to the BAC clones of the region of interest prior to making a control 3C template. Second, 5C primers are designed for the region of the internal control. Typically, an alternating format is used for the design of these 5C primers.

References

1. ENCODE-consortium. (2007) Identification and analysis of functional elements in 1% of the human genome by the ENCODE pilot project. *Nature* **447**, 799–816.

2. Kleinjan, D. A. and van Heyningen, V. (2005) Long-range control of gene expression: emerging mechanisms and disruption in disease. *Am. J. Hum. Genet.* **76**, 8–32.

3. Dekker, J. (2008) Gene regulation in the third dimension. *Science* **319**, 1793–1794.
4. Tolhuis, B., Palstra, R. J., Splinter, E., Grosveld, F. and de Laat, W. (2002) Looping and interaction between hypersensitive sites in the active beta-globin locus. *Mol. Cell.* **10**, 1453–1465.
5. Spilianakis, C. G. and Flavell, R. A. (2004) Long-range intrachromosomal interactions in the T helper type 2 cytokine locus. *Nat. Immunol.* **5**, 1017–1027.
6. Cremer, T. and Cremer C. (2001) Chromosome territories, nuclear architecture and gene regulation in mammalian cells. *Nat. Rev. Genet.* **2**, 292–301.
7. Sexton, T., Schober, H., Fraser, P. and Gasser, S. M. (2007) Gene regulation through nuclear organization. *Nat. Struct. Mol. Biol.* **14**, 1049–1055.
8. Dekker, J., Rippe, K., Dekker, M. and Kleckner, N. (2002) Capturing chromosome conformation. *Science* **295**, 1306–1311.
9. Splinter, E., Grosveld, F. and de Laat, W. (2004) 3C technology: analyzing the spatial organization of genomic loci in vivo. *Meth. Enzymol.* **375**, 493–507.
10. Miele, A., Gheldof, N., Tabuchi, T. M., Dostie, J. and Dekker, J. Mapping chromatin interactions by Chromosome Conformation Capture (3C). (2006) In: Ausubel, F. M., Brent, R., Kingston, R. E., Moore, D. D., Seidman, J. G., Smith, J. A. and Struhl, K., eds. *Current Protocols in Molecular Biology.* Vol. Supplement 74. Hoboken, NJ: John Wiley & Sons, 21.11.1–21.11.20.
11. Lomvardas, S., Barnea, G., Pisapia, D. J., Mendelsohn, M., Kirkland, J. and Axel, R. (2006) Interchromosomal interactions and olfactory receptor choice. *Cell* **126**, 403–413.
12. Spilianakis, C.G., Lalioti, M. D., Town, T., Lee, G. R. and Flavell, R. A. (2005) Interchromosomal associations between alternatively expressed loci. *Nature* **435**, 637–645.
13. Bacher, C. P., Guggiari, M. and Brors, B., et al. (2006) Transient colocalization of X-inactivation centres accompanies the initiation of X inactivation. *Nat Cell Biol.* **8**, 293–299.
14. Xu, N., Tsai, C. L. and Lee, J. T. (2006) Transient homologous chromosome pairing marks the onset of X inactivation. *Science* **311**, 1149–1152.
15. Dostie, J., Richmond, T. A. and Arnaout, R. A., et al. (2006) Chromosome Conformation Capture Carbon Copy (5C): A massively parallel solution for mapping interactions between genomic elements. *Genome Res.* **16**, 1299–1309.
16. Dostie, J. and Dekker, J. (2007) Mapping networks of physical interactions between genomic elements using 5C technology. *Nat. Protoc.* **2**, 988–1002.
17. Dekker, J. (2006) The 3 C's of Chromosome Conformation Capture: controls, controls, controls. *Nat. Methods* **3**, 17–21.

Chapter 14

Analysis of Nascent RNA Transcripts by Chromatin RNA Immunoprecipitation

Ales Obrdlik and Piergiorgio Percipalle

Abstract

Biochemical methods to analyze co-transcriptional recruitment of co-activators to nascent RNA molecules have lagged behind for many years. Most of the information on co-transcriptional regulation of nascent RNA came from invaluable in situ studies using single-cell model systems. More recently, the chromatin RNA immunoprecipitation technique has been developed to evaluate at the molecular level the association of proteins with nascent RNA which is still coupled to chromatin. Similar to chromatin immunoprecipitation, the chromatin RNA immunoprecipitation method is suitable to study events along specific genes, and it has been successfully used in numerous applications to demonstrate the cross-talk between transcription and RNA processing. This technique has a considerable margin of technological development especially in high-throughput screening experiments in combination with microarrays. In this chapter, we describe a RIP protocol optimized in our laboratory to study association of RNA binding proteins with specific nascent mRNA transcripts.

Key words: Chromatin, nascent RNA transcripts, pre-mRNA, ribonucleoprotein complexes, mRNA processing, immunofluorescence, confocal microscopy, in vivo cross-linking, immunoprecipitation.

1. Introduction

The co-transcriptional interplay between elongating RNA polymerase and nascent RNA molecules is important for recruitment of multiple regulatory factors to active genes (1, 2). The RNA polymerase (pol) II carboxy-terminal domain mediates association of chromatin remodeling complexes, histone modifying enzymes, and mRNA processing factors (3), while nascent mRNA molecules assemble into pre-messenger ribonucleoprotein (pre-mRNPs) particles. This process occurs through direct association with specific RNA binding proteins, including heterogeneous nuclear ribonucleoproteins (hnRNPs) (**Fig. 14.1**), splicing factors and repressors,

Philippe Collas (ed.), *Chromatin Immunoprecipitation Assays,* Methods in Molecular Biology 567,
DOI 10.1007/978-1-60327-414-2_14, © Humana Press, a part of Springer Science+Business Media, LLC 2009

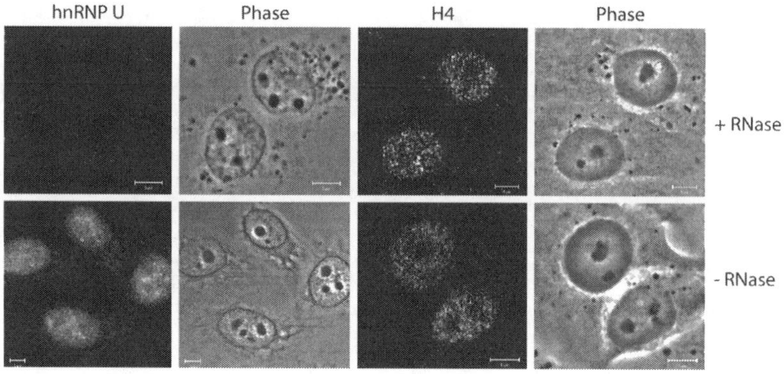

Fig. 14.1. The intranuclear distribution of hnRNP U is sensitive to RNase treatment. HeLa cells are permeabilized with TritonX-100 (0.1%, 10 min, RT) and treated with RNase A (1 mg/mL, 10 min, RT) prior to fixation. After treatment cells are immunostained with a mouse monoclonal antibody against hnRNP U or with a rabbit polyclonal antibody against histone H4. The distribution of endogenous proteins is analyzed by confocal microscopy (scale bar 5 μm).

transport facilitators and mediators as well as regulators of mRNA translation (4). hnRNPs comprise a large number of proteins, classified into several families based on structural and functional motifs. In mammals, there are more than 20 major and a large number of minor protein species, designated A1 to U hnRNPs. hnRNPs display a general role in transcription elongation by facilitating mRNA packaging and have specialized functions in different aspects of gene expression (5). In this complex scenario, it is therefore important to be able to monitor co-transcriptional binding of proteins to nascent RNA.

In situ analyses on active genes in single-cell model systems such as *Xenopus laevis* oocytes, *Drosophila melanogaster*, and *Chironomus tentans* provided considerable contributions to our understanding of co-transcriptional regulatory events leading to efficient mRNA synthesis. For instance, electron microscopy studies carried out on the Balbiani ring genes in the *C. tentans* polytene chromosomes revealed co-transcriptional association of the spliceosome with elongating RNA polymerase and nascent transcripts (6, 7). However, these studies did not provide molecular insights.

Recent developments in chromatin immunoprecipitation (ChIP) identified a new method – chromatin RNA immunoprecipitation (RIP) – to analyze at the molecular level the interactions between pre-mRNA and regulatory factors occurring during elongation and RNA processing events. The RIP method was originally devised to study association of the histone acetyl transferase (HAT) elongator with nascent RNA emanating from the elongating pol II (8). Since then several RIP applications have been described. For instance, RIP assays contributed to establish molecular cross-talks between transcription and RNA processing via Brm, a subunit of the SWI/SNF chromatin remodeling complex, and to conclude that recognition of trimethylated H3Lys4 facilitates recruitment of transcription post-initiation factors and pre-mRNA splicing (9, 10). The RIP technique involves in vivo crosslinking of proteins to RNA, cell lysis, fragmentation, and preparation of soluble chromatin followed by immunoprecipitation and DNase treatment. This extra step which is not performed in ChIP experiments allows analysis of RNA segments specifically associated with a protein of interest. The immunoprecipitated RNA is reverse-transcribed and subsequently identified by PCR amplification with gene-specific primers. As preliminary assay accompanying the RIP protocol, we also perform a short RNase treatment on HeLa cells and we analyze the distributions of candidate RNA binding proteins by immunofluorescence and confocal microscopy (11). If alterations in the intracellular distributions are observed, it is likely that the protein of interest associates with RNA and therefore we proceed with the RIP assay.

Here, we describe the steps to perform in situ RNase treatment on HeLa cells and a detailed RIP protocol optimized in our lab to study the specific in vivo association of certain hnRNPs and RNA binding proteins with a nascent pol II transcript.

2. Materials

2.1. Cell Culture

1. 175 cm^2 cell culture flask with vented cap or equivalent tissue cell culture dish.
2. Dulbecco's Modified Eagle Medium (DMEM), 10% fetal calf serum (FCS), penicillin (0.01 mg/mL), streptomycin (0.01 mg/mL).
3. Trypsin 2.5 mg/mL, cell culture grade. Aliquot and store at −20°C.
4. Low-passage HeLa cells.
5. Water-jacketed incubator or equivalent tissue cell culture incubator with maintained standard temperature, humidity, and CO_2 level.
6. Sterile 10X phosphate buffered saline (PBS). Dilute 1:10 with sterile water when required and keep refrigerated on ice.
7. Sterile Teflon cell scrapers (Fisher).

2.2. RNase Treatment and Confocal Immunofluorescence

1. RNase A in powder (Sigma-Aldrich). Prepare a 10 mg/mL stock solution in 1X PBS and store in aliquots at −20°C.
2. Microscope cover slips (22 × 40 × 0.15 mm) and glass slides.

3. 10X PBS buffer. Dilute 1:10 with bi-distilled water. Keep at room temperature (RT).
4. 37% formaldehyde stock solution. This reagent is light sensitive and must be kept in the dark. For cell fixation, dilute 1:10 with 1X PBS. Protect from light and keep at RT.
5. Permeabilization solution: 0.1% (v/v) Triton X-100 in 1X PBS.
6. Blocking solution: 1% milk powder dissolved in 1X PBS. Keep refrigerated at 4°C.
7. Primary and secondary antibodies dilution buffer: 1% milk in 1X PBS.
8. Primary mouse monoclonal antibody against hnRNP U (3G6, Abcam) (12)and rabbit polyclonal antibody against histone H4 (Upstate Technology).
9. Secondary: anti mouse and anti-rabbit IgGs conjugated to Alexa488 (Invitrogen).
10. Nuclear stain: 300 nM DAPI (4,6-diamino-2-phenylindole) in water. The solution is light sensitive and must be kept in the dark. Store at 4°C.
11. Mounting medium: Mowiol (Mowiol 4.88, Calbiochem). Prepare stock solutions and store at−20°C (*see* **Note 1**).

2.3. In Vivo *Cross-Linking and Cell Lysis*

1. 250 mL DMEM cell culture medium. Store at−20°C and thaw at RT well in advance prior to use.
2. 37% formaldehyde stock solution. Prepare fresh solutions every time. This reagent is toxic and volatile and all procedures must be performed under a chemical hood.
3. 25 mL 2 M glycine solution in 1X PBS.
4. Diethyl-pyrocarbonate (DEPC).
5. 1 L 1X PBS pH 7.5, sterile filtered.
6. 50 mL lysis buffer including 1X PBS, pH 7.5, 1 mM phenylmethylsulfonylfluoride (PMSF), 0.2% nonylphenolethoxylate-40 (NP40), Complete Mini protease inhibitor Mix (Roche), 20 U/mL RNase Guard^{TM} (GE-Healthcare). Prepare fresh solutions before use and keep on ice.
7. 50 mL 1X PBS, pH 7.5, 1 mM PMSF, Complete Mini Inhibitor Mix, 20 U/mL RNaseGuard^{TM}. Prepare fresh solutions and keep on ice.
8. Glass tissue grinder.

2.4. Immunoprecipitation

1. Sepharose G (Invitrogen).
2. Primary mouse monoclonal antibodies against hnRNP U (12) and rabbit polyclonal antibodies against CBP20 (13), pol II CTD (3). Prepare fresh antibody solutions before use.

3. Primary rabbit polyclonal antibody against the TATA Box binding protein TBP (Abcam) and unspecific polyclonal rabbit IgGs (Dianova) as an alternative negative control. Fresh solutions should be prepared before use.
4. RIPA wash buffer including 1% NP40, 0.1% sodium deoxycholate, 1 mM PMSF, 0.05% sodiumdodecylsulfate (SDS), RNaseGuard™(5 U/mL), in 1X PBS. Prepare as fresh solution before use.
5. RIPA-1000 wash buffer including 1 M NaCl, 1 mM PMSF, 1% NP40, 0.1% sodium deoxycholate, 0.05% SDS, RNaseGuard (5 U/mL), in 1X PBS. Prepare before use.
6. LiCl wash buffer (optional) including 250 mM LiCl, 1% NP40, 0.1% sodium deoxycholate, 0.05% SDS, RNaseGuard (5 U/mL) in 1X PBS.
7. Elution buffer including 10 mM dithiothreitol (DTT), 1% SDS, 1 mM PMSF, and 1X PBS. Prepare before use. In case of prolonged storage, keep buffer at −20°C because DTT is prone to degradation at RT.

2.5. RNA Preparation and DNase I Treatment

1. Tri Reagent© (Sigma-Aldrich) (*see* **Note 2**).
2. Chloroform p.a.
3. Isopropanol p.a.
4. 70% ethanol aqueous solution.
5. Amplification grade RNase free DNase I (1 U/μL) (Invitrogen).
6. 10X DNase I reaction buffer including 200 mM Tris-HCl, pH 8.5, 20 mM $MgCl_2$or supplied reaction buffer by the manufacturer (see above).
7. RNaseGuard™. Store at −20°C.
8. 50 mM EDTA, pH 8.5.

2.6. cDNA Synthesis and mRNA-Specific PCR Amplification

1. 100 mM dNTP mix (Roche). Prepare 10 mM dNTP mix in prepared in sterile DEPC-treated bi-distilled water. Store at −20°C until further use.
2. RNaseGuard™.
3. Reverse transcriptase kit Superscript II (Invitrogen). Store at −20°C.
4. N6-random oligonucleotide primers (250 ng/μL).
5. Gene-specific primers: 100 μM stock solutions of S19 mRNA-specific forward and reverse primers and tRNA primers for as negative controls (**Figs. 14.2** and **14.3**). Store at −20°C.
6. Conventional Taq DNA polymerase (5 U/μL) kit with corresponding reaction buffers (Invitrogen). Keep at −20°C.

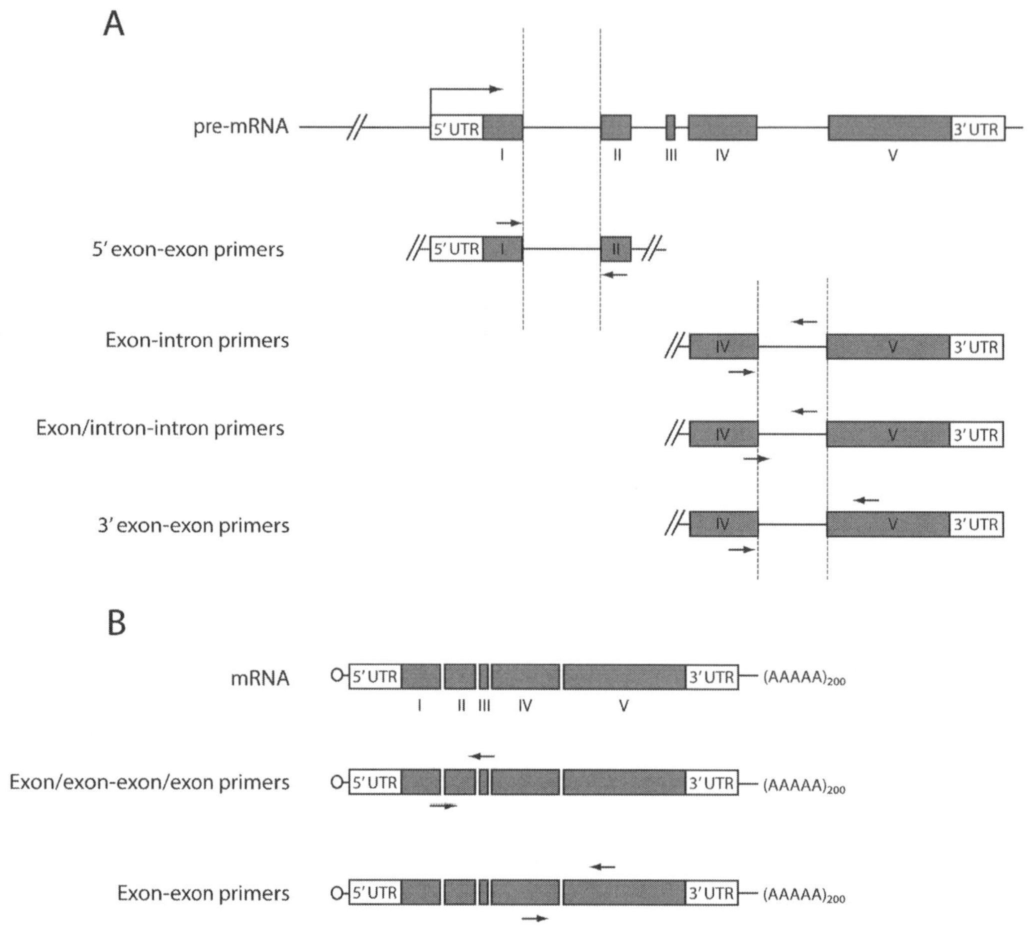

Fig. 14.2. Schematic illustration of gene specific primer design to study hypothetical pre-mRNA and mRNA transcripts. (**A**) Exon–exon and exon/intron–intron primer combinations are preferred to screen pre-mRNA. In this case, primers should be complementary to internal exonic or intronic sequences. (**B**) For mRNA analysis, exon/exon–exon/exon primer combinations should be designed spanning exon–exon borders. This type of combination is especially recommended when the focus of the experiment is on the mechanisms underlying mRNA processing.

7. 100% dimethylsulfonyloxid (DMSO) (its use depends on the primer features). Store at RT. If kept at 4°C or at −20°C, DMSO has the tendency to crystallize.

2.7. Databases for Primer Design

1. Entrez Gene: http://www.ncbi.nlm.nih.gov/sites/entrez.
2. Blastn: http://www.ncbi.nlm.nih.gov/blast/Blast.cgi.
3. Primer3Input: http://frodo.wi.mit.edu/cgi-bin/primer3/primer3_www.cgi.
4. OligoCalc, "Oligonucleotide Properties Calculator": http://www.basic.northwestern.edu/biotools/oligocalc.html.

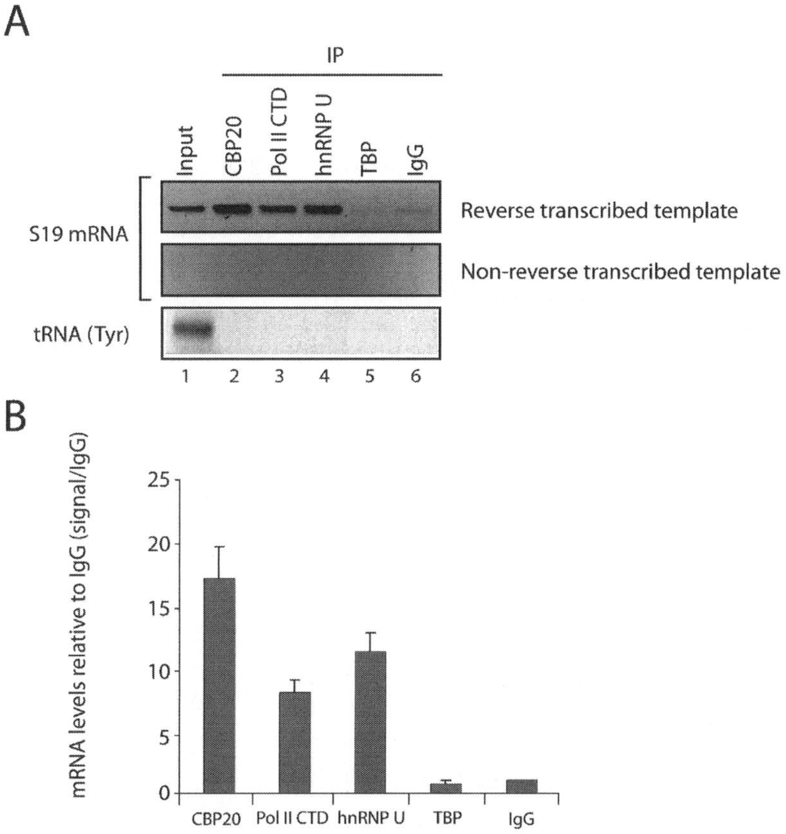

Fig. 14.3. Analysis of the RNA binding activity of CBP20, hnRNP U, pol II CTD, and TBP. (**A**) CBP20, hnRNP U, and pol II CTD are associated with nascent S19 mRNA. For RIP analysis, nuclear extracts from cross-linked HeLa cells were incubated with the indicated antibodies and co-precipitated RNA was subjected to RT-PCR using primers that amplify S19 mRNA and tRNA. PCR reactions were also performed on non-reverse transcribed templates as control for DNA contamination. Lane 1, PCR amplification of 1% input RNA. (**B**) The bar diagram shows the relative amounts of different RNAs precipitated with antibodies to the indicated proteins determined in three independent RIP experiments. Error bars represent standard deviations.

2.8. Quantitative Analysis of Specific PCR Products

1. 50X TAE electrophoresis buffer containing 2 M Tris-acetate, pH 8.5, 0.05 M Na-EDTA. Prepare before the electrophoresis run.
2. 6X loading buffer. Store in small aliquots at −20°C.
3. Gene Ruler, DNA Ladder Mix (Fermentas).
4. Electrophoresis grade agarose.
5. Ethidium bromide, stock solution (5.25 mg/mL). This solution is extremely harmful. Furthermore it is light sensitive and must be stored in the dark (*see* **Note 15**).
6. Horizontal gel electrophoresis system with 8 cm migration range.
7. Software for quantification of digital gel slab images: ImageJ (http://rsb.info.nih.gov/ij/download.html).
8. Statistical software such as Statistica®, Microsoft Excel®.

3. Methods

As general guidelines through the entire RIP procedure, we recommend that all buffers are prepared with DEPC-treated water and sterile-filtered when handling RNA samples. Work bench as well as all pipettes must be cleaned properly using alkaline detergents and 70% ethanol to reduce RNase contamination. We also recommend the use of pre-sterilized filter pipette tips in all protocol steps as well as originally sealed reaction tubes. In our applications, the RIP procedure is combined with an RNase treatment in living cells and confocal immunofluorescence analysis to evaluate whether our protein of interest is integrated in RNPs.

3.1. RNase Treatment and Immunofluorescence Analysis

1. Grow HeLa cells on cover slips to approximately 50% confluence.
2. Prior to RNase treatment, cells are detergent-permeabilized with PBS containing 0.1% Triton X100 for 10 min at RT. For RNase treatment, incubate cells with 1X PBS containing 1.0 mg/mL RNase A for 10 min at RT. Incubate control HeLa cells with PBS alone.
3. Fix cells with 3.7% formaldehyde for 10 min and block with a 1% milk solution in 1X PBS, 20 min to 1 h at RT, or overnight at 4°C.
4. Wash cells three times 10 min with 1X PBS and individually incubate for 1 h at RT with solutions of primary antibodies against hnRNP U and histone H4 supplemented with 1% milk in 1X PBS.
5. After three times 10 min washes for detection of the primary antibodies, incubate cells for 1 h at RT with Alexa488-conjugated mouse and rabbit secondary antibodies diluted according to the manufacturer instruction leaflet in the same solution as the primary antibodies. At the end of the procedure, stain cells with DAPI (0.1 μg/mL) for 3 min before mounting the cover slips on glass slides.
5. In the meantime prepare the mounting medium Mowiol. Put 6 g glycerol in a 50 mL centrifuge tube, add 2.4 g Mowiol, and stir to mix. While stirring, add 6 mL distilled water and leave 2 h at RT. Add 12 mL 2 M Tris-HCl, pH 8.5, and NaN_3(sodium azide) to a final concentration of 0.02% (optional). Incubate the tube in a hot water bath (50–60°C) for 10 min to dissolve the Mowiol. This can be repeated over several hours if necessary. Centrifuge at 5,000*g* for 15 min to remove any particulate. For storage, keep Mowiol as 1 mL aliquots at −20°C. Before use, warm tubes to room

temperature. Opened tubes can be stored at 4°C for approximately 1 month. Discard if any crystalline material is seen in the tube or on the slides.

6. For mounting, carefully turn cover slips upside down (cells facing down) with tweezers and place on a drop of Mowiol pre-spotted on the glass slide. Mowiol has the advantage of polymerizing at room temperature and does not require extra sealing of the cover slip on the glass slide (*see* **Note 1**). After mounting cover slips on the glass slides, leave in the dark overnight to harden before oil immersion lenses are used.
7. After Mowiol polymerization, the cells are ready to be viewed by microscopy. We recommend analysis at the phase contrast followed by confocal microscopy. Laser excitation at 364 nm induces DAPI fluorescence. The use of other excitation wavelengths depends on the fluorochromes conjugated to the secondary antibodies. For Alexa488 the excitation wavelength is 488 nm. Alexa conjugates are recommended in this application since they are considerably more stable (less bleaching under the laser beam) than the standard fluorescein or rhodamine fluorochromes.
8. After data collection, images can be analyzed using the software which accompanies the microscope.

3.2. Preparation of HeLa Cells for RIP

1. Grow low passage HeLa cells in DMEM medium supplemented with 10% FCS in 175 cm^2cell culture flasks to approximately 90% confluence. Remove medium and wash cell layer with 1X PBS. Repeat procedure 2–3 times.
2. Add 2 mL Trypsin (2% w/v stock solution) to the cell layer. Incubate for 5 min at RT. Re-suspend detached cells in 20 mL DMEM, 10% FCS containing penicillin and streptomycin.
3. Centrifuge cells, discard supernatant, and take up pellets in 120–150 mL DMEM supplemented with 10% FCS, penicillin, and streptomycin. Split the cell suspension into 4–5 new 175 cm^2cell culture flasks. Grow cells to 80–90% confluence and then subjected to cross-linking (*see* **Note 3**). All of the above procedures require sterile conditions and are carried out under a laminar flow hood in a laboratory equipped for sterile tissue cell culture work.

3.3. Cross-Linking and Cell Lysis

1. All materials for the protocol are prepared in advance. For each flask prepare 50 mL of DMEM medium without FCS and antibiotics, but supplemented with 1% v/v formaldehyde (final concentration). Remove cell culture flasks from incubator and replace culture medium with 50 mL DMEM, 1% formaldehyde. Place flasks on a slowly shaking platform and allow the cross-linking reaction for 10 min at RT. Sterile conditions are not required.

2. Quench the cross-linking reaction by adding 5 mL of a 2 M glycine solution to reach a final concentration of 0.2 M glycine. Incubate for an additional 10 min period at RT.
3. After cross-linking, rinse the cells twice with 20 mL 1X PBS and subsequently scrape them using a cell scraper or equivalent to detach them from the bottom of the flasks. Immediately transfer the cell suspension to a cold pre-chilled 50 mL reaction tube. Repeat the procedure for all flasks, in the meantime keeping each 50 mL tube on ice until further processing (*see* **Note 4**).
4. Centrifuge cell suspensions for 10 min at 800*g* at 4°C, discard the supernatants, re-suspend cell pellets in 4 mL lysis buffer (*see* **Section 2**), and incubate them on ice for 20 min.
5. For isolation of intact nuclei, transfer the suspension into a chilled glass tissue grinder with a tight fitting bulb-pestle. Break open cells by smoothly moving the pestle up and down in the grinder avoiding formation of foam. To keep the suspension cold, perform the grinding in an ice box. Repeat the movement of the pestle about 20 times.
6. Transfer the lysed cells in a 15 mL reaction tube and spin at 500*g*, 4°C for 10 min. The resulting pellet contains intact nuclei. Take them up in 3 mL ice-cold 1X PBS and centrifuged again at 500*g*, 4°C for 10 min. To remove any remaining traces of buffer, after centrifugation discard supernatant using a micropipette. At this stage the pellet of highly pure cross-linked nuclei can be flash-frozen in liquid nitrogen and stored for up to three months at −70°C or even up to a year in liquid nitrogen (*see* **Note 5**).

3.4. Chromatin Fragmentation and Preparation of Soluble Chromatin

1. To isolate soluble chromatin, re-suspend the nuclear pellet in 1 mL cold 1X PBS. Sample the nuclear suspension à 250 μL in 0.5 μL reaction tubes. Place tubes in a pre-chilled Diagenode Bioruptor water-bath sonifier (or equivalent). We set the pulse at 30 s and energy input on "H" high levels. Shear chromatin for an initial cycle of 10 min. Refill water bath with ice and subject the suspension to another cycle of 5 min. After shearing, pool aliquots in 1.5 mL reaction tube(s) and spin for 35 min at 18,000–20,000*g* at 4°C. Collect the supernatant at the end of the centrifugation run.
2. During the centrifugation run, prepare the material required for further processing of the solubilized chromatin. For equilibration of sepharose G beads (*see* **Section2**), adjust 50 μL bed volume with 1 mL 1X PBS and centrifuge for 5 min at 700*g* at 4°C. Discard supernatant and repeat procedure twice.

3. Load clarified nuclear extract containing soluble chromatin (obtained from Step 1) on the pre-equilibrated sepharose G beads (from Step 2). Close tightly the reaction tube and place it on a rotating device at 4°C for 1 h. This step is referred to as pre-clearing of the nuclear extract containing soluble chromatin. It is important because it represents a rather reliable way to get rid of those nuclear extract components which bind to the resin in a non-specific manner. Centrifuge samples at 700*g*, at 4°C for 5 min. Collect pre-cleared nuclear extract containing fragmented soluble chromatin and proceed to immunoprecipitation (*see* **Note 6**).

3.5. Immunoprecipitation

1. Incubate 200–250 μL aliquots of pre-cleared chromatin with primary antibodies against CBP20, pol II CTD, hnRNP U, and TBP (**Fig. 14.3**). In parallel, incubate an aliquot of pre-cleared chromatin with non-specific rabbit IgGs used as controls for the specificity of the immunoprecipitation reactions. Normally antibodies are added to a final concentration of 2–8 μg/mL. Incubation with specific antibodies is performed for 6–18 h under continuous agitation in a rotating platform placed at 4°C. Each aliquot is also supplemented with 20 U of RNaseGuard. Prior to incubation with the antibody of interest, keep 25–50 μL of pre-cleared chromatin stored at −70°C as input control for RNA extraction and PCR analysis discussed in the next steps (*see* **Note 7**).
2. For each immunoprecipitation reaction, pre-equilibrate 25 μL bed volume of sepharose G with 1X PBS using clean 1.5 mL reaction tubes. Add the antibody chromatin mix from Step 1 and place reaction tubes on a rotating device at 4°C. Allow immunoprecipitations for 1–1.5 h. We have experienced that before incubation with equilibrated sepharose G beads it is good practice to perform a short 5 min centrifugation run in a pre-cooled microcentrifuge at 18,000–20,000*g* to remove eventual protein aggregates that may in turn interfere with the efficiency of the immunoprecipitation.
3. After antibody incubation, spin samples at 700*g* at 4°C for 5 min and discard the supernatant. Re-suspend beads in 1 mL RIPA-1000 buffer (*see* **Section 2**) and place the reaction tubes on a rotating device at 4°C for 5 min. Centrifuge beads at 700*g*, at 4°C for 5 min. At this point, discard supernatant and re-suspend beads in 1 mL RIPA buffer (*see* **Section 2**). Place the tubes on a rotating device at 4°C and incubate for 5 min. Repeat this washing procedure at least three times.

4. As optional step, beads can be further washed by re-suspending them in 1 mL LiCl–wash buffer (*see* **Section 2**), incubated at 4°C on a rotating device for 5 min and centrifuged at 700*g* for 5 min at 4°C. Discard the supernatant at the end of the procedure.
5. Finally, re-suspend both sepharose G beads and input (from Step 1) in 100 μL elution buffer and place the tubes in a pre-heated block for 1–1.5 h. This step is required to reverse the cross-linked RNA–protein complexes (*see* **Note 8**).

3.6. RNA Preparation and DNase I Treatment

Even though there are many commercially available kits for RNA preparation from cell extracts or lysates, we almost entirely rely on the use of the phenol-based TRI Reagent™ for RNA extraction. The TRI Reagent allows for a reliable and fast method which is also suitable for high-throughput extractions on multiple samples (*see* **Note 2**). Here we propose the protocol which is used in our laboratory and adapted from Chomczynski (1993) (14).

1. After reversion of the cross-linking, add 500 μL of TRI Reagent™. Vortex shortly and incubate at RT under permanent agitation for 5 min. Add 100 μL chloroform to each sample, vortex for 15–20 s, and incubate samples at RT under permanent agitation for 15 min.
2. After incubation, centrifuge samples at 18,000–20,000*g*, 4°C for 15 min. Transfer the aqueous supernatant containing RNA to new, clean 1.5 mL reaction tubes. Add 250 μL isopropanol to each tube, vortex, and allow samples to stand for 10 min at RT.
3. Centrifuge samples at 18,000–20,000*g*, 4°C for 10 min. Carefully remove the supernatant, add three volumes of ice cold 100% ethanol, and place the reaction tubes in dry ice for 15–30 min to precipitate RNA. The majority of the DNA fraction remains at the interphase (*see* **Note 2**).
4. Centrifuge samples at 18,000–20,000*g*, 4°C for 10 min and re-suspend pellets in 1 mL 70% ethanol by vortexing. Samples can be stored in 70% ethanol at −70°C for up to 3 months.
5. To remove the 70% ethanol solution, centrifuge samples at 18,000–20,000*g*, 4°C for 10 min. Carefully remove the supernatant and dry samples in a speed-vac at maximum vacuum and rotating speed without applying additional heat.
6. Perform DNase I treatment to get rid of the contaminating DNA fraction in the above samples. Prepare a DNase I reaction master-mix at a final concentration of 0.08 U/μL in 20 mM Tris-HCl, pH 8.5, 2 mM $MgCl_2$ and supplement with RNaseGuard (0.02 U/mL).

7. Re-suspend pellets in 15 µL DNase I reaction mix and allow enzymatic digestion for 20 min at RT.
8. For DNase I inactivation, add 1.5 µL 50 mM EDTA to each sample, vortex shortly, and place in a heating block with the temperature set at 70°C for 10 min.
9. If samples are not utilized immediately, freeze, and store them at –70°C for maximum 3 months. Otherwise proceed to cDNA synthesis (*see* **Note 9**).

3.7. cDNA Synthesis

1. Assemble a typical reaction mix in a total volume of 20 µL. Prepare annealing mix in 0.2 mL PCR reaction tubes using 5 µL RNA sample, 1 µL N_6random oligo primer (12.5 ng/µL final concentration), 1 µL from a 10 mM dNTP mix, 1 µL gelatine and bring to volume with DEPC-treated bi-distilled water.
2. Pre-heat the above mixture in the PCR thermocycler machine at 65°C for 5 min and immediately chill on ice. Add pre-mixed solutions for reverse transcription using 4 µL 5X First Strand Buffer (provided by the manufacturer), 2 µL from a stock solution of 0.1 M DTT, 1 µL RNaseGuard™ and 1 µL Superscript™ II reverse transcriptase.
3. Once the reaction has been assembled, incubate samples at 25°C for 12 min (annealing step), at 42°C for 50 min (elongation step), and finally leave at 70°C for 15 min (inactivation step).
4. If screening for pre-mRNA, add 2 U of RNase H to each reaction mix and incubate for an additional 20 min at 37°C.
5. At the end of the procedure, store samples at –20 or –70°C for a longer period of time, otherwise proceed to PCR amplification (*see* **Note 10**).

3.8. Primer Design and PCR Analysis of cDNA

Preliminary considerations are required for gene-specific primers design. It is important to be aware of the type of transcript that we want to analyze, whether it is unprocessed pre-mRNA or processed mRNA (**Fig. 14.2**). In addition, certain physical parameters such as GC content and length must be optimized (*see* **Note 11**).

We recommend using open source tools which search in the sequence template of interest for the best fitting primer combinations. In the following protocol steps, we propose the use of online accessible *in silico* tools which ensure the use of proper nucleic acid sequence templates and design of oligonucleotides with recommended physical parameters.

3.8.1. Choice of Amplification Regions Within a Specific mRNA Sequence

1. To analyze whether CBP20, hnRNP U, pol II CTD, and TBP are associated with nascent mRNA, use the mRNA encoding ribosomal protein S19 (15)as model gene (NG007080.2/ NM0011022.3). For the design of S19 mRNA-specific primers utilize exon–exon junctions as targets (*see* **Note 12**). Choose potential primer sequences using the tool Primer3Input which is found at http://frodo.wi.mit.edu/cgi-bin/primer3/primer3_www.cgi. Paste in the amplification region of interest and adjust the physical parameters as follows:
   - mispriming library: *Homo sapiens*
   - optimal GC content: 50%
   - optimal T_m: 59°C
   - optimal sequence length: 20 bp
   - GC clamp: 1, poly X: 3
   - erase all ranges from the product size window except of 150–250 and 100–300 bp.
2. To cross-validate physical parameters of the chosen S19 mRNA primer sequences use the program Oligonucleotide Properties Calculator found at http://www.basic.northwestern.edu/biotools/oligocalc.html. Paste in the potential oligonucleotide sequences and check for the calculated T_m and for self-complementarity.
3. If no self-complementarity is detected proceed with the "blastn" search at http://www.ncbi.nlm.nih.gov/blast/Blast.cgi. Choose option "nucleotide blast" and paste in the oligonucleotide sequence. Set species parameters and perform blast search. We recommend using only those sequences which exhibit 80–100% complementarity as primers for the target mRNA species (*see* **Note 13**).

3.8.2. Specific PCR Analysis of cDNA

1. Analyze the cDNA prepared as above described using the following S19 mRNA specific primers: forward primer 5′-ACGCGAGCTGCTTCCACAG and reverse primer 5′-AGCTGCCACCTGTCCGGC. As control for the template specificity, use non-reverse transcribed material as template for the PCR reaction. As control for the specificity of the immunoprecipitation experiments, analyze the cDNA also with primers specific for $tRNA_{Tyr}$, forward primer 5′-CCTTCGATAGCTCAGCTGGTAGAGCGGAGG and reverse primer, 5′-CGGAATTGAACCAGCGACCTAAGGATGTCC (**Fig. 14.2A**).
2. The materials for the PCR protocol are made ready in advance. First, transfer 0.6 μL from each cDNA sample into a thermocycler-approved 0.2 mL PCR reaction tubes. Second, assemble a reaction master mix for all samples according to the example reported below in a final volume of 25 μL:

1 μL	dNTP-mix (10 mM stock solution)
0.1 μL	S19 mRNA forward primer (100 μM stock solution)
0.1 μL	S19 mRNA reverse primer (100 μM stock solution)
1.25 μL	DMSO
2.5 μL	10X PCR buffer (*see* **Section 2**)
0.75 μL	$MgCl_2$(50 mM stock solution)
0.5 μL	Taq polymerase (5 U/μL stock solution)
17.95 μL	double-distilled water.

3. Transfer 24.4 μL of the above master mix to each PCR tube containing 0.6 μL of cDNA template and place tubes in a pre-programmed PCR thermocycler.
4. For the PCR reaction, an initial period of 4 min at 95°C is required for efficient sample denaturation. Each sample is then subjected to a cycle that consists of 30 s incubation at 95°C (denaturation step), followed by 30 s at a temperature between 58 and 60°C which depends on the primer sequences used (annealing step) and 30 s at 72°C (elongation step). Reiterate each cycle 25–29 times. At the end of the PCR reaction allow a longer incubation at 72°C for 7 min (*see* **Note 14**).

3.8.3. Visualization of PCR Products for Densitometry Measurement

1. Prepare a 2% (w/v) agarose gel in 1X TAE buffer. Heat up the solution until cooking point is reached and all agarose powder is melted.
2. Add ethidium bromide from a stock solution to a final 1/10,000 dilution (*see* **Section 2**).
3. Prepare gel chamber with appropriate combs (20–30 μL sample volume) and cast the gel. Allow enough time for the agarose to cool down and polymerize. Fill chamber completely with 1X TAE buffer until the gel slab is covered and remove the combs.
4. Mix PCR reaction with appropriate amount of 6X DNA loading dye (*see* **Section 2**).
5. Load 20 μL of each sample in individual wells. Load 2.5–4 μL of ready-to-use DNA ladder in the last well (*see* **Section 2**).
6. Run gel electrophoresis at constant voltage, 15 V/cm of gel slab, approximately corresponding to 121 V for one agarose gel slab with 8 cm migration length. Run gel electrophoresis until the migration lane of the xylene and bromophenol blue dyes reach a distance of 1.5 cm (approximately 15 min of run time).

7. Remove gel slab from chamber and place it in a gel scanner. If quantification of band is not possible with the software provided, export gel scans to tif (tagged image file) format in 16 bit mode (*see* **Note 15**).

3.9. Densitometry and Statistical Evaluation of the PCR Data

1. Analyze conventional PCR experiments in a semi-quantitative manner by densitometry measurements, performed on the PCR products separated by agarose gel electrophoresis (*see* **Note 16**). A number of different softwares are available for quantitative analysis of the scanned gels. We commonly use ImageJ which is user friendly and allows several options in combination with Microsoft Excel.
2. Open tif version of the scanned gel using Image J. Measure band and general background signal of the agarose gel. Repeat for each sample lane. Export measurements to a Microsoft Excel work sheet.
3. Open the Excel document and subtract background from signal values obtained from the densitometry measurement of each band. Repeat measurements on each gel at least three times.
4. Compare signals obtained from specific immunoprecipitations against those signals obtained from control immunoprecipitations. Plot results in a graph, including standard deviations to determine the significance of the measured values (**Fig. 14.3B**).

4. Notes

1. Mowiol (Mowiol 4.88, Calbiochem cat. no. 475904) is a solution of polyvinyl alcohol which normally hardens overnight after slide preparation, and does not require the cover slips to be sealed with nail polish. Do not use so much mounting solution that the cover slips are floating. Normally, 15–20 μL is sufficient for 22 × 22 mm cover slips. 22 × 50 mm cover slips require about 40–50 μL. The addition of PPD (*p*-phenylenediamine or 1,4-benzenediamine hydrochloride, Sigma) is recommended to reduce bleaching of fluorescent probes. Note that PPD is carcinogenic and should be handled with care. Make up a 0.1% aqueous solution of PPD, aliquot, and freeze in 1 mL Eppendorf tubes wrapped in aluminum foil. To use, thaw at RT and add 1 part to 9 parts of Mowiol. Refreeze immediately after use. Discard any solution that becomes pink-brown in colour.

2. The Tri Reagent is a complete and ready-to-use solution for the isolation of total RNA or the simultaneous isolation of RNA, DNA, and protein from diverse biological material, including samples of human, animal, plant, yeast, bacterial, and viral origin. This highly reliable technique performs well with samples larger than 5 mg tissue or 5×10^5 cultured cells. The Tri Reagent solution combines phenol and guanidine thiocyanate in a monophasic solution to rapidly inhibit RNase activity. A biological sample can be homogenized or lysed in Tri Reagent solution, and the homogenates or lysates are then separated into aqueous and organic phases by adding chloroform and centrifuging. RNA partitions to the aqueous phase, DNA to the interphase, and proteins to the organic phase. Next the RNA is precipitated from the aqueous phase with isopropanol, and finally it is washed with ethanol and re-suspended in the buffer of choice.
3. It is crucial for the reproducibility of a RIP experiment that all cell cultures are started from the same batch of cells and have the same level of confluence. We recommend amplification of low-passage HeLa cell batches and freeze several aliquots to be used again as starting cultures for RIP experiments.
4. From this step on, we strongly recommend the use of DEPC-treated water for all aqueous solutions and buffers. For DEPC-treatment, add 1 mL of DEPC per 1000 mL bi-distilled water and perform standard autoclaving at 121°C for 15 min.
5. For cross-linking, the incubation time with formaldehyde is strictly dependent on the room temperature. We have observed season-specific fluctuations in the cross-linking efficiency when working in non-air-conditioned laboratory spaces. As mentioned in **Section 2**, formaldehyde is highly volatile and very toxic upon inhalation. Therefore, it is crucial to perform all steps which entail the use of formaldehyde under a chemical hood and the cell culture flasks remained closed during the actual cross-linking procedure. As a general tip, the cross-linking step has to be optimized for each cell type. Pilot experiments using different cell lines, settings, and incubation times should be performed. In each case, after reversion of the cross-linking and phenol extractions, the size of the DNA should be evaluated by agarose gel electrophoresis.
6. In general, the sonication conditions including those presented in this chapter are empirically estimated depending on the instrument available in the laboratory. In any case at the end of the sheering process, the experimental setup must yield a chromatin population consisting of 500–700 bp chromatin fragments. Treatment with DNase I can help to improve chromatin fragmentation. Nevertheless we do not

recommend the use of DNase I to samples prior to the immunoprecipitation step because this usually leads to increased levels of unspecific stickiness. Another major argument against the use of DNase I at this stage is the high affinity for actin. Since actin is an essential regulator of gene transcription and component of pre-mRNP particles (16), the use of DNase I at this stage of the protocol may interfere with the entire procedure through unwanted depletion and disruption of actin associated RNA-binding factors within the RNP.

7. We recommend that the best primary antibody dilution should be empirically estimated prior to the RIP assay. For ChIP experiments, one can find information about recommended antibody concentrations on the Websites of commercial manufacturers. In general, these dilutions can be applied also in RIP experiments. If the antibody is raised in the lab we recommend to use it 10-fold more concentrated in comparison to Western blotting.

8. It is crucial to perform the sequential wash with RIPA-1000 and RIPA at least three times to increase the signal-to-noise ratio. In our hands the additional LiCl–wash step does not improve the signal-to-noise ratio, suggesting that this step may not be a general requirement. It may be needed for certain antibodies or, alternatively LiCl washes may improve the experiment when analyzing abundant RNA transcripts derived from genes transcribed at high rates. It is also advisable that at the end of the immunoprecipitation, during the washing steps, one should avoid high-speed centrifugation because beads may be sensitive to mechanical shock.

9. It is crucial to use DNase I with high degree of purity. The conventional DNAse I as lyophylized powder is cheaper but retains a high degree of RNAse and protease activity. Therefore, we recommend the use of the commercially available RNase-free DNase I. Moreover, before proceeding to cDNA synthesis, we also recommend splitting the RNA samples in 3X 5 μL aliquots and freeze the remaining samples which are not utilized.

10. cDNA synthesis is an important step not only to reverse-transcribe RNA to DNA and, therefore, being able to run PCR analysis, but also for creating stable templates which can be stored at –70°C for several years.

    When planning cDNA synthesis, two types of mRNA molecules can be analyzed. Pre-mRNA represents the primary gene transcript which is not yet fully processed and, therefore, still contains exon and intron sequences. Alternatively, mRNA represents the fully processed transcript which does not contain introns. The cDNA synthesis procedure has to be designed depending on whether we analyze pre-mRNA or mRNA transcripts. For instance, the use of oligo-d(T) primers during cDNA synthesis a priori excludes the possibility to analyze pre-mRNA because this primer only anneals with the mRNA poly (A)-tails (**Fig. 14.2**). On the other hand, the use of random N6 hexamer primers allows PCR analysis of both pre-mRNA and mRNA transcripts and moreover, allows intrinsic control for the specificity of the RNA precipitations.

11. If screening for pre-mRNA, consideration should be given to the fact that all important maturation processes take place co-transcriptionally with a high degree of processivity. Because of this transient nature, it is more difficult to analyze pre-mRNA transcripts in comparison to mature mRNA molecules, which persist in the cell nucleus for a longer time and are, therefore, easier to detect. In addition, it is also important to be aware of possible naturally occurring splicing variants for the transcript of interest. Finally, it is necessary to identify whether within the genome analyzed, the gene of interest exhibits intron-less variants for instance resulting from ancient retroviral activity.

    For specific mRNA, amplification primers must be designed with the correct physical parameters, including a GC content of around 50% of the entire nucleotide sequence and 18–21 bp in length. For the application discussed in this chapter, we recommend the use of primer sequences which have a slightly higher melting temperature (T_m= 59–63°C) in comparison to genome-specific oligonucleotides (T_m= 55–58°C). Generally, higher melting temperatures allow higher degree of sensitivity and can be adjusted to standard PCR conditions by running the reactions in the presence of DMSO. In our experience, DMSO allows to decrease the primers annealing temperature by approximately 2°C.

12. For pre-mRNA analysis, design at least two types of primer pairs. The schematic illustration in **Fig. 14.2A** provides simple guidelines, where forward primers are complementary to an exon or an exon–intron junction and reverse primers are complementary to the adjacent intronic sequence (**Fig. 14.2A**). For mRNA analysis, use exon–exon junctions as targets for the design of primers (**Fig. 14.2B**). For simultaneous screening of pre-mRNA with respect to mRNA, we recommend the use of exclusively exon-binding primer sequences. The primer combination identified to simultaneously analyze pre-mRNA with respect to mRNA is suitable only if the exon linking intron sequence is short and does not result in large amplification products which are longer than 500 bp.

13. When designing primers, if the gene sequence information is not available, go to the site http://www.ncbi.nlm.nih.gov/sites/entrez?db=pubmed to obtain information about gene sequence and corresponding mRNA. Choose the "Gene" database and type in the search window "gene-name" AND "Organism". For already sequenced and characterized genes for instance in the human- and mouse genomes, one normally obtains information about gene sequence with its 5′ and 3′ untranslated region (UTR) as well as its intron–exon positions. One can also extract the corresponding mRNA sequence as EST annotation.
14. The number of PCR cycles has to be empirically determined. Moreover, primer-annealing temperatures are dependent on the physical parameters of the corresponding primer oligonucleotides and can be evaluated with publicly available algorithms using the link provided in **Section 2**. Finally, for densitometric measurements the final PCR products separated by agarose gel electrophoresis should not be saturated after ethidium bromide detection.
15. As mentioned in **Section 2**, ethidium bromide is an extremely toxic mutagen. Handle with care, use gloves, wear lab coat, and perform agarose gel electrophoresis in a place reserved for ethidium bromide work only. In addition, to visualize the DNA fragments a UV transilluminator is needed. In this situation, the operator risks to be exposed to UV light when monitoring the DNA migration on the gel. Therefore, to avoid serious eye problems, it is mandatory to use goggles when analyzing the DNA run under a UV light source.
16. Quantitative real-time PCR should be considered as a method of choice to analyze PCR data. Even though rather expensive, the method is extremely sensitive and allows unbiased quantification. If using quantitative real-time PCR, simply skip the PCR screening reported in Step 3.6.2 and directly perform real-time PCR reactions on the cDNA following the manufacturer's instruction manual. Compare the amplification curves of cDNA obtained from specific immunoprecipitations with the curves obtained in control immunoprecipitations using the software provided.

Acknowledgments

Our work is supported by grants from the Swedish Research Council and Cancerfonden to PP.

References

1. Granneman, S. and Baserga, S. J. (2005) Crosstalk in gene expression: coupling and co-regulation of rDNA transcription, pre-ribosome assembly and pre-rRNA processing. *Curr. Op. Cell Biol.* **17**, 281–286.
2. Eissenberg, J. C. and Shilatifard, A. (2006) Leaving a mark: the many footprints of the elongating RNA polymerase II. *Curr. Opin. Genet. Dev.* **16**, 184–190.
3. Hirose, Y. and Ohkuma, Y. (2007) Phosphorylation of the C-terminal domain of RNA Polymerase II plays central roles in the integrated events of Eucaryotic gene expression. *J. Biochem.* **141**, 601–608.
4. Daneholt, B. (2001) Assembly and transport of a pre-messenger RNP particle. *Proc. Natl. Acad. Sci. U. S. A.* **98**, 7012–7017.
5. Dreyfuss G., Kim, V. N. and Kataoka, N. (2002) Messenger-RNA-binding proteins and the messages they carry. *Nat. Rev. Mol. Cell Biol.* **3**, 195–205.
6. Wetterberg, I., Zhao, J., Masich, S., Wieslander, L. and Skoglund, U. (2001) In situ transcription and splicing in the Balbiani ring 3 gene *EMBO J.* **20**, 2564–2574.
7. Daneholt, B. (2001) Packing and delivery of a genetic message. *Chromosoma* **110**, 173–185.
8. Gilbert, C., Kristjuhan, A., Winkler, G. S. and Svejstrup, J. Q. (2004) Elongator interactions with nascent mRNA revealed by RNA immunoprecipitation. *Mol. Cell* **14**, 457–464.
9. Batsche, E., Yaniv, M. and Muchardt, C. (2006) The human SWI/SNF subunit Brm is a regulator of alternative splicing. *Nat. Struct. Mol. Biol.* **13**, 22–29.
10. Sims III, R. J., Millhouse, S., Chen, C.-F., Lewis, B. A., Erdjument-Bromage, H., Tempst, P., Manley, J. L. and Reinberg, D. (2007) Recognition of trimethylated histone H3 lysine 4 facilitates the recruitment of transcription postinitiation factors and pre-mRNA splicing. *Mol. Cell* **28**, 665–676.
11. Fomproix, N. and Percipalle, P. (2004) An actin–myosin complex on actively transcribing genes. *Exp. Cell Res.* **294**, 140–148.
12. Kiledjian, M. and Dreyfuss, G. (1992) Primary structure and binding activity of the hnRNP U protein: binding RNA through RGG box. *EMBO J.* **11**, 2655–2664.
13. Visa, N., Izaurralde, E., Ferreira, J., Daneholt, B. and Mattaj, I. W. (1996) A nuclear cap-binding complex binds Balbiani ring pre-mRNA cotranscriptionally and accompanies the ribonucleoprotein particle during nuclear export. *J. Cell Biol.* **133**, 5–14.
14. Chomczynski, P. (1993) A reagent for the single-step simultaneous isolation of RNA, DNA and proteins from cell and tissue samples. *Biotechniques* **15**, 532–534, 536–537.
15. Wilson, D. N. and Nierhaus, K. H. (2005) Ribosomal proteins in the spotlight. *Critic. Rev. Biochem. Mol. Biol.* **40**, 243–267.
16. Percipalle, P. and Visa, N. (2006) Molecular functions of nuclear actin in transcription. *J. Cell Biol.* **172**, 967–971.

Chapter 15

Methyl DNA Immunoprecipitation

Jean-Jacques Goval and Juana Magdalena

Abstract

Epigenetics is the study of heritable changes in gene expression. Chromatin immunoprecipitation (ChIP) and methylation status analysis of genes have been applied to the study of epigenetic modifications, often perturbed in human cancer. ChIP is a technique allowing the analysis of the protein association with specific genomic regions in the context of intact cells. ChIP and immunoprecipitation (IP) of methylated DNA, both rely on the use of well-characterized specific antibodies. The first is described in Chapter 2 and the second is shown here. At Diagenode, a novel METHYL kit has been designed to immunoprecipitate methylated DNA (Methyl DNA IP). This kit allows you to perform DNA methylation analysis of your sample together with optimized internal IP controls, all in one tube. This brand new Methyl DNA IP method provides methylated DNA (meDNA) and unmethylated DNA (unDNA) controls to be used together with your DNA sample, allowing direct correlation between immunoprecipitated material and methylation status. Such methylation analysis is highly specific and each IP is quality controlled, two essential keys for reliable results. In addition, the kit protocol is fast and user-friendly.

Key words: Immunoprecipitation, methylated DNA, MeDIP, internal controls.

1. Introduction

Current methods used for the detection of methylated DNA are methods based on methylation-sensitive enzymatic digestion (1), bisulfite treatment (2–4), enrichment by binding to methylated binding domain (MBD) (5–8), and enrichment by immunoprecipitation of methylated DNA (9). DNA methylation detection assays using methylation-sensitive restriction enzymes to digest unmethylated DNA while leaving methylated DNA intact are DNA-sequence dependent and not compatible with high-throughput (HTP) analysis. The use of sodium bisulfite to deaminate cytosine to uracil while leaving 5-methylcytosine intact is a cumbersome

Philippe Collas (ed.), *Chromatin Immunoprecipitation Assays,* Methods in Molecular Biology 567,
DOI 10.1007/978-1-60327-414-2_15, © Humana Press, a part of Springer Science+Business Media, LLC 2009

technique, which involves subsequent optimization of specific PCR or sequencing, involving time- and labor-intensive chemical treatments that damage DNA and limit throughput. It was shown that the methyl-binding domain (MBD) of MeCP2 has some sequence preference besides its recognition of methyl-CpGs (10) and that the MBD method does require relatively high methyl-CpG density (11). Methylated DNA enrichment by binding to MBDs can, therefore, be sequence dependent and can also show low specificity and potential for false-positive results due to capture of unmethylated DNA. The MBD method combined with enzymatic restriction of the DNA has been adapted to HTP (8, 12).

Yet another way to enrich for methylated DNA is by Methyl DNA IP. Methyl DNA IP uses bead-immobilized anti-5-methyl cytosine antibodies to isolate the methylated DNA, which allows highly efficient enrichment of methylated DNA dose-dependent and sequence-independent, with high specificity. Moreover, the use of the Methyl DNA IP technique is compatible with high-throughput platforms (13) and can directly give reliable qualitative results as well as semi-quantitative data. The use of antibody instead of MBD to pull-down methylated sequences presents, therefore, several advantages. In addition, it is important to point out that monoclonal antibody is produced with far less lot-to-lot variation in comparison with a fusion protein expressed in *E. coli*.

Our novel METHYL kit is designed to immunoprecipitate methylated DNA. This kit allows you to perform DNA methylation analysis of your sample including optimized internal IP controls.

Table 15.1
qPCR module following Methyl DNA IP

Primer pairs (10 μM each)	Specificity (size of amplified DNA)	Input DNA sample (which includes Ctrls) amplification:	Methyl DNA IP (which includes Ctrls) amplification:
hum meDNA primer pair (AlphaX1)	Human DNA (81 bp)	Yes (*if sample is*)	Yes
hum unDNA primer pair (GAPDH)	Human DNA (102 bp)	Yes (*human DNA*)	No
meDNA pos control primer pair #1	Kit Positive Ctrl (81 bp)	Yes	Yes
meDNA pos control primer pair #2	Kit Positive Ctrl (87 bp)	Yes	Yes
unDNA neg control primer pair #1	Kit Negative Ctrl (84 bp)	Yes	No
unDNA neg control primer pair #2	Kit Negative Ctrl (92 bp)	Yes	No

The internal IP controls consist of methylated DNA (meDNA) and unmethylated DNA (unDNA) added to your DNA sample, such that a direct correlation between immunoprecipitated material and methylation status can be done. This methylation analysis is highly specific due to the use of a well-characterized monoclonal antibody and each IP is directly quality controlled: two essential keys for reliable results. In addition, the kit protocol is fast and user-friendly. The METHYL kit includes three modules; they are used sequentially as follows for genomic DNA preparation (*see* **Section 3.1**), immunoprecipitation of methylated DNA (*see* **Section 3.2**), and qPCR analysis of the immunoprecipitated DNA (*see* **Section 3.3** and **Table 15.1**). Each module is provided with adapted buffers and detailed protocols.

2. Materials

2.1. DNA Preparation and Shearing

1. Cultured cells and Trypsin–EDTA.
2. GenDNA module (cat. no. mc-green-002, Diagenode).
3. Phosphate buffered saline (PBS).
4. Phenol:chloroform:isoamyl alcohol (25:24:1), chloroform:isoamyl alcohol (24:1), 100% ethanol, 70% ethanol. Fume hood. Vortex.
5. Agarose and TAE buffer, DNA molecular weight marker.
6. Bioruptor (cat. no. UCD-200, Diagenode).

2.2. Methylated DNA Immunoprecipitation and Analysis of Immunoprecipitated DNA

1. METHYL kit (cat. no. mc-green-003, Diagenode), which includes the Methyl DNA IP module (cat. no. mc-green-001) and the GenDNA and qPCR modules (cat. no. mc-green-002).
2. Autoclaved tips.
3. Rotating wheel.
4. Thermomixer (50 and 65°C).
5. Incubator (37°C).
6. Quantitative PCR facilities and reagents.

3. Methods

3.1. DNA Preparation and Module

The GenDNA module from Diagenode has been optimized for the preparation of genomic DNA from cultured cells to be then used in Methyl DNA IP (*see* **Note 1**). The goal of this first step is to get high molecular weight genomic DNA. Cell culturing is the starting point before cell collection and lysis described below.

3.1.1. Cell Collection and Lysis

1. Pellet suspension culture out of its serum-containing medium. Trypsinize adherent cells and collect cells from the flask. Centrifuge at 300*g* for 5 min at 4°C.
2. Discard the supernatant. Resuspend cells in 5–10 mL ice-cold PBS. Centrifuge at 500*g* for 5 min. Discard the supernatant. Repeat this resuspension and centrifugation step once more. This step is to wash the cells.
3. Meanwhile, place the GenDNA digestion buffer at room temperature (RT) and the GenDNA proteinase K on ice.
4. Add GenDNA proteinase K to the GenDNA digestion buffer before use. The stock of provided proteinase K is 200X; e.g., add 5 μL per 1 mL of digestion buffer, i.e., the freshly prepared complete digestion buffer to be used directly.
5. Resuspend cells in complete digestion buffer (1 volume). For 3 million cells, use 300 μL complete digestion buffer. For 10 million cells, use 500 μL complete digestion buffer. It might be necessary to use more buffer to avoid problems when performing the extractions below. If necessary, for 3 million cells, use up to 600 μL of buffer. For 10 million cells, use up to 1,000 μL of buffer.
6. Cell lysis: Incubate the samples with shaking at 50°C for 12–18 h in tightly capped tubes. That is the cell lysis step. At this stage, samples are viscous. After 12 h incubation the tissue should be almost indiscernible, a sludge should be apparent from the organ samples, and tissue culture cells should be relatively clear.

3.1.2. Extraction of Nucleic Acids and DNA Purification

1. Thoroughly extract the samples with an equal volume of phenol:chloroform:isoamyl alcohol. Add 1 volume of phenol:chloroform:isoamyl alcohol (25:24:1). One volume is about 500 μL. It is possible to incubate the samples at RT for 10 min on a rotating wheel before centrifugation. Use gentle rotation and do not vortex. Work under a fume hood.
2. Centrifuge at 1,700*g* for 10 min in a swinging bucket rotor.
3. Transfer the aqueous (top) layer to a new tube. Increase volume if necessary (see above) and pipette slowly.
4. Add 1/2 volume of GenDNA precipitant and 2 volumes of 100% ethanol (*see* **Note 2**). That is to purify the DNA. One volume is about 500 μL and corresponds to the original amount of top layer. Add therefore 250 μL of precipitant and 1,000 μL of 100% ethanol. The DNA should immediately form a stringy precipitate.

5. Recover DNA by centrifugation at 1,700*g* for 2 min. This brief precipitation in the presence of an optimized high salt precipitant (GenDNA precipitant) reduces the amount of RNA in the DNA sample. For long-term storage, it is convenient to leave the DNA in the presence of ethanol.
6. Rinse the pellet with 70% ethanol. Decant ethanol and air-dry the pellet. It is important to rinse extensively to remove any residual of salt and phenol.
7. Resuspend the pellet of DNA at ~1 mg/mL in GenDNA TE until dissolved. Shake gently at room temperature or at 65°C for several hours to facilitate solubilization. Store at 4°C. From 3 million cells, ~20–30 μg of DNA can be expected in a volume of 20–30 μL. From 10 million cells, ~50–100 μg of DNA can be expected (in a volume of 200–300 μL). If possible, it is recommended to get at least 30 μg of DNA (when enough material is available) to be able to work with 30 μg of DNA (*see* **Section 3.1.3**).
8. If necessary, residual RNA can be removed at this step by adding 2 μL of GenDNA RNase (DNase-free) per milliliter of DNA sample and incubating 1 h at 37°C, followed by phenol: chloroform extraction and ethanol precipitation (similar to above).
9. For DNA analysis, run samples in a 1% agarose gel along with DNA size marker to visualize the DNA preparation efficiency.

3.1.3. DNA Shearing

1. In a 1.5 mL tube, dissolve the DNA sample in TE to reach 0.1 μg/μL.
2. Use a final volume of 300 μL of DNA sample in 1.5 mL tubes.
3. Shear the DNA using the Bioruptor: at "LOW" power using the following cycles: (15 s "ON" and 15 s "OFF") for a total time of 10 min.
4. Sheared DNA can be analyzed on agarose gel.

3.2. Methylated DNA Immunoprecipitation and Methyl DNA IP Kit

In the Methyl DNA IP module, our antibody directed against 5-methyl cytidine is provided as well as meDNA and unDNA internal IP controls. The IP has been optimized to specifically select and precipitate the methylated DNA, using our antibody, buffers and protocol. The IP efficiency can be demonstrated for the IP internal controls and can serve as normalization purposes from IP to IP.

3.2.1. Immunoprecipitation of Methylated DNA and Washes

When preparing the IP incubation mix at this stage, the IP incubation mix contains the methylated and unmethylated DNA internal IP controls, your DNA sample being added at a future stage. When possible it is best to perform the IP at least in duplicate. Keep an input sample equivalent to 20% of the IP sample, which includes the two DNA internal IP controls and your DNA sample. Take into account that such input sample does not undergo IP, but is used as input control next to each IP when DNA is purified (*see* **Section 3.2.2**) and analyzed by PCR (*see* **Section 3.3**).

1. Prepare the IP incubation mix w/o DNA sample as follows. For one IP: 20.00 μL buffer A, 5.00 μL buffer B, 1.25 μL of positive meDNA control and 1.25 μL negative unDNA control and 37.5 μL water.
2. Label new 1.5 mL tubes. Add the DNA sample to the IP incubation mix.
3. Add per labeled "IP" tube: the IP incubation mix. Then, add 1 μg of DNA sample per tube. Using DNA samples at a concentration of 0.1 μg/μL: add 65 μL of IP incubation mix and 10 μL of DNA per tube. The total volume per IP is 75 μL. When using DNA samples at a concentration that is not 0.1 μg/μL, adjust the volumes.
4. Add per "input sample" tube: 20% of what is used per IP above. Using DNA samples at a concentration of 0.1 μg/μL, add 13 μL of IP incubation mix and 2 μL of DNA per tube. The total volume for 20% input is 15 μL. When using DNA samples at a concentration that is not 0.1 μg/μL, adjust the volume (use 20% of the volumes used per IP).
5. Incubate at 95°C for 3 min.
6. Quickly chill on ice (it is best to use ice-water).
7. Quickly perform a short spin at 4°C.
8. Label new 1.5 mL tubes: one per IP. Add beads to all tubes: 20 μL meDNA-IP blocked beads (50% suspension). Keep on ice. To be used in Step 11 below.
9. In a new tube, prepare the diluted antibody mix. For one IP: prepare a 1:10 antibody dilution as follows: (0.3 μL antibody, 0.6 μL buffer A, and 2.1 μL water). Then, add 2 μL of buffer C. Final volume is 5 μL. Scale the volumes accordingly based on the number of IPs that are performed on the day.
10. Add 5 μL of diluted antibody mix per IP tube (Step 3 above). Antibody is added to the IP tubes, which contain IP incubation mix and your DNA sample.
11. Mix and add to the labeled tubes containing beads prepared earlier (Step 8 above). That is the IP incubation which comprises the IP samples, the diluted antibody mix, and the beads. The final volume in each tube is 100 μL.
12. Place on a rotating wheel at 4°C for 4 h or overnight.
13. The Methyl DNA IP samples are then washed as follows: add 450 μL of ice-cold wash buffer to each IP tube, starting with wash buffer-1. Place the four wash buffers on ice and perform the washes in a cold room.

14. Rotate for 5 min at 4°C.
15. Centrifuge at 4,000*g* for 1 min at 4°C.
16. Discard the supernatant. Do not disturb the pellet. Keep the pellet.
17. Wash the pellet again (as described above: Steps 12–15) as follows: perform one more wash with wash buffer-1, then one wash with wash buffer-2, and one wash with wash buffer-3. Finally perform two more washes using the wash buffer-4.
18. After the last wash, discard the last traces of wash buffer (using a P200 pipette). Keep the bead pellets. These are the Methyl DNA IP samples. To the beads, the immunoprecipitated methylated DNA is bound.

3.2.2. DNA Elution and Purification

1. Take the input samples, centrifuge briefly, and from now onwards treat the input DNA samples and IP samples in parallel.
2. Prepare the complete elution buffer by mixing buffers D, E, and F as follows. For one IP: 360 μL of buffer D, 40 μL of buffer E, 16 μL of buffer F. The total volume is of 416 μL.
3. Add 416 μL of freshly prepared complete elution buffer to the bead pellets (the Methyl DNA IP samples).
4. Add 416 μL of freshly prepared complete elution buffer to the input samples.
5. Incubate in a thermo-shaker for 10 min at 65°C at 1,000–1,300 rpm.
6. Cool down samples to room temperature, add 1 volume of phenol:chloroform:isoamyl alcohol (25:24:1).
7. Centrifuge for 2 min at 14,000*g* at RT. Transfer the top aqueous phase into a new 1.5 mL tube.
8. Add 1 volume of chloroform:isoamyl alcohol (24:1).
9. Centrifuge for 2 min at 14,000*g* at RT. Transfer the top aqueous phase into a new 1.5 mL tube.
10. Thaw on ice the DNA co-precipitant.
11. Per tube: add 5 μL of the provided meDNA-IP co-precipitant and 40 μL of the meDNA-IP precipitant. Then, add 1 mL of ice-cold 100% ethanol. Mix well. Leave at -20°C for 30 min.
12. Centrifuge for 25 min at 14,000*g* at 4°C. Carefully remove the supernatant and add 500 μL of ice-cold 70% ethanol to the pellet.
13. Centrifuge for 10 min at 14,000*g* at 4°C. Carefully remove the supernatant, and leave tubes opened for 30 min at RT to evaporate the remaining ethanol. The pellets are (i) DNA that was purified from the sheared DNA (input sample(s)) and (ii) DNA that was isolated by IP (Methyl DNA IP samples).

14. Add of 50 μL TE to the IP and input samples. Suspend the DNA evenly; place the tubes in a shaker for 30 min at 14,000*g* at RT to dissolve the pellets.

3.3. qPCR Analysis of Immunoprecipitated DNA

This step consists in the analysis by qPCR of the purified DNA obtained from the sheared DNA (input sample(s)) and the DNA that was isolated by Methyl DNA IP (Methyl DNA IP sample(s)). The qPCR module is used to analyze input and Methyl DNA IP samples, and also internal kit DNA controls. The module includes indeed validated primer pairs specific to four types of DNA: *a:* the methylated DNA control (meDNA positive ctrls #1 and #2), *b:* the unmethylated DNA control (unDNA negative ctrls #1 and #2), *c:* one methylated human DNA region (X-linked alpha satellites), and *d:* one unmethylated human DNA region (GAPDH promoter). The description of the primer pairs provided in the METHYL kit is shown in **Table 15.1**: names, specificity, size of amplified DNA, and expected amplification with input DNA and with immunoprecipitated DNA are given. The results obtained are also shown (**Fig. 15.1**).

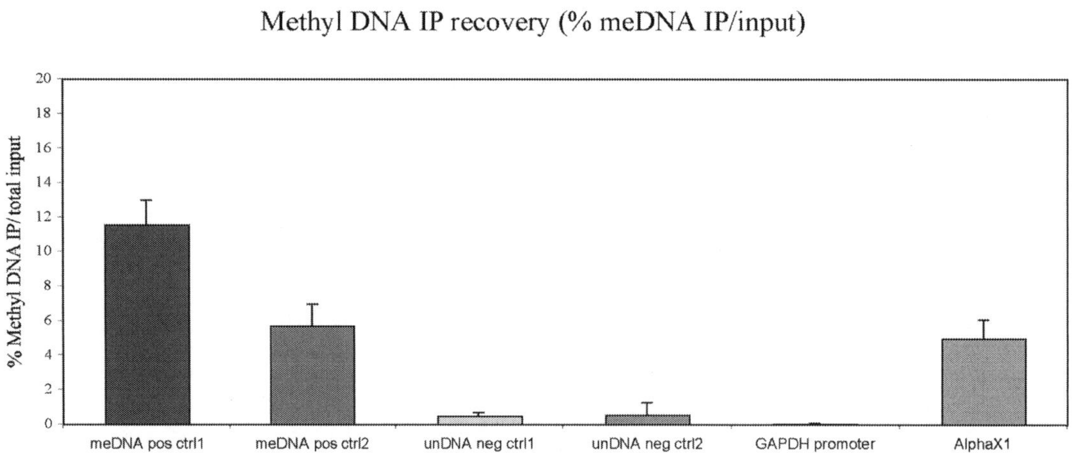

Fig. 15.1. Methyl DNA IP results obtained with the Diagenode METHYL kit (cat. no. mc-green-03). Methyl DNA IP assays were performed using DNA from MCF7 cells, the Diagenode antibody directed against 5-methyl cytidine and optimized PCR primer pairs for qPCR. The DNA was prepared with the GenDNA module. The IP was performed including the kit internal controls together with the human DNA sample. The internal positive and negative DNA controls included in the IP assay are methylated DNA (meDNA) and unmethylated DNA (unDNA). The DNA is then purified from the immunoprecipitated material and analysed by PCR using the primer pairs included in the kit (see below). Data shown are taken from three independent experiments (mean±SD). Each "primer pair" targets a specific DNA and expected results are as follows: *Internal DNA controls:* "meDNA pos ctrl1" and "meDNA pos ctrl2", both primer pairs target the meDNA control and positive signals are obtained as the methylated DNA should be immunoprecipitated; "unDNA neg ctrl1" and "unDNA neg ctrl2" primer pairs target unDNA control and no signal is obtained as non-methylated DNA should not be immunoprecipitated (0% methylation). *Human DNA sample:* "GAPDH promoter" – no signal is expected as this region is not methylated; "AlphaX1 satellite" – a signal is expected as it is a methylated region. The Methyl DNA IP controls reveal IP efficiency.

1. Make aliquots of the purified DNA and prepare dilutions. Use the purified DNA from Methyl DNA IPs and DNA input(s) (*see* **Section 3.2.2**). From 50 μL of purified DNA, transfer 10 μL into a new tube (keep 40 μL for Methyl DNA IP-on-chip analysis (*see* **Note 3**) or further PCR analysis,). For the first PCR analysis, dilute 10 μL of each purified DNA sample as follows: to 10 μL of purified DNA sample (from IP and input), add 35 μL of water. Final volume is 45 μL. Use 5 μL per PCR (see below). Note: when testing the hum meDNA primer pair (AlphaX1), dilute the DNA sample 1:1,000.
2. Prepare your qPCR mix using SYBR PCR Green master mix and qPCR. qPCR mix (total volume of 25 μL/reaction): 1.0 μL of provided primer pair (stock: 10 μM each: reverse and forward), 12.5 μL of master mix (e.g., iQ SYBR Green supermix), 5.0 μL of diluted purified DNA sample (see above for DNA dilutions), and 6.5 μL of water.
3. PCR cycles: amplification: 1X 95°C for 7 min, 40 cycles of (95°C for 15 s, 60°C for 1 min and 95°C for 1 min).
4. When the PCR is done, analyze the results.

Our first MeDIP kit as described here has been optimized using sepharose beads in IP and traditional IP wash methods. About a year later, we then launched the magnetic kit version, simplifying the protocol and reducing the number of buffers needed per experiment. Moreover, we recently automated the Methyl DNA IP, which allows standardization and ensures high reproducibility. Our methods allow the targeted analysis of methylation and has also been plugged to genome wide analysis and to multiple sample analysis (such as the new generation sequencing, as it is increasingly used in the research field).

We also designed and tested novel positive and negative primer pairs to analyse from human, mouse and rat DNA samples the IP'd material obtained after Methyl DNA IP. Positive primer pairs target a specific methylated DNA region and the negative primer pairs target an unmethylated DNA sequence.

4. Notes

1. The GenDNA module provides you with a high excess of buffer for the preparation of DNA. Sufficient buffer is given for the preparation of four genomic DNA batches, each obtained from 3 to 10 $\times$ 10^6 cultured cells (scale accordingly based on your starting material). From about 3 million cells, 20–30 μg of DNA can be expected. From about 10 million

cells, 50–100 μg of DNA can be expected. It is also possible to start with less cells, keeping in mind that 1 μg of DNA is needed per IP. Note that the protocol can be adapted to mammalian tissues.

2. When preparing the DNA (*see* **Section 3.1.2**), it is possible to omit Steps 4–8 and perform a dialysis instead. Although dialysis is time consuming, it is a good alternative and allows the prevention of possible shearing of high molecular weight DNA. In brief, to dialyze, remove organic solvents and salt from the DNA by at least two dialysis steps against a minimum of 100 volumes of TE buffer. Because of the high viscosity of the DNA, it is necessary to dialyze for a total of at least 24 h.

3. The DNA obtained following a Methyl DNA IP must be submitted to amplification before hybridization to DNA array. The method of choice is T7 based linear amplification described by in (14).

References

1. Ushijima, T., Morimura, K., Hosoya, Y., Okonogi, H., Tatematsu, M., Sugimura, T. and Nagao, M. (1997) Establishment of methylation-sensitive-representational difference analysis and isolation of hypo- and hypermethylated genomic fragments in mouse liver tumors. *Proc. Natl. Acad. Sci. U.S.A.* **94**, 2284–2289.
2. Clark, S. J., Harrison, J., Paul, C. L. and Frommer, M. (1994) High sensitivity mapping of methylated cytosines. *Nucleic Acids Res.* **22**, 2990–2997.
3. Fraga, M. F. and Esteller, M. (2002) DNA methylation: a profile of methods and applications. *Biotechniques* **33**, 636–649.
4. Yang, H. J., Liu, V. W., Wang, Y., Chan, K. Y., Tsang, P. C., Khoo, U. S., Cheung, A. N. and Ngan, H. Y. (2004) Detection of hypermethylated genes in tumor and plasma of cervical cancer patients. *Gynecol. Oncol.* **93**, 435–440.
5. Cross, S. H., Charlton, J. A., Nan, X. and Bird, A. P. (1994) Purification of CpG islands using a methylated DNA binding column. *Nat. Genet.* **6**, 236–244.
6. Yegnasubramanian, S., Lin, X., Haffner, M. C., DeMarzo, A. M. and Nelson, W. G. (2006) Combination of methylated-DNA precipitation and methylation-sensitive restriction enzymes (COMPARE-MS) for the rapid, sensitive and quantitative detection of DNA methylation. *Nucleic Acids Res.* **34**, e19.
7. Gebhard, C., Schwarzfischer, L., Pham, T. H., Andreesen, R., Mackensen, A. and Rehli, M. (2006) Rapid and sensitive detection of CpG-methylation using methyl-binding (MB)-PCR. *Nucleic Acids Res.* **34**, e82.
8. Rauch, T., Li, H., Wu, X. and Pfeifer, G. P. (2006) MIRA-assisted microarray analysis, a new technology for the determination of DNA methylation patterns, identifies frequent methylation of homeodomain-containing genes in lung cancer cells. *Cancer Res.* **66**, 7939–7947.
9. Weber, M., Davies, J. J., Wittig, D., Oakeley, E. J., Haase, M., Lam, W. L. and Schübeler, D. (2005) Chromosome-wide and promoter-specific analyses identify sites of differential DNA methylation in normal and transformed human cells. *Nat. Genet.* **37**, 853–862.
10. Klose, R. J., Sarraf, S. A., Schmiedeberg, L., McDermott, S. M., Stancheva, I. and Bird, A. P. (2005) DNA binding selectivity of MeCP2 due to a requirement for A/T sequences adjacent to methyl-CpG. *Mol. Cell* **19**, 667–678.
11. Zhang, X., Yazaki, J., Sundaresan, A., Cokus, S., Chan, S. W., Chen, H., Henderson, I. R., Shinn, P., Pellegrini, M., Jacobsen, S. E. and Ecker, J. R. (2006) Genome-wide high-resolution mapping and functional analysis of DNA methylation in arabidopsis. *Cell* **126**, 1189–1201.
12. Rauch, T. A., Zhong, X., Wu, X., Wang, M., Kernstine, K. H., Wang, Z., Riggs, A. D.

and Pfeifer, G. P. (2008) High-resolution mapping of DNA hypermethylation and hypomethylation in lung cancer. *Proc. Natl. Acad. Sci. U.S.A.* **105**, 252–257.

13. Weber, M. and Schübeler, D. (2007) Genomic patterns of DNA methylation: targets and function of an epigenetic mark. *Curr. Opin. Cell Biol.* **19**, 273–280.
14. Liu, C. L., Schreiber, S. L. and Bernstein, B. E. (2003) Development and validation of a T7 based linear amplification for genomic DNA. *BMC Genomics* **4**, 19.

Chapter 16

Immunoprecipitation of Methylated DNA

Anita L. Sørensen and Philippe Collas

Abstract

DNA methylation contributes to the regulation of long-term gene repression by enabling the recruitment of transcriptional repressor complexes to methylated cytosines. Several methods for detecting DNA methylation at the gene-specific and genome-wide levels have been developed. Methylated DNA immunoprecipitation, or MeDIP, consists of the selective immunoprecipitation of methylated DNA fragments using antibodies to 5-methylcytosine. The genomic site of interest can be detected by PCR, hybridization to DNA arrays, or by direct sequencing. This chapter describes the MeDIP protocol and quality control tests that should be performed throughout the procedure.

Key words: DNA methylation, immunoprecipitation, 5-methylcytosine antibody, microarray.

1. Introduction

DNA methylation consists in the post-replicative addition of a methyl group to the 5 position of a cytosine in a cytosine-phosphate-guanine (CpG) dinucleotide (**Fig. 16.1A**). CpG methylation alters the interaction of DNA with proteins which in turn may modulate transcription – it is either impaired, by methylation of activator sites, or enhanced, by methylation of insulators and silencers (1). CpG methylation in vertebrates is symmetrical (it occurs on both DNA strands) (**Fig. 16.1B**) and targets isolated or clustered CpGs. In plants, cytosines are methylated both symmetrically (CpG or CpNpG) and asymmetrically (CpNpNp), where N is any nucleotide. *Drosophila melanogaster* only exhibits DNA methylation in early stages of development, while*Saccharomyces cerevisiae* shows no DNA methylation.

Philippe Collas (ed.), *Chromatin Immunoprecipitation Assays,* Methods in Molecular Biology 567,
DOI 10.1007/978-1-60327-414-2_16, © Humana Press, a part of Springer Science+Business Media, LLC 2009

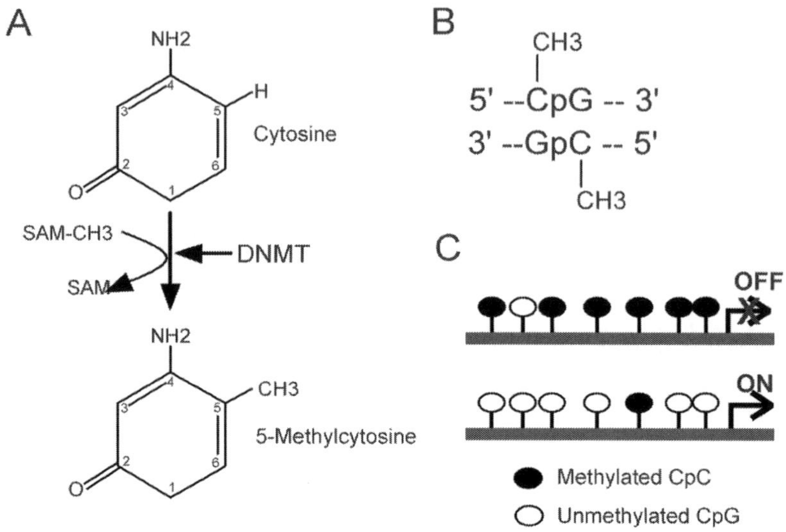

Fig. 16.1. Principles of DNA methylation. (**A**) Mechanism of DNA methylation. (**B**) CpG methylation is symmetrical and occurs on both DNA strands. (**C**) DNA methylation correlates with long-term gene repression.

CpG methylation is catalyzed by DNA methyltransferases (DNMTs). The maintenance DNA methyltransferase DNMT1 specifically recognizes hemi-methylated DNA after replication and methylates the daughter strand, ensuring fidelity in the methylation profile after replication (2). In contrast to DNMT1, DNMT3a and DNMT3b are implicated in de novo DNA methylation that takes place during embryonic development and cell differentiation (3), as a means of shutting down genes whose activity is no longer required as cells differentiate (e.g., that of pluripotency-associated genes). The fourth DNMT, DNMT2, has to date no clear ascribed function in DNA methylation but has been shown to have cytoplasmic transfer RNA methyltransferase activity (4, 5).

DNA methylation is a hallmark of long-term gene silencing (6, 7) (**Fig. 16.1C**). The methyl groups create target sites for methyl-binding proteins which induce transcriptional repression by recruiting co-repressor complexes, histone deacetylases, or histone methyltransferases (7). DNA methylation is essential for development (8–11), X chromosome inactivation (12), and genomic imprinting (13–15). The relationship between DNA methylation and gene expression is complex (1) and recent evidence based on genome-wide CpG methylation profiling highlights promoter CpG content as a component of this complexity (16).

Several approaches have been developed to analyze DNA methylation profiles. Protocols relying on bisulfite conversion of DNA have been recently reviewed and the widely used bisulfite genomic sequencing approach has been extensively improved and described (17). An alternative to bisulfite sequencing is the immunoprecipitation of methylated DNA, referred to as methylated DNA immunoprecipitation, or MeDIP (18). The principle of MeDIP is simple; genomic DNA is randomly fragmented by sonication and methylated fragments are selectively immunoprecipitated using an antibody to 5-methyl cytosine (5mC). Detection of a gene of interest in the methylated DNA fraction can be done by polymerase chain reaction (PCR), hybridization to genomic arrays (MeDIP-chip), or high-throughput sequencing (MeDIP-seq) (19, 20). We have used MeDIP-chip for the analysis of DNA methylation profiles in various mesenchymal stem cell (MSC) types. This chapter describes the MeDIP assay as it is performed in our laboratory, including control tests that can be performed along the way (**Fig. 16.2**). The protocol is derived from that established in Dirk Schübeler's laboratory (16, 18).

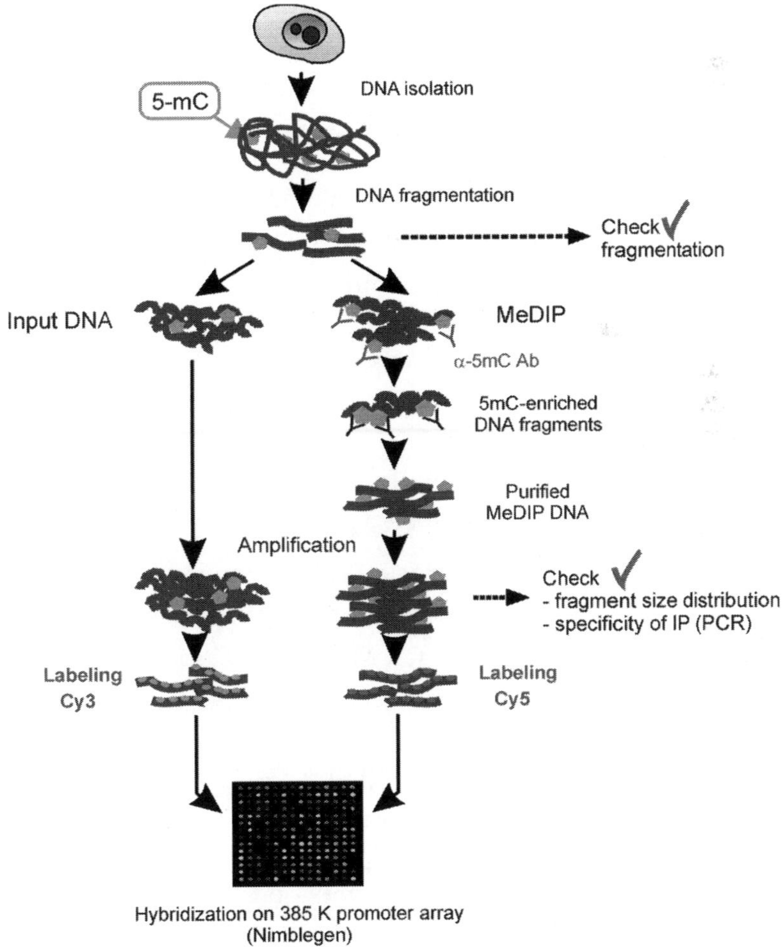

Fig. 16.2. The MeDIP assay. Genomic DNA are purified from cells, fragmented to ~300–1,000 bp by sonication, and 5-mC enriched fragments are immunoprecipitated using an anti-5mC antibody. Precipitated and input DNA are amplified. For array-based analysis, input and MeDIP DNA samples are differentially fluorescently labeled.

2. Materials

2.1. Laboratory Equipment

1. 1.5 mL centrifuge tubes.
2. Magnetic rack for 1.5 mL tubes.
3. Probe sonicator (Sartorius Labsonic M sonicator fitted with 3 mm diameter probe, or similar).
4. Thermomixer (Eppendorf).
5. Table top centrifuge.
6. Minicentrifuge.
7. Vortex.
8. Rotator.
9. Thermal cycler with accessories.

2.2. Reagents

1. Anti-5mC antibody (Diagenode, cat. no. Mab-5MECYT-500).
2. Dynabeads® M-280 sheep anti-mouse IgG (Invitrogen).
3. GenomePlex Complete Whole Genome Amplification Kit (Sigma-Aldrich, WGA2-50RXN).
4. MinElute PCR Purification Kit (Qiagen).
5. 500 mM EDTA.
6. 1 M Tris-HCl, pH 8.0.
7. 5 M NaCl.
8. Glycogen at 20 mg/mL.
9. Proteinase K stock solution: 20 mg/mL proteinase K in MilliQ water.
10. 3 M NaAc, pH 7.0.
11. Phenol:chloroform:isoamylalcohol (25:24:1).
12. Chloroform:isoamylalcohol (24:1).
13. PCR Master Mix (Promega).
14. MilliQ deionized water.
15. Crushed ice.

2.3. Buffers

1. Cell lysis buffer: 10 mM Tris-HCl, pH 8.0, 0.1 M EDTA, pH 8.0, 0.5% SDS.
2. 10X immunoprecipitation (IP) buffer: 1.4 M NaCl, 100 mM Na-phosphate, pH 7.0, 0.5% Triton X-100.
3. Na-phosphate: 39 mL of 2 M NaH_2PO_4 (276 mg/mL), 61 mL 2 M Na_2HPO_4 (284 mg/mL), 100 mL MilliQ water. This should make a 1 M Na-phosphate solution at pH 7.0. Measure pH.

4. PBS–BSA solution: 0.05 g BSA in 50 mL PBS (i.e., 0.1% BSA).
5. TE buffer: 10 mM Tris-HCl, pH 8.0, 1 mM EDTA.
6. Proteinase K digestion buffer: 50 mM Tris-HCl pH 8.0, 10 mM EDTA, 0.5% SDS.

3. Methods

3.1. Purification of Genomic DNA

The procedure described here is for DNA purification from $\leq 10^6$ cells harvested and washed by your standard protocol to obtain enough DNA for a duplicate MeDIP (*see* **Note 1**).

1. Suspend the cell pellet (10^6 cells) in 400 μL of cell lysis buffer in a 1.5 mL centrifuge tube.
2. Add 1.2 μL proteinase K solution (20 mg/mL stock) per 100 μL cell lysis buffer.
3. Incubate at 55°C for 1 h.
4. Add another 0.6 μL of proteinase K solution per 100 μL cell lysis buffer and incubate at 37°C overnight.
5. Add 1 volume of phenol:chloroform:isoamylalcohol, mix the contents by inversion of the tube 10–20 times, centrifuge at 15,000*g* for 5 min, and transfer the aqueous phase to a clean tube.
6. Repeat Step 5.
7. Add 1 volume of chloroform:isoamylalcohol, mix by inversion, and centrifuge at 15,000*g* for 5 min and transfer the aqueous phase to a clean tube.
8. Precipitate the DNA by adding 0.1 volumes of 3 M NaAc, pH 7.0, and 2.5 volumes of 96% ethanol at −20°C; mix and incubate for 1 h at −20°C.
9. Centrifuge at 20,000*g* for 15 min at 4°C and remove the supernatant.
10. Add 0.5 mL of 70% ethanol to wash the pellet, vortex, and centrifuge at 20,000*g* for 10 min at 4°C. Remove the ethanol.
11. If starting material were more than 10^6 cells, collect DNA pellets (from each tube) into one tube.
12. Dissolve the DNA in 100 μL TE buffer per 10^6cells (~50 μg DNA). Make sure the DNA is properly dissolved before RNAse treatment.

3.2. RNAse Treatment of Genomic DNA

It is imperative to treat the genomic DNA with RNase because the antibody also recognizes 5-methylcytidine in the context of RNA.

1. Place 100 μL genomic DNA (~50 μg DNA) into a 1.5 mL tube. The amount of DNA to be RNase-treated should be that needed to continue with MeDIP.
2. Add 6 μL RNase solution (final concentration, 30 μg/mL) and incubate for 2 h at 37°C.

3.3. Fragmentation of Genomic DNA

1. Dilute the RNase-treated DNA in a total of 200 μL in TE, pH 8.0, in a 1.5 mL tube placed on ice.
2. Sonicate on ice for three times 30 s, with 30 s pauses on ice between each sonication session, using the probe sonicator. With the Labsonic M sonicator, we use the following settings: cycle 0.5, 30% power (*see* **Note 2**).
3. Repeat for each DNA sample (if relevant) while leaving the sonicated samples on ice.
4. To assess fragmentation, resolve 4 μL of sonicated DNA by 1.6% agarose gel electrophoresis and ethidium bromide staining (**Fig. 16.3A**).

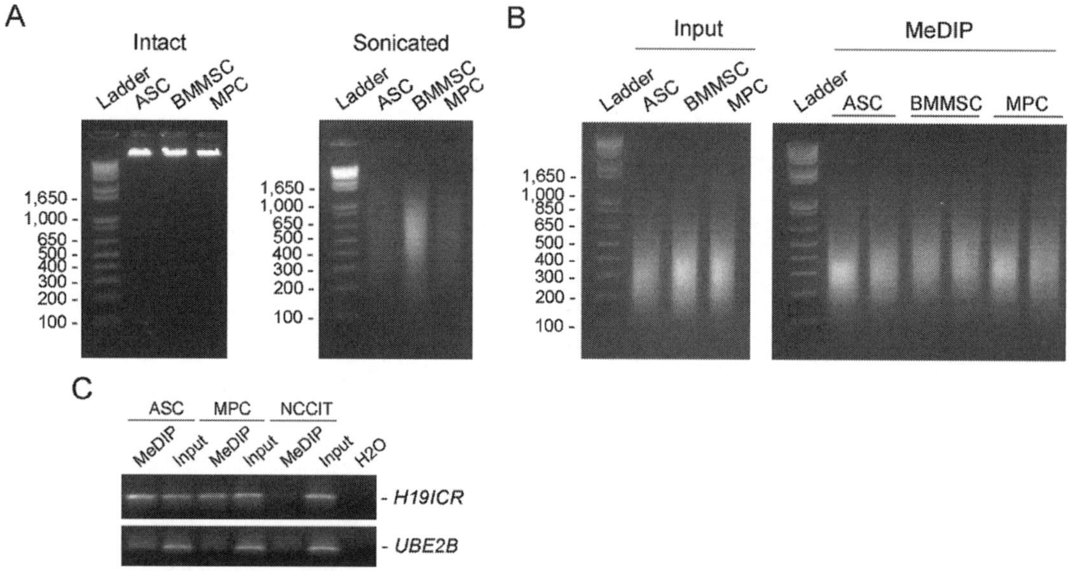

Fig. 16.3. Quality assessments of DNA during the MeDIP assay. (**A**) Assessment of fragmentation by sonication. Both intact (non-fragmented) and sonicated DNA samples from adipose stem cells (ASCs), bone marrow mesenchymal stem cells (BMMSCs), and muscle progenitor cells (MPCs) were analyzed by electrophoresis in 1.2% agarose and ethidium bromide staining. (**B**) Assessment of input and MeDIP DNA fragment size distribution and uniformity after amplification. Input and MeDIP DNA samples from ASCs, BMMSCs, and MPCs were amplified, purified, and resolved by electrophoresis in 1.2% agarose and ethidium bromide staining. Note the uniformity of the fragment sizes. Such samples are ready for processing for labeling and microarray hybridization. (**C**) PCR analysis of specificity of the MeDIP assay. Input and MeDIP DNA were analyzed by PCR using primers specific for the human *H19* Imprinting Control Region (*H19* ICR), which is methylated in somatic cells but not in male germ-cell-derived embryonal carcinoma NCCIT cells, and for the human *UBE2B* (ubiquitin-conjugating enzyme E2B) locus, which is unmethylated. DNA from ASCs, BMMSCs, and NCCIT cells were used in this analysis. PCR products were resolved by agarose gel electrophoresis and stained with ethidium bromide.

5. If necessary, continue with sonication until the desired DNA fragment length is achieved.
6. Precipitate the sonicated DNA by adding 1 μL of glycogen, 16 μL 5 M NaCl stock (400 mM final concentration), and 400 μL of 100% ethanol; mix and incubate at −80°C for 1 h. Thaw the tubes and centrifuge at 20,000*g* for 15 min at 4°C. It is important to remove all the ethanol. The pellet may be left to dry at room temperature for 15 min.
7. Dissolve the DNA in 60 μL MilliQ water and measure DNA concentration.

3.4. Immunoprecipitation of Methylated DNA

1. Dilute 4 μg sonicated DNA in TE buffer to a total volume of 450 μL. Remember to store the rest of the DNA for input.
2. Denature for 10 min in boiling water and immediately chill on ice for 10 min.
3. Add 51 μL of 10X IP buffer.
4. Add 10 μL of anti-5mC antibody and incubate for 2 h at 4°C on a rotator set to 40 rpm.
5. Pre-wash 40 μL of Dynabeads with 800 μL PBS–BSA for 5 min at room temperature with shaking at 800 rpm on a thermomixer.
6. Place on magnetic rack to collect the beads; remove the PBS–BSA and repeat the wash (Step 5) with 800 μL PBS–BSA.
7. Collect the beads with a magnetic rack and resuspend in 40 μL of 1X IP buffer.
8. Add Dynabeads to the sample and incubate for 2 h at 4°C on a rotator set at 40 rpm.
9. Place the tube on a magnetic rack to collect the beads and wash with 700 μL 1X IP buffer for 10 min at room temperature on a thermomixer at 950 rpm.
10. Repeat wash Step 9 once.
11. Transfer the contents of the tube to a clean 1.5 mL tube. This tube shift step eliminates any non-specifically bound DNA stuck on the tube wall, which may give rise to background in the analysis.
12. Place the tube on the magnetic rack, collect the beads, and wash once with 700 μL 1X IP buffer at room temperature on a thermomixer at 950 rpm.
13. Collect the beads with the magnetic rack and resuspend in 250 μL proteinase K digestion buffer.
14. Add 3.5 μL proteinase K solution (20 mg/mL stock).
15. Incubate for 3 h at 50°C on a thermomixer at 950 rpm.

16. Place the tube on the magnetic rack, transfer the content of the tube to a clean 1.5 mL tube.
17. Extract DNA once with 250 μL phenol:chloroform:isoamylalcohol and once with 250 μL chloroform:isoamylalcohol. It is critical not to harvest proteins together with the extracted DNA; carefully collect the upper (water) phase after each extraction, leaving 10–15 μL of the water phase behind.
18. Precipitate DNA by adding 20 μL of 5 M NaCl stock (400 mM final concentration), 1 μL glycogen, and 500 μL 100% ethanol; mix and incubate at −80°C for 1 h. Centrifuge at 20,000*g* for 15 min at 4°C. Make sure to remove all ethanol after centrifugation.
19. Dissolve the DNA in 15 μL MilliQ water overnight at 4°C.
20. Measure DNA concentration with a Nanodrop. The sample can be stored at −20°C.

3.5. Amplification of the Immunoprecipitated DNA

Following immunoprecipitation, the yield of MeDIP DNA is low (300–450 ng in our hands) and incompatible with hybridization to microarrays (Nimblegen promoter arrays require 4 μg DNA per array). A genomic DNA amplification step, followed by a clean up, is therefore introduced in the protocol. For MeDIP and input DNA amplification, we use the Sigma WGA2 GenomePlex Complete Whole Genome Amplification Kit, but omit the DNA fragmentation step (*see* **Note 3**).

3.5.1. Amplification of Precipitated and Input DNA

1. Place 11 μL (i.e., 100 ng) of MeDIP DNA into a 0.2 mL tube, and add 2 μL of 1X library preparation buffer provided with the WGA2 kit, and 1 μL of library stabilization solution.
2. Dilute input DNA with MilliQ water similar to concentration of MeDIP DNA and repeat Step 1 with input DNA.
3. Vortex thoroughly, centrifuge briefly, and denature by placing in a thermal cycler at 95°C for 2 min.
4. Place the samples on ice, centrifuge briefly, and return the samples on ice.
5. Add 1 μL of library preparation enzyme solution provided with the WGA2 kit, vortex well, and centrifuge briefly.
6. Place samples in a thermal cycler and run the following program: 16°C for 20 min, 24°C for 20 min, 37°C for 20 min, 75°C for 5 min and hold at 4°C.
7. Remove samples from the thermal cycler, centrifuge briefly, and either freeze and store at −20°C (for up to 3 days) or proceed with amplification.

8. Prepare a PCR master mix by adding to the 15 μL reaction from Step 7: 7.5 μL of 10X amplification master mix, 47.5 μL nuclease-free water, and 5 μL of WGA DNA polymerase (provided with the WGA2 kit).
9. Vortex thoroughly and centrifuge briefly.
10. Incubate samples in a thermal cycler with the following program: 95°C for 3 min and 14 cycles of 94°C for 15s (denaturation) and 65°C for 5 min (annealing/extension). Hold the reaction at 4°C when completed. The amplified sample can be stored at −20°C, similarly to genomic DNA.

3.5.2. Clean Up of the Amplified DNA

The amplified DNA needs to be cleaned up regardless of the mode of analysis that follows. We use the MinElute PCR Purification Kit protocol from Qiagen as per manufacturer's instructions. The kit is designed for purification of DNA fragments ranging from 70 bp to 4 kb and thus is well suited for fragmented MeDIP and input DNA fragments.

1. Divide the 75 μL amplified DNA sample in **Section 3.5.1** in two tubes and add 5 volumes of PB buffer provided with the kit and mix; 200 μL buffer to 40 μL sample and 175 μL to 35 μL sample.
2. Place a MinElute column in a provided 2 mL collection tube, add the sample from Step 1, and centrifuge for 1 min to bind DNA to the membrane in the column.
3. Discard the flow-through and return the column back into the same tube.
4. Add 750 μL buffer PE provided with the kit (wash buffer) and centrifuge for 1 min.
5. Discard the flow-through, return the column to the tube, and centrifuge again for 1 min at full speed.
6. Place the MinElute column into a clean 1.5 mL centrifuge tube.
7. Add 10 μL MilliQ water to the center of the membrane, let the column stand for 1 min, and centrifuge for 1 min. The eluate volume should be ~9 μL starting from 10 μL elution solution (*see* **Notes 4** and **5**).

3.6. Analysis of the Precipitated DNA

Analysis of DNA methylation can be performed by PCR, hybridization to genomic arrays, or by high-throughput sequencing (16, 18, 21–23).

3.6.1. Assessment of DNA Fragment Size Distribution

Before any analysis of MeDIP DNA, we routinely check that the range of fragment sizes is conserved between multiple samples. This is done by 1.6% agarose gel electrophoresis of an aliquot of

2 μL amplified and cleaned-up DNA. **Figure 16.3B** shows that all input and MeDIP DNA samples display fragment sizes uniformly ranging from 200 to ~850 bp, with most fragments around 300 bp.

3.6.2. PCR Analysis

It is recommended to verify the specificity of the MeDIP by PCR analysis of immunoprecipitated and input DNA samples prior to fluorescent labeling and array hybridization, primarily due to the labor and costs involved. Both input (positive control) and MeDIP DNA samples should be analyzed using primer pairs to loci known to be methylated or unmethylated (such as housekeeping genes). As methylated control locus, we use the *H19* imprinted control region (*H19* ICR) with the following primer pair: 5'-GAGCCGCACCAGATCTTCAG-3' and 5'-TTGGTG GAACACACTGTGATCA-3' (annealing temperature 60°C). As unmethylated control locus, we use primers to the promoter of the housekeeping *UBE2B*gene with the following primer pair: 5'-CTCAGGGGTGGATTGTTGAC-3' and 5'-TGTGGA TTCAAAGACCACGA-3' (annealing temperature 60°C). **Figure 16.3C** illustrates the result of MeDIP and input DNA sample PCR analysis using the above primers for three different cell types. Note that in NCCIT embryonal carcinoma cells, the *H19*ICR is unmethylated (**Fig. 16.3C**). A list of candidate methylated and unmethylated loci and respective PCR tests after MeDIP has been recently reported for human primary fibroblasts (16).

3.6.3. Microarray-Based Analysis

Several commercial platforms exist for hybridization of MeDIP samples. Choice of platform depends on the experimental objective (e.g., investigation of CpG islands specifically, or promoters), array design, probe density, previous experience, and cost. We have used Nimblegen human HG18 RefSeq Promoter arrays (www.nimblegen.com) in an investigation of methylation profiles in the promoters of various cell types (*see also* Ref. (16) for an earlier version of these arrays). **Figure 16.4** shows the confirmation that MeDIP-array results match the PCR data (*see* **Fig. 16.3C** for the *UBE2B* promoter methylation) and the PCR and array data of Weber and colleagues (16). In addition, **Fig. 16.5** shows the MeDIP methylation profiles of several adipogenic, myogenic, and endothelial gene promoters in human adipose tissue stem cells. These profiles are in agreement with CpG methylation patterns reported by bisulfite genomic sequencing (24–26).

3.6.4. High-Throughput Sequencing

By similarity to chromatin immunoprecipitation (ChIP)-sequencing approaches (27), it is possible to analyze MeDIP DNA in an unbiased manner by direct quantitative high-throughput sequencing (23).

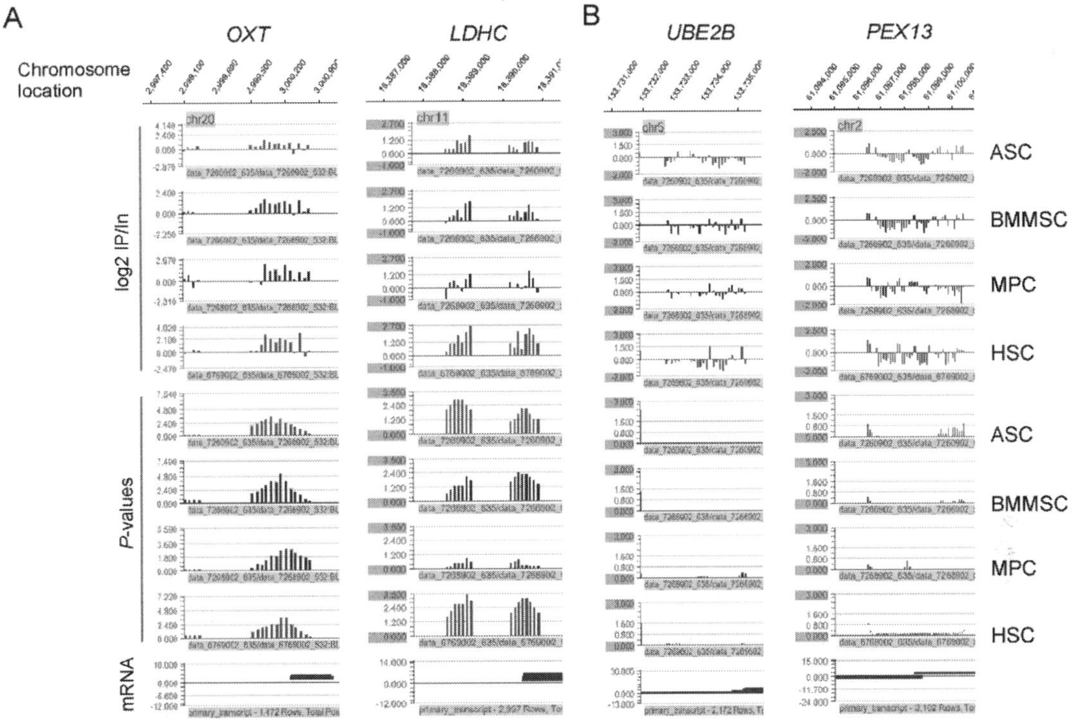

Fig. 16.4. Validation of the MeDIP assay: MeDIP methylation profile of genes known to be methylated and unmethylated, in adipose stem cells (ASCs), bone marrow MSCs (BMMSCs), muscle progenitor cells (MPCs), and hematopoietic stem cells (HSCs). (**A**) Methylation profiles of two known methylated promoters (*OXT, LDHC*). (**B**) Methylation profiles of two known unmethylated housekeeping gene promoters (*UBE2B, PEX13*). Log_2 IP/input ratios, *P*-values, and transcripts are shown.

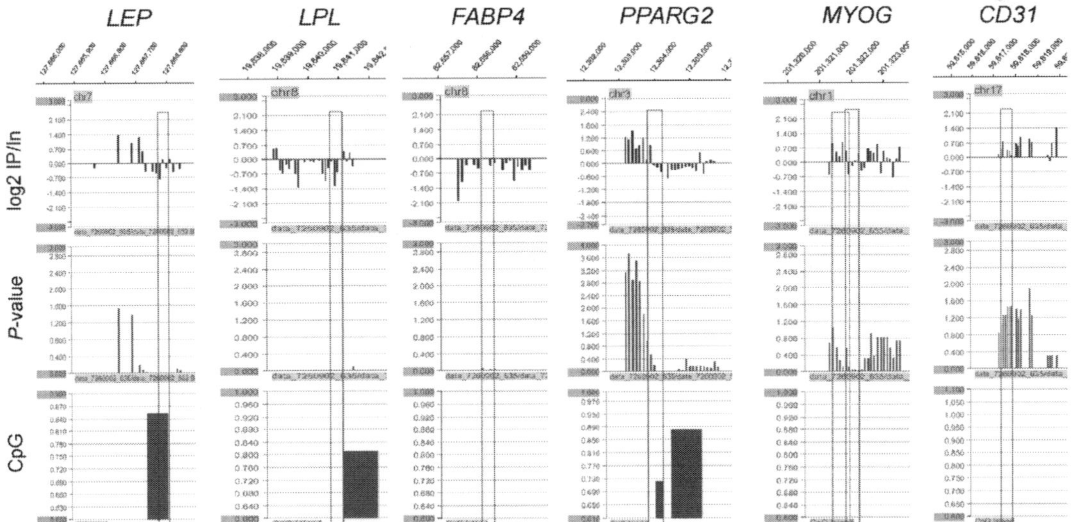

Fig. 16.5. MeDIP methylation profiles of the *LEP, LPL, FABP4, PPARG2, MYOG,* and *CD31* promoters in adipose stem cells. Rectangle boxes delineate genomic regions analyzed earlier by bisulfite sequencing (24–26). A CpG density track (CpG) is also shown.

4. Notes

1. We find that 10^6 cells are usually sufficient for one MeDIP in duplicate. Note that the extent of DNA recovery may vary between cell types.
2. Sonication should produce DNA fragments of ~300–1,000 bp (**Fig. 16.3A**). DNA fragments of less than 200 bp will be difficult to label when analyzed by microarray. The sonication protocol reported here is suitable for a variety of cell lines and primary cell types such as NCCIT cells, 293T cells, skin fibroblasts, keratinocytes, adipose-, bone marrow- and muscle-derived MSCs, or hematopoietic stem cells. Optimization of sonication conditions may be required for a different cell type and other sonicator models. Samples should not foam as this reduces sonication efficiency.
3. DNA amplification: The Sigma WGA2 DNA amplification procedure includes a three-step process: DNA fragmentation, library generation, and PCR amplification. The first two steps, fragmentation and library generation, are recommended by the manufacturer to be carried out without interruption. However, in the MeDIP assay, the DNA to be amplified is already fragmented, so we omit the fragmentation step of the WGA2 protocol and start at the library preparation step. The WGA2 kit recommends starting with a minimum of 10 ng DNA; however, we consistently start with 100 ng DNA. At the end of the amplification procedure as described in **Section 3.5.1**, we obtain 20 μL of DNA at ~500 ng/μL, or a total of approximately 10 μg DNA.
4. We elute DNA with MilliQ water, as performed by Farnham and colleagues (28). However, it should be possible to use the Qiagen elution buffer (EB) provided with the clean-up kit. We do not recommend eluting with TE buffer because EDTA may inhibit further enzymatic reaction, notably during sample labeling for array hybridization.
5. Using this protocol, the MeDIP assay generally yields ~5% of the initial DNA amount from the cell types we have investigated. This is similar to what is reported from the Schübeler laboratory (www.epigenome-noe.net/researchtools/protocol.php).

Acknowledgments

The basis for this MeDIP protocol has been the procedure established in Dirk Schübeler's laboratory (Friedrich Miescher Institute for Biomedical Research, Basel, Switzerland) by Michaël Weber and Dirk Schübeler and posted on the Epigenome Network of Excellence website (http://www.epigenome-noe.net/research-tools/protocol.php?protid=33). We are also grateful to Dirk Schübeler for discussion and advice. Our work is supported by the Research Council of Norway.

References

1. Jones, P. A. and Takai, D. (2001) The role of DNA methylation in mammalian epigenetics. *Science* **293**, 1068–1070.
2. Jaenisch, R. and Bird, A. (2003) Epigenetic regulation of gene expression: how the genome integrates intrinsic and environmental signals. *Nat. Genet.* **33 Suppl**, 245–254.
3. Turek-Plewa, J. and Jagodzinski, P. P. (2005) The role of mammalian DNA methyltransferases in the regulation of gene expression. *Cell Mol. Biol. Lett.* **10**, 631–647.
4. Rai, K., Chidester, S., Zavala, C. V., Manos, E. J., James, S. R., Karpf, A. R., Jones, D. A. and Cairns, B. R. (2007) Dnmt2 functions in the cytoplasm to promote liver, brain, and retina development in zebrafish. *Genes Dev.* **21**, 261–266.
5. Goll, M. G., Kirpekar, F., Maggert, K. A., Yoder, J. A., Hsieh, C. L., Zhang, X., Golic, K. G., Jacobsen, S. E. and Bestor, T. H. (2006) Methylation of tRNAAsp by the DNA methyltransferase homolog Dnmt2. *Science* **311**, 395–398.
6. Hoffman, A. R. and Hu, J. F. (2006) Directing DNA methylation to inhibit gene expression. *Cell Mol. Neurobiol.* **26**, 425–438.
7. Klose, R. J. and Bird, A. P. (2006) Genomic DNA methylation: the mark and its mediators. *Trends Biochem. Sci.* **31**, 89–97.
8. Morgan, H. D., Santos, F., Green, K., Dean, W. and Reik, W. (2005) Epigenetic reprogramming in mammals. *Hum. Mol. Genet.* **14**, R47–R58.
9. Young, L. E. and Beaujean, N. (2004) DNA methylation in the preimplantation embryo: the differing stories of the mouse and sheep. *Anim. Reprod. Sci.* **82**, 61–78.
10. Mann, J. R. (2001) Imprinting in the germ line. *Stem Cells* **19**, 287–294.
11. Razin, A. and Shemer, R. (1995) DNA methylation in early development. *Hum. Mol. Genet.* **4**, 1751–1755.
12. Hellman, A. and Chess, A. (2007) Gene body-specific methylation on the active X chromosome. *Science* **315**, 1141–1143.
13. Tremblay, K. D., Saam, J. R., Ingram, R. S., Tilghman, S. M. and Bartolomei, M. S. (1995) A paternal-specific methylation imprint marks the alleles of the mouse H19 gene. *Nat. Genet.* **9**, 407–413.
14. Sapienza, C., Peterson, A. C., Rossant, J. and Balling, R. (1987) Degree of methylation of transgenes is dependent on gamete of origin. *Nature* **328**, 251–254.
15. Reik, W., Collick, A., Norris, M. L., Barton, S. C. and Surani, M. A. (1987) Genomic imprinting determines methylation of parental alleles in transgenic mice. *Nature* **328**, 248–251.
16. Weber, M., Hellmann, I., Stadler, M. B., Ramos, L., Paabo, S., Rebhan, M. and Schubeler, D. (2007) Distribution, silencing potential and evolutionary impact of promoter DNA methylation in the human genome. *Nat. Genet.* **39**, 457–466.
17. Clark, S. J., Statham, A., Stirzaker, C., Molloy, P. L. and Frommer, M. (2006) DNA methylation: bisulphite modification and analysis. *Nat. Protoc.* **1**, 2353–2364.
18. Weber, M., Davies, J. J., Wittig, D., Oakeley, E. J., Haase, M., Lam, W. L. and Schubeler, D. (2005) Chromosome-wide and promoter-specific analyses identify sites of differential DNA methylation in normal and transformed human cells. *Nat. Genet.* **37**, 853–862.
19. Zilberman, D. and Henikoff, S. (2007) Genome-wide analysis of DNA methylation patterns. *Development* **134**, 3959–3965.
20. Jacinto, F. V., Ballestar, E. and Esteller, M. (2008) Methyl-DNA immunoprecipitation (MeDIP): hunting down the DNA methylome. *Biotechniques* **44**, 35, 37, 39.
21. Penterman, J., Zilberman, D., Huh, J. H., Ballinger, T., Henikoff, S. and Fischer, R. L. (2007) DNA demethylation in the Arabidopsis genome. *Proc. Natl. Acad. Sci U. S. A* **104**, 6752–6757.
22. Zilberman, D., Gehring, M., Tran, R. K., Ballinger, T. and Henikoff, S. (2007) Genome-wide analysis of *Arabidopsis thaliana* DNA methylation uncovers an interdependence between methylation and transcription. *Nat. Genet.* **39**, 61–69.

23. Jacinto, F. V., Ballestar, E., Ropero, S. and Esteller, M. (2007) Discovery of epigenetically silenced genes by methylated DNA immunoprecipitation in colon cancer cells. *Cancer Res.* **67**, 11481–11486.
24. Boquest, A. C., Noer, A., Sorensen, A. L., Vekterud, K. and Collas, P. (2007) CpG methylation profiles of endothelial cell-specific gene promoter regions in adipose tissue stem cells suggest limited differentiation potential toward the endothelial cell lineage. *Stem Cells* **25**, 852–861.
25. Noer, A., Sørensen, A. L., Boquest, A. C. and Collas, P. (2006) Stable CpG hypomethylation of adipogenic promoters in freshly isolated, cultured and differentiated mesenchymal stem cells from adipose tissue. *Mol. Biol. Cell* **17**, 3543–3556.
26. Noer, A., Boquest, A. C. and Collas, P. (2007) Dynamics of adipogenic promoter DNA methylation during clonal culture of human adipose stem cells to senescence. *BMC Cell Biol.* **8**, 18–29.
27. Mikkelsen, T. S., Ku, M., Jaffe, D. B., Issac, B., Lieberman, E., Giannoukos, G., Alvarez, P., Brockman, W., Kim, T. K., Koche, R. P., Lee, W., Mendenhall, E., O'Donovan, A., Presser, A., Russ, C., Xie, X., Meissner, A., Wernig, M., Jaenisch, R., Nusbaum, C., Lander, E. S. and Bernstein, B. E. (2007) Genome-wide maps of chromatin state in pluripotent and lineage-committed cells. *Nature* **448**, 553–560.
28. Acevedo, L. G., Iniguez, A. L., Holster, H. L., Zhang, X., Green, R. and Farnham, P. J. (2007) Genome-scale ChIP-chip analysis using 10,000 human cells. *Biotechniques* **43**, 791–797.

INDEX

Note: locators with the letter 'f' denote a figure on that page.

N

O

P

Q

R

S